# ALIPHATIC COMPOUNDS

(SULPHUR, PHOSPHORUS, SILICON & BORON)

# ALIPHATIC COMPOUNDS

(SULPHUR, PHOSPHORUS, SILICON & BORON)

*By*

**Amit Arora**

**DISCOVERY PUBLISHING HOUSE**

**NEW DELHI-110002**

First Published-2006

ISBN 81-8356-179-9



*Published by*

**DISCOVERY PUBLISHING HOUSE**
4831/24, Ansari Road, Prahlad Street,
Darya Ganj, New Delhi-110002 (India)
Phone: 23279245 • Fax: 91-11-23253475
E-mail:dphtemp@indiatimes.com

*Printed at:*

Amit Enterprises
Delhi-53

# Preface

This book is written for B.Sc., B.Sc. (Hons.) and M.Sc. students of various universities of India. In this book my aim has been describe the fundamental principles of organic chemistry. Since I do not consider the chemistry of natural products to be fundamental chemistry but rather the application of fundamental principles. The subject matter described in this book covers much of the basic organic chemistry that is needed by a student who wish to study chemistry as a main subject at degree level. The arrangement of the subject-matter is based on homologous series and in general, descriptions of reactions are followed by discussion of their mechanisms and these includes an elementary account of the sort of evidence that led workers to suggest mechanisms that are acceptable at the present time.

We are sure you will appreciate the thought and care that has gone into making of this text. This book has everything you want to have in your advanced organic chemistry text.

The author wish to thank all those who helped in the preparation of the text and illustration, suggestion for improvement and criticism will be gratefully received and acknowledge.

**Author**

# CONTENTS

# 1

# ALIPHATIC COMPOUNDS OF SULPHUR, PHOSPHORUS, SILICON AND BORON

## INTRODUCTION

Sulphur lies directly below oxygen in Group VI of the Periodic Table. Therefore, there two elements have similar outer shell configuration.

| | | | | | | |
|---|---|---|---|---|---|---|
| O (at no = 8) | $1s^2$ | $2s^2$ | $2p^4$ | | | |
| S (at no = 16) | $1s^2$ | $2s^2$ | $2p^6$ | $3s^2$ | $3p^4$ | $3d^0$ |

Hence, sulphur is capable of forming several classes of organic compounds parallel to those containing oxygen, such as alcohols, ethers and esters.

### Mercaptans, Thiols or Thioalcohols, RSH

These compounds occur in petroleum and give rise to 'sour petrol'.

### Nomenclature

One method is to name them as alkyl *mercaptans,* the—SH group being known as the *mercapto* or *sulph-hydryl group.* On the other hand, according to the I.U.P.A.C. system of nomenclature, the—SH group is known as the *thiol* group, and the suffix of the series is *thiol.*

This method of naming the mercaptans arises from the fact that they are the sulphur analogues of the alcohols, the usual procedure of showing that an oxygen atom has been replaced by a sulphur atom being indicated by the prefix *thio.*

| | | |
|---|---|---|
| $CH_3SH$ | methyl mercaptan | methanethiol |
| $(CH_3)_2CHCH_2SH$ | isobutyl mercaptan | 2-methylpropanethiol |

## General Methods of Preparation

Thiols may be prepared by heating potassium or sodium hydrogen sulphide with an alkyl halide, a sodium alkyl sulphate, or with a tosylate.

$$RSH \xleftarrow{RX} KSH \xrightarrow[\text{or RsOR}]{ROSO_2ONa} RSH$$

Oxygen in an alcohol may be directly replaced by sulphur, *e.g.,* with phosphorus pentasulphide, but a more satisfactory method is to pass a mixture of alcohol vapour and hydrogen sulphide over a thoria catalyst at 400°C:

$$ROH + H_2S \xrightarrow{ThO_2} RSH + H_2O \ (\textit{f.-g.})$$

Alternatively, alkenes may be used instead of alcohols, and the catalyst is nickel sulphide, *e.g.,*

$$MeCH{=}CH_2 + H_2S \xrightarrow[300°C]{NiS} MeCH(SH)Me + MeCH_2CH_2SH$$

The major product is the s-thiol (Markownikoff product; polar mechanism). Addition of hydrogen sulphide can also be induced by ultraviolet light to give predominantly the anti-Markownikoff product (free-radical mechanism).

Compounds containing the thiol group are known as *thiyl compounds,* and the addition of these to multiple bonds is called *thiylotion* (see above examples).

The best method for preparing thiols is to decompose an *S-alkylisothiouronium salt* with alkali; these salts may be prepared by interaction of an alkyl halide (preferably the bromide or iodide) and thiourea (cyanamide polymers are also produced):

$$RBr + S{=}C(NH_2)_2 \longrightarrow [RSC({=}NH_2)NH_2]^+Br^- \xrightarrow{NaOH}$$
$$RSH + NH_2CN + NaBr + H_2O$$

## General Properties

The thiols, except methanethiol, which is a gas, are colourless volatile liquids with disagreeable smells. Their boiling points are lower than those of the corresponding alcohols; this is due to the fact that they are very little associated, hydrogen bonding not readily taking place between hydrogen and sulphur, which Is less electronegative than oxygen. The thiols are also less soluble in water than the corresponding alcohols, again, no doubt, due to their inability to form hydrogen bonds with water.

Thiols are stronger acids than alcohols, *e.g.*, EtSH, $pK_a$ 10.5; EtOH, $pK_a$ 15.5. This is unexpected in view of the fact that sulphur is less electronegative than oxygen. One possible explanation is that the S—H bond (347.3 kJ) is weaker than the O—H bond (464.4 kJ) and, because of its larger size, the sulphide ion can accommodate the negative charge more easily than can the smaller oxygen atom in the hydroxide ion.

The infra-red absorption region of the thiol group (S—H stretch) is 2590–2550cm$^{-1}$(w), and that of the C—S (stretch) is 705–570cm$^{-1}$ (w),

**Reactions**

The thiols resemble the alcohols in many ways (sulphur and oxygen occur in the same periodic group); the main difference is their behaviour towards oxidising agents.

1. Thiols form mercaptides with the evolution of hydrogen when treated with alkali metals.

$$2RSH + 2Na \longrightarrow 2RS^-Na^+ + H_2$$

These alkali mercaptides are salts, and are decomposed by water.

$$R\bar{S}Na^+ + H_2O \longrightarrow RSH + NaOH$$

2. Thiols precipitate mercaptides when treated with an aqueous solution of the salt of a heavy metal, *e.g.*,

$$2RSH + (CH_3CO_2)_2Pb \longrightarrow (RS)_2Pb + 2CH_3CO_2H$$

Thiols also attack mercuric oxide in aqueous solution to form the mercury mercaptide.

$$2RSH + HgO \longrightarrow (RS)_2Hg + H_2O$$

It is this reaction which is the origin of the name mercaptans (*mercurius,* mercury; *captans,* seizing). These heavy metal mercaptides are *covalent* compounds.

3. Thiols react with carboxylic acids, preferably in the presence of inorganic acid, to form a thiolester.

$$R^1CO_2H + R^2SH \xrightleftharpoons{H^+} R^1COSR^2 + H_2O$$

The equilibrium lies mainly to the left, but a good yield of ester may be obtained by using an acid chloride instead of acid.

4. Thiols may be oxidised, the nature of the product depending on the oxidising agent used. Mild oxidation with, *e.g.*, air, hydrogen peroxide, cupric chloride or sodium hypochloric results in the formation of *dialkyl disulphides*.

$$2RSH + H_2O_2 \rightarrow R—S—S—R + 2H_2O$$

An alkaline solution of sodium plumbite containing sulphur (doctor solution) converts thiols into disulphides.

$$2RSH + Na_2PbO_2 + S \longrightarrow R—S—S—R + PbS + 2NaOH$$

Dialkyl disulphides may also be prepared by the action of iodine on sodium mercaptides.

$$2R\bar{S}Na^+ I_2 \longrightarrow R_2S_2 + 2NaI$$

These disulphides have an unpleasant smell (not as unpleasant as that of thiols). Diallyl disulphide, $(CH_2 = CHCH_2 — )_2S_2$, occurs in garlic. Disulphides are reduced to thiols by lithium aluminium hydride or by zinc and acid.

When oxidised with, *e.g.*, nitric or periodic acid, thiols are converted into *xulphonic acids*.

$$RSH + 3[O] \xrightarrow{HNO_3} RSO_3H$$

5. Thiols readily undergo desulphurisation with Raney nickel.

$$RSH \xrightarrow{Ni} RH + NiS$$

Desulphurisation, in general, involves the breaking of a C—S bond in an organic compound, and the source of the hydrogen is usually the hydrogen adsorbed by the Raney nickel prepared in the ordinary way.

The first step is believed to be chemisorption of the S atom on the surface of the catalyst, and this is then followed by fission of the C—S bond to give free radicals and finally hydrogenation.

6. Thiols readily combine with aldehydes and ketones, in the presence of hydrochloric acid, to form *mercaptals* and *mercaptols* respectively: *e.g.*, ethanethiol forms diethylmethyl mercaptal with acetaldehyde, and diethyldimethyl mercaptol with acetone (see also sulphones, below).

$$CH_3CHO + 2C_2H_5SH \xrightarrow{HCl} CH_3CH(SC_2H_5)_2 + H_2O$$

Refluxing mercaptals (and mercaptols) with ethanol in the presence of freshly prepared Raney nickel replaces the thiol group by hydrogen (Wolfrom *et al.*, 1944).

$$>C(SR)_2 + 4[4] \longrightarrow >CH_2 + 2RSH$$

Thus a carbonyl group can be converted into a methylene group *(cf.* the Clemmensen and Wolff-Kishner reductions).

The aldehyde group may be 'protected' in *acid* solution by conversion into a mercaptal, and can be regenerated by treatment of the latter with mercuric chloride in the presence of cadmium carbonate.

An alternative way of protecting a carbonyl group makes use of β-mercaptoethanol.

$$>C{=}O + HSCH_2CH_2OH \xrightarrow{TaOH} \left[>C<\begin{matrix}S\\O\end{matrix}\right] \xrightarrow{Ni} >C{=}O + C_2H_4 + S$$

*Thioethers of alkyl sulphides, $R_2S$ :* These are the sulphur analogues of the ethers, from which they differ considerably in a number of ways.

## General Methods of Preparation

1. By heating potassium sulphide with an alkyl halide, potassium alkyl sulphate, or a tosylate, *e.g.*,

$$S^{=} \; R{-}X \longrightarrow X^{-} + R{-}S^{-}$$

$$R{-}S^{-} \; R{-}X \longrightarrow R_2S + X^{-} \text{ (f.g. -g.)}$$

2. Heating an ether with phosphorus pentasulphide.

$$5R_2O + P_2S_5 \longrightarrow 5R_2S + P_2O_5 \text{ (p).}$$

3. By heating an alkyl halide with a sodium mercaptide (cf. Williamson's synthesis).

$$R^1X + R^2\bar{S}Na^+ \longrightarrow R^1{-}S{-}R^2 + NaX$$

4. By passing a thiol over a mixture of alumina and zinc sulphide at 300°C

$$2RSH \longrightarrow R_2S + H_2S$$

5. By the addition of a thiol to an alkene in the presence of peroxides, in the absence of the latter very little reaction occurs (Kharasch *et al.*, 1939).

$$(R^1CO_2)_2 \longrightarrow 2R^1. + 2CO_2$$
$$R^1. + R^2SH \longrightarrow R^1H + R^2S.$$
$$R^3CH{=}CH_2 + R^2S. \longrightarrow R^3\dot{C}HCH_2SR^2 \xrightarrow{R^2SH} R^3CH_2CH_2SR^2 + R^2S.$$

## General Properties and Reactions

The thioethers are unpleasant-smelling oils, insoluble in water but soluble in organic solvents. Like the ethers, they behave as weak bases, *e.g.*, they dissolve in 100 per cent sulphuric acid to form sulphonium salts. $[R_2SH]^+SHO_4^-$. They may be oxidised to *sulphoxides* which, on further oxidation, are converted into *sulphones, e.g.* ethyl sulphide, on oxidation with hydrogen peroxide in glacial acetic acid, gives first diethyl sulphoxide and then diethyl sulphone.

$$(C_2H_5)_2S \xrightarrow{H_2O_2} (C_2H_5)_2S{=}O \xrightarrow{H_2O_2} (C_2H_5)_2S{\overset{\nearrow O}{\underset{\searrow O}{}}}$$

Only sulphoxide is obtained with periodic acid at 0°C; at higher temperatures, the sulphone is produced (Leonard *et al.*, 1962). Sulphoxide is also obtained by oxidation of sulphides with 1-chlorobenzotriazole (Johnson *et al.*, 1969).

Originally, sulphur was thought to be quadrivalent in sulphoxides and sexavalent in sulphones. Then, to conform with the 'octet' theory, the valencies were changed to 3 and 4, respectively, the oxygen atoms now being linked by co-ordinate bonds. More recently, however, bond length measurements of the S—O bond in sulphoxides and sulphones indicate that these bonds are almost all double bonds, M.O. calculations have also shown that the S—O bond is largely double in character and that the *3d* orbitals of the sulphur atom are involved in the formation of this bond (Moffitt, 1950).

The valency shell of sulphur is $(3s)^2(3/P)^4$. Also the nature of hybridisation depends on the number of σ-electrons and lone pairs in the atom in the bonded state. Unless one wishes to consider multiple bonds as 'bent' bonds, electrons in w-bonds (double or triple) are not counted when deciding the type of hybridisation. In sulphides, each R group supplies one electron, and since the S atom has 6 valency electrons, the bond orbitals are $sp^3$, two being occupied by bonding pairs, and two by lone pairs, *i.e.,* $R_2S$. In sulphonium salts, the halide ion takes its bonding pair, and the R group uses one of the lone pairs of the sulphur atom, *i.e.*, $R3\ddot{S}^+X^-$. In sulphoxides, $R_2\ddot{S} = O$, the orbitals are still $sp^3$, but here the R groups are joined by π-bonds, the oxygen atom by a σ and $\pi_{d\text{-}p}$ bond (one of the sulphur electrons has been promoted to 3*d)*, and a lone pair remains in the fourth $sp^3$ orbital. In sulphones, $R_2S\begin{smallmatrix}\nearrow O\\ \searrow O\end{smallmatrix}$ there are four σ-bonds and two $\pi_{d\text{-}p}$ bonds (one to each oxygen atom).

Alkyl sulphides readily undergo desulphurisation with Raney nickel.

$$R^1-S-R^2 + H_2 \xrightarrow{Ni} R^1H + R^2H + NiS$$

The alkyl sulphides form various addition products, *e.g.*, with bromine the alkyl sulphide dibromide is formed, $R_2SBr_2$. Alkyl sulphides also combine with a molecule of alkyl halide to form *sulphonium salts,* in which the sulphur is tercovalent unielectrovalent. When a sulphonium salt is heated, it decomposes into alkyl sulphide and alkyl halide *(cf.* quaternary ammonium salts):

$$[R_3S]^+ \longrightarrow R_2S + RI$$

When they are treated with moist silver oxide, the *sulphonium hydroxide* is formed.

$$[R_3S]^+I^- + \text{'AgOH'} \longrightarrow [R_3S]^+\ OH^-\ AgI$$

Sulphonium hydroxides are strongly basic, and on heating form alkyl sulphide and alkene, *e.g.*,

$$[(C_2H_5)_3S]^+OH^- \longrightarrow (C_2H_5)_2S + C_2H_4 + H_2O$$

This is believed to occur by the E2 mechanism.

$$HO^-\ H-CH_2-CH_2-S^+(C_2H_5)_2 \longrightarrow H_2O + CH_2{=}CH_2 + (C_2H_5)_2S$$

This reaction can also occur by the El mechanism provided the alkyl groups are of a suitable type, *e.g.*,

$$MeC-S^+Me_2 \xrightarrow{OH^-} Me_2S + S + Me_3C^+ \longrightarrow MeC{=}CH_2 + H_2O$$

*Sulphur ylids* : Many bases remove α — H from sulphonium salts to form ylids, *e.g.*, dimethylsulphonium methylide

$$Me_3S^+X^- \xrightarrow{RLi} RH + LiX + Me_2\overset{+}{S} - CH_2 \longleftrightarrow Me_2S{=}CH_2$$

The most satisfactory base appears to be the methylsulphinyl carbanion, $MeSOCH_2^{\ -}$, produced by the action of NaH, $NaNH_2$, etc., on dimethyl sulphoxide (see also below.)

$$MeSOMe \xrightarrow{NaH} MeSOCH_2^- \xrightarrow{Me_3S^+I^-} MeSOMe + Me_2S{=}CH_2$$

Sulphonium ylids react with ketones to form epoxides.

$$Me_2S{=}CH_2 \;\; CR_2{=}O \longrightarrow Me_2\overset{+}{S}{-}CH_2{-}\underset{}{\overset{^-O}{|}}CR_2 \longrightarrow MeS + CH_2{-}CR_2 \text{ (ring with } CH_2\text{)}$$

Compounds containing activated double bonds readily form cyclopropanes, *e.g.*,

$$Ph_2C{=}CH_2 \;\; CH_2{=}SMe_2 \longrightarrow Ph_2C{-}CH_2 \text{ (with } CH_2{-}\overset{+}{S}Me_2\text{)} \longrightarrow Ph_2C{-}CH_2 \text{ (ring with } CH_2\text{)} + Me_2SO$$

Dimethyl sulphoxide also forms an ylid, *dimethylsulphoxonium methylide,* which reacts with ketones to form epoxides.

$$Me_2SO \xrightarrow{MeI} MeSOI^- \xrightarrow{NaNH_2} Me\overset{O}{\overset{\|}{S}}{=}CH_2 \xrightarrow{R_2CO} R_2C{-}CH_2 \text{ (epoxide, O bridge)} + Me_2SO$$

*Mustard gas,* 2,2'. *dichlorodiethyl sulphide, bis (2-chloroethyl) sulphide,* b.p. 215°C, may be prepared by the action of sulphur monochloride on ethylene.

$$2C_2H_4 + S_2Cl_2 \longrightarrow (ClCH_2CH_2-)_2S + S$$

It may also be prepared, in purer state, by heating ethylene chlorohydrin with sodium sulphide, and treating the product with hydrochloric acid.

$$2HOCH_2CH_2Cl + Na_2S \longrightarrow 2NaCl + (HOCH_2CH_2-)_2S$$

$$\xrightarrow{HCl} (ClCH_2CH_2-)_2S$$

Mustard gas is a poison and a vesicant.

*Sulphinium salts :* Diaryl sulphides or sulphoxides produce sulphur radical ions when dissolved in concentrated sulphuric acid.

$$Ar_2S \xrightarrow{-e} Ar_2S.^+ HSO_4^-$$

Solutions of this radical ion are coloured (red or blue), and sulphinium salts have been isolated as their hexachlorostannates.

*Thiocyanic acid, isothiocyanic acid and their derivatives :* Thiocyanic acid was believed to be a tautomeric equilibrium mixture of thicyanic acid, HSCN, and isothiocyanic acid. HNCS. Spectro-scopic studies of thiocyanic acid, however, indicate the structure HNCS (Bread *et al.*, 1947), and this has been confirmed by microwave studies (Dousmanis *et al.*, 1953)

Alkyl derivatives of both forms are known (cf. cyanic acid).

Thiocyanic acid may be prepared by heating a mixture of potassium thiocyanate and potassium hydrogen sulphate.

$$KNCS + KHSO_4 \longrightarrow HNCS + K_2SO_4$$

Its dilute aqueous solutions are fairly stable, the concentrated solutions decomposing to form carbonyl sulphide and ammonia.

$$HNCS + H_2O \longrightarrow COS + NH_3$$

*Alkyl thiocyanates* may be prepared by heating potassium thiocyanated with an alkyl halide or tosyl ester.

$$RI + KNCS \longrightarrow RSCN + KI$$

Since the thiocyanate is obtained rather than the isothiocyanate, this indicates that sulphur is a stronger nucleophile than nitrogen.

Thiocyanates are very conveniently prepared form sodium thiosulphate.

$$NaSSO_3Na \xrightarrow{RI} RSSO_3Na \xrightarrow{NaCN} RSCN + Na_2SO_3$$

The alkyl thiocyanates are fairly stable volatile oils, with a slight odour of garlic. They are oxidised to sulphonic acids by concentrated nitric acid, and reduced to thiols by, *e.g.*, zinc and sulphuric acid.

$$RSO_3H \xleftarrow{HNO_3} RSCN \xrightarrow{Zn/H_2SO_4} RSH$$

They are converted into sulphonyl chlorides by chlorine water.

$$RSCN + Cl_2 \xrightarrow{H_2O} RSO_2Cl + ClCN$$

Alkyl thiocyanates rearrange on heating to the more thermodynamically stable isothiocyanate.

$$R-SCN \rightleftharpoons [R^+ + SCN^-] \rightleftharpoons RNCS$$

*Alkyl isothiocyanates* or *mustard oils* may be prepared by heating alkyl thiocyanates, but a much more satisfactory method is the *Hofmann mustard-oil reaction* (1868), which is carried out by heating a mixture of a primary amine, carbon disulphide and mercuric chloride.

Another convenient method is to add an aqueous solution of a primary amine to carbon disulphide in sodium hydroxide solution, and then ethyl chloroformate.

$$RNH_2 + CS_2 + NaOH \longrightarrow RNHCS_2Na + H_2O \xrightarrow{ClCO_2C_2H_5}$$
$$NaCl + RNHCS_2CO_2C_2H_5 \longrightarrow RNCS + COS + C_2H_5OH \quad (60\text{-}70\%)$$

The alkyl isothiocyanates are liquids with a powerful mustard smell; they are lachrymatory and vesicatory. They are hydrolysed to primary amines when heated with hydrochloric acid, and reduced to primary amines and thioformaldehyde by, *e.g.,* zinc and sulphuric acid.

$$RNH_2 + CO_2 + H_2S \xleftarrow[2H_2O]{HCl} RNCS \xrightarrow{Zn/H_2SO_4} RNH_2 + HCHS$$

*Allyl isothiocyanate (allyl mustard oil),* $CH_2 = CHCH_2NCS$. When the glucoside *sinigrin* (which occurs in mustard seed) is hydrolysed by acid or by the enzyme *my rosin* (which is found in mustard seeds), allyl isothiocyanate, glucose and potassium hydrogen sulphate are obtained.

Allyl isothiocyanate is a colourless oil, b.p. 151°C, and is the substance which gives mustard its characteristic odour and taste. It is lachrymatory and vesicatory, and is a convenient starting material for the preparation of allylamine.

*Thiocyanogen,* $(SCN)_2$, may be prepared by treating lead thiocyanate with bromine in ethereal solution at 0°C.

$$Pb(SCN)_2 + Br_2 \longrightarrow MSCN)_2 + PbBr_2$$

It is a gas and resembles the halogens in that it adds on to-double bonds *(thiocyanation)* to form dithiocyanates.

*Alkyl sulphoxides (alkylsulphinylalkanes)* may be prepared by oxidising alkyl sulphides with the *theoretical* amount of hydrogen peroxide (in acetic acid) or with *dilute* nitric acid.

$$R_2S + [O] \longrightarrow R_2S = O$$

The sulphoxides are odourless, relatively unstable solids, soluble in water, ethanol and ether, and are feebly basic, *e.g.,* they form salts with hydrochloric acid, $[R_2S - OH]^+Cl^-$. Sulphoxides are reduced to sulphides by zinc and acetic acid or by LAH.

The infra-red absorption region of the sulphoxide group (S = O stretch) is $1070\text{–}1030 cm^{-1}$(s).

*Dimethyl sulphoxide,* DMSO, b.p. 188°C, is prepared by the oxidation of dimethyl sulphide by air. It is an extremely valuable solvent and may be used as an oxidising agent for certain compounds, *e.g.,* phenacyl bromide. A solution of sodium dimethyl sulphoxide appears to be very useful for the estimation of all forms of active hydrogen. Titration is carried out in the presence of triphenylmethane as indicator; the blood-red colour of the triphenyl-methyl anion is formed immediately in the presence of excess of the reagent.

$$MeSOCH_3 \xrightarrow{NaH} MeSOCH_2^{-\cdot} \xrightarrow{Ph_3CH} MeSOCH_3 + Ph_3C^-$$

*Alkyl sulphones* may be prepared by oxidising alkyl sulphides with hydrogen peroxide *in excess* (in acetic acid) or with *concentrated* nitric acid.

$$R_2S + 2[O] \longrightarrow R_2SO_2$$

They are colourless, odourless, very stable solids, soluble in water, they absorb in the region 1160-M20cm$^{-1}$ (vs). They are very resistant to reduction, but some are reduced to sulphide by lithium aluminium hydride. The a-hydrogen atom in sulphones is very reactive *(cf.* the active methylene group) and so sulphones can enter into the Claisen-type of condensation. Many sulphones produce sulphinic acids when fused with potassium hydroxide at 200°C, *e.g.,*

$$(C_2H_5)_2SO_2 + KOH \longrightarrow C_2H_4 + C_2H_5SO_2K + H_2O \quad (60\%)$$

Aliphatic sulphinic acids are viscous oils unstable to heat, but their salts are stable.

*Sulphenic acids* (RSOH) are known only as their derivatives, *e.g.,* sulphenyl halides, RSX.

*Sulphens,* $R^1R^2C=SO_2$. These may be generated by various methods, but because of their reactivity, they polymerise *(cf.* ketens), or they may be trapped. Conclusive evidence for the intermediate formation of sulphens has been obtained by Truce *et al.,* (1964).

$$CH_3SO_3Me \xleftarrow{MeOD} CH_3SO_2Cl + Et_3N \longrightarrow Et_3N.HCl$$

$$+ CH_2=SO_2 \xrightarrow{MeOD} CH_2DSO_3Me$$

The methyl methanesulphonate was shown, by NMR spectroscopy and mass spectrometry, to be a mixture of monodeuterated and undeuterated ester.

*Sulphines,* $R^1R^2C = S = O$, have been obtained, *e.g.*, by the action of triethylamine on sulphinyl chlorides, RSOCl. Sulphines are stable enough to be isolated.

*Sulphonic acids :* The aliphatic sulphonic acids are named either as alkylsulphonic acids or as alkanesulphonic acids; in the latter case the sulphonic acid group is considered as a substituent group, *e.g.*,

$CH_3SO_3H$ methylsulphonic acid or methanesulphonic acid

$(CH_3)_2CHSO_3H$ isopropylsulphonic acid or propane-2-sulphonic acid

**General Methods of Preparation**

1. By the action of sulphuric acid, oleum, or chlorosulphonic acid on an alkane. The sulphonic acid group enters the chain, probably at the second carbon atom, *e.g.*,

$$CH_3(CH_2)_4CH_3 + SO_3 \xrightarrow{H_2SO_4} CH_3(CH_2)_3CH(SO_3H)CH_3 \quad \text{(p.-f.)}$$

2. By the action of sulphuryl chloride, or a mixture of chlorine and sulphur dioxide, on a hydrocarbon in the presence of light or peroxide at 40–60°C. The sulphonyl chloride is obtained, often in high yield; the sulphonyl chloride group appears to enter mainly at the second carbon atom, but varying amounts of product with this group at the first carbon are also obtained.

$$RH + SO_2Cl_2 \longrightarrow RSO_2Cl + HCl$$

3. By the oxidation of a thiol with nitric acid, permanganate, sodium hypobromite, etc., *e.g.*,

$$RSH + 3[O] \xrightarrow{HNO_3} RSO_3H \quad \textit{(g.-v.g.)}$$

4. The strecker reaction (1868). This is carried out by heating an alkyl halide with sodium sulphite.

$$RX + Na_2SO_3 \longrightarrow RSO_3Na + NaX \quad \textit{(g.-v.g.)}$$

5. Sodium hydrogen sulphite adds on to alkenes in the presence of peroxides to form sulphonic acids (Kharasch *et al.,* 1939).

$$RCH = CH_2 + NaHSO_3 \longrightarrow RCH_2CH_2SO_3Na \text{ (ex).}$$

6. By the oxidation of S-alkylisothiouronium salts.

**General Properties and Reactions**

The sulphonic acids are generally thick liquids, soluble in water. They are isomeric with the alkyl hydrogen sulphites.

$$\underset{\text{alkanesulphonic acid}}{ROSO_2OH} \qquad \underset{\text{alkyl hydrogen sulphite}}{ROSOOH}$$

Sulphonic acids have two absorption regions (S = O stretch), 1260–1150 (s) and 1080 – 1010 $cm^{-1}$(s), whereas sulphites (S–O stretch) have only one absorption region, 1220 – 1170 $cm^{-1}$ (s)

The sulphonic acids are strong acids, forming salts with metallic hydroxides or carbonates, the lead and barium salts are very soluble in water. The form the acid chloride, the sulphonyl chloride, when treated with phosphorus pentachloride.

$$RSO_3H + PCl_5 \longrightarrow RSO_2Cl + HCl + POCl_3$$

These sulphonyl chlorides are only very slowly hydrolysed by water (cf. acyl chlorides), they react readily with concentrated aqueous ammonia to form *sulphonamides,* and with alkoxides to form esters.

$$RSO_2NH_2 \xleftarrow{NH_3} RSO_2Cl \xrightarrow{RONa} RSO_3R$$

Sulphonyl chlorides are reduced by lithium aluminium hydride to thiols (Marvel *et al.,* 1950).

Thioacids may be prepared by the action of phosphorus pentasulphide on a carboxylic acid.

$$5RCO_2H + P_2S_5 \longrightarrow 5RCOSH + P_2O_5$$

The thioacids have a most disagreeable odour, and slowly decompose in air. In most of their reactions, the thioacids and their salts behave as if they contained the mercapto-group, but in a few reactions they behave as if they contained a hydroxyl group. This may be accounted for by tautomerism for the acid and resonance for the salts.

*ithioacid* $R{-}C({=}O){-}S{-}H \rightleftharpoons R{-}C(=S){-}O{-}H$

*salts* $R{-}C(=O){-}S{:}^- \longleftrightarrow R{-}C(=S){-}O^-$

**Nomenclature**

The methods of nomenclature are illustrated by the following example.

$$\underset{(I)}{CH_3COSH} \quad \underset{(II)}{CH_3CSOH}$$

According to the trivial system of nomenclature, both (I) and (II) are named as thioacetic acid. According to the I.U.P.A.C. rules, the suffixoic of the corresponding oxygen acid is changed to thiolic. Thus both (I) and (II) are named either as ethanethioic acid or methanecarbothioic acid. On the other hand, the suffixthiolic is used if it is certain that the oxygen of the hydroxyl group is replaced by sulphur, and the suffix thionic if it is the oxygen of the carbonyl group. Thus (I) is ethanethiolic acid or methane-carbothiolic acid, and (II) is ethanethionic or methanecarbothionic acid.

The most characteristic reaction of the thioacids is their extreme readiness -to acylate alcohols and amines.

$$R^1COOR^2 \xleftarrow[(-H_2S)]{R^2OH} R^1COSH \xrightarrow[(-H_2S)]{R^2NH_2} R^1CONHR^2$$

*Dithioacids* may be prepared by the action of a Grignard reagent on carbon disulphide.

$$RMgX + S{=}C{=}S \longrightarrow R{-}C(=S){-}SMgX \xrightarrow{\text{acid}} RCS_2H$$

**Nomenclature**

The methods of nomenclature are illustrated by the following example $CH_3CS_2H$ is named as dithioacetic acid, ethanethionthiolic acid or methanecarbodithioic acid. A very important dithioacid is *dithiocarbonic acid,* $HOCS_2H$. The free acid is unknown, but many of its derivatives have been prepared, *e.g., potassium xanthate* may be prepared by the reaction between potassium hydroxide, ethanol and carbon disulphide.

$$EtOH \xrightarrow{KOH} EtO^-K^+ \xrightarrow{CS_2} EtOC(=S)S^-K^+$$

*Thiocarbonyl compounds :* Aliphatic thioaldehydes *(thials)* are unknown; attempts to prepare them result in polymers. On the other hand, some thioketones *(thiones)* have been isolated as monomers.

The reason for this instability is believed to be due to the difficulty with which sulphur forms a $\pi_{d-p}$ bond with carbon (overlapping of the $S3d_\pi$, with the $C2p_\pi$ orbital is very difficult), and the result is ready polymerisation (to dimer, trimer, etc.), *e.g.* (6-ring).

$$R_2C{=}S \quad CR_2{=}S \quad CR_2{=}S \longrightarrow R_2C{-}C{-}S{-}CR_2{-}S{-}CR_2{-}S$$

*Vulcanisation of rubber.* This process is carried out by heating crude rubber with 4-5 per cent sulphur and certain organic compounds which accelerate the reaction between the rubber and sulphur. These organic compounds are known as *accelerators,* and all contain sulphur or nitrogen, or both. Vulcanising rubber causes the rubber to lose its stickiness, makes it no longer sensitive to temperature changes, causes it to retain its elasticity over a wide temperature range, and increases its tensile strength. The function of the sulphur appears to be to cross-link the long hydrocarbon chains in crude rubber.

## PHOSPHORUS COMPOUNDS

Both nitrogen and phosphorus occur in Group V, and their valence shells are similar. N, $(2s)^2$ $(2p)^3$, P, $(3s)^2$ $(3p)^3$. The two elements, however, differ in some important respects. Nitrogen exhibits a maximum covalency of 4 (there are no available d-orbitals), whereas phosphorus can exhibit a covalency of 5, *e.g.,* $PCl_5$ ($sp^3d$; trigonal bipyramid), and 6, *e.g.,* $PF_6^-$ ($sp^3d^2$; octahedron).

The majority of P compounds are $sp^3$ hybridised, $PX_4^+$ (*cf.* N), but although both $PX_3$ and $NH_3$ are pyramidal structures, the angle in P compounds is less than that in N compounds, *e.g.,* ∠HPH ($PH_3$) is 93°; ∠HNH ($NH_3$) is 107°. It therefore appears that the lone pair ($3s^2$) is less involved in tricovalent P compounds than the lone pair ($2s^2$) in the corresponding N compounds.

### Nomenclature of Organophosphorus Compounds

There is a very large variety of organophosphorus compounds; som are discussed here. For compounds containing *one* phosphorus atom, three parent structures have been used (whether they have an independent existence or not. *N.B.* In the following structures, *PO is P=O).*

| **Hydrides** | *Examples* |
|---|---|
| Phosphine, $H_3P$ | $Me_2PH$, dimethylphosphine |
| Phosphine oxide, $H_3PO$ | $Et_3PO$, triethylphosphine oxide |
| Phosphorane, $H_5P$ | $Ph_5P$, pentaphenylphosphorus |
| **Tervalent acids** | *Examples* |
| Phosphorous' acid, $H_3PO_3$ | $(EtO)_3P$, triethyl phosphite |
| Phosphonous acid, $HP(OH)_2$ | HP(OH)OMe, methyl hydrogen phosphonite |
| Phosphinous acid, $H_2POH$ | $H_2POEt$, ethyl phosphinite. |

Structures with carbon-phosphorus bonds formed by replacement of H joined to P by an alkyl, aryl, of a heterocyclic group, are named by prefixing to the parent name the name of the replacing group. Compounds with P—X or P—N bonds are named respectively as acid halides or amides if *all* OH groups have been replaced. If, however, at least one OH group is still present, then affixes are used and the name acid is retained, *e.g.*,

| **Parent** | *Examples* |
|---|---|
| $HP(OH)_2$ | $MePCl_2$, methylphosphonous dichloride |
| | $HP(OH)NH_2$, phosphonamidous acid |
| $(HO)_3P$ | $HOPCl_2$, phosphorodichloridous acid |
| | $P(NH_2)_3$, phosphorous triamide |
| **Quinquevalent acids** | *Examples* |
| Phosphoric acid, $(HO)_3PO$ | $(EtO)_3PO$, triethyl phosphate |
| | $HOPO(NH_2)_2$, phosphorodiamidic acid |
| | $(HO)_2PO(Cl)$, phosphorochloridic acid |
| Phosphonic acid, $HPO(OH)_2$ | $HPO(Cl_2)$, phosphonic dichloride |
| | $MePO(OH)NH_2$, methylphosphonamidic acid |
| | MePO(Cl)(NH2), methylphosphonamidic chloride |
| Phosphinic acid, $H_2PO(OH)$ | MeHPO(Cl), methylphosphinic chloride $H_2PO(OEt)$, ethyl phosphinate |

## Preparation and Reactions of Some Organophosphorus Compounds

Since the phosphorus atom in tercovalent compounds has a lone pair of electrons, these compounds behave as nucleophiles in many reactions. It is because of this lone pair that phosphines are good reducing agents. Also, the C—P bond is very strong, and is resistant to oxidation and hydrolysis.

A number of organophosphorus compounds are used as nerve gases, insecticides and fungicides.

## Phosphines and Phosphonium Salts

Trialkylphosphines are prepared most conveniently by the action of excess of Grignard reagent on phosphorus trihalide.

$$3RMgBr + PCl_3 \longrightarrow R_3P + 3MgClBr \quad (g.\ -v.g.)$$

This method may also be used to prepare triarylphosphines (from ArMgBr), and another useful method is.

$$3ArBr + PCl_3 + 6Na \longrightarrow Ar_3P + 3NaCl + 3NaBr$$

The three classes of phosphines may be prepared .by interaction of alkyl halide and metal phosphides or metal hydrogen phosphides, *e.g.*,

$$3Na \xrightarrow[\text{liq.}NH_3]{PH_3} Na_3P \xrightarrow{3RX} R_3P + 3NaX$$

$$Na \xrightarrow[\text{liq.}NH_3]{PH_3} NaPH_2 \xrightarrow{RX} RPH_2 + NaX$$

$$Na \xrightarrow[\text{liq.}NH_3]{R^1PH_2} NaPHR^1 \xrightarrow{R^2X} R^1R^2PH + NaX$$

Mono and di-alkylphosphines are formed when an alkyl halide is heated with phosphonium iodide in the presence of zinc oxide (to neutralise the HI formed in the reaction).

$$PH_4^+ I^- + EtI + ZnO \longrightarrow EtPH_2 + ZnI_2 + H_2O$$
$$2EtPH_2 + 2EtI + ZnO \longrightarrow Et_2PH + ZnI_2 + H_2O$$

**General Properties and Reactions**

Except for methylphosphine (a gas), all the alkylphosphines are colourless, unpleasant-smelling liquids or low-melting solids. One of their most characteristic reactions is the ease with which they combine with atmospheric oxygen, often undergoing spontaneous combustion in the process. On the other hand, controlled oxidation is readily carried out with nitric acid.

$$EtPH_2 \xrightarrow{HNO_3} EtPO(OH)_2 \quad \text{ethylphosphonic acid}$$

$$Et_2PH \xrightarrow{HNO_3} Et_2PO(OH) \quad \text{diethylphosphinic acid}$$

$$Et_3P \xrightarrow{HNO_3} Et_3PO \quad \text{triethylphosphine oxide}$$

t-Phosphines are also oxidised to phosphine oxide by ozone.

$$R_3P + O_3 \longrightarrow R_3PO + O_2$$

All the alkylphosphines are reducing agents, but trialkylphosphines appear to be the most useful, they react with a variety of oxygen-containing compounds.

$$R_3P + ZO \longrightarrow R_3PO + Z$$

Thus, *e.g.*, triphenylphosphine reduces epoxides to alkenes and amine oxides to amines. The major driving force in these reactions is the unusually large strength of the P = O bond (502.1.627.6kJ $mol^{-1}$).

Phosphines are weaker bases than the corresponding amines, and the order of basicity in phosphines is t > s > prim. This is not quite parallel with that of the amines, and a possible explanation is that solvation effects are less important in phosphines than in amines because of the larger size of the phosphorus atom (and consequently the charge is more diffuse than in amines).

Trialkylphosphines readily form phosphonium salts with alkyl halides (preferably iodides); the mechanism is $S_N2$

$$R^1_3P: \; R_2—X \longrightarrow R^1_3PR2X^-$$

(*cf.* the Menshutkin reaction). t-Phosphines are stronger nucleophiles than t-amines; this is illustrated by the fact that aminophosphines combine with alkyl halides to give phosphonium salts and not ammonium salts.

$$R^1_2NPR^1_2 + R^2X \longrightarrow R^1_3N\overset{+}{P}R^1_2R^2X^-$$

This greater nucleophilicity has been explained on the basis that the lone pair of electrons of phosphorus has a greater polarisability than that of nitrogen, due to the larger size of the phosphorus atom. However, according to Henderson *et al.*, (1960), this greater nucleophilicity, *i.e.*, greater rate of reaction, is accounted for by the entropy of activation term in the two reactions.

When phosphonium halides are treated with moist silver oxide, the phosphonium hydroxide is produced.

$$(C_2H_5)_4P^+I^- + \text{'AgOH'} \longrightarrow (C_2H_5)_4P^+OH^- + AgI$$

The quaternary phosphonium hydroxides are strongly basic, comparable in strength with the quaternary ammonium hydroxides and sodium hydroxide. When heated they form the trialkylphosphine oxide and a hydrocarbon (cf $R_4NOH$).

$$R_4P^+OH^- \longrightarrow R_3PO + RH$$

Some alkene, however, is formed when R contains 0-hydrogen (with respect to the phosphorus).

Pentaphenylphosphorus (m.p. 124°C) has been prepared, and by means of X-ray analysis and dipole moment measurements, has been shown to have a trigonal bipyramidal structure.

*Phosphine oxides :* Phosphine oxides may be prepared by oxidation of a trialkylphosphine with, *e.g.,* nitric acid or hydrogen peroxide; by heating a phosphonium hydroxide (see above); or by reaction between a Grignard reagent and phosphoryl chloride.

$$3RMgBr + POCl_3 \longrightarrow R_3PO + 3MgBrCl$$

Tertiary phosphine oxides are usually solids, and form one of the most stable groups of organophosphorus compounds.

Primary ($RH_2PO$) and secondary ($R_2HPO$) phosphine oxides are also known, and may be prepared by oxidation of the corresponding phosphine with hydrogen peroxide.

*Phosphorus ylids : Phosphoranes* may be prepared from phosphonium halides by reaction with a strong base. Triphenylphosphine is usually the starting point; it is commercially available because of its stability in air; *e.g.,* methylenetriphenylphosphorane.

$$Ph_3P + CH_3Br \longrightarrow Ph_3PC\,CH_3^+\,Br^- \xrightarrow{PhLi} Ph_3P$$

$$= CH_2 \leftrightarrow Ph_3\overset{+}{P}-CH_2^- + PhH + LiBr$$

In general, the bases commonly used are butyl- and phenyl-lithium in ether or tetrahydrofuran, sodium or lithium ethoxide in ethanol or dimethylformamide, and dimethyl sulphoxide and sodium hydride. The last reagent is the best.

These alkylidenephosphoranes are ylids, and because they readily react with oxygen (see later), reactions involving them are carried out in an atmosphere of nitrogen. Their reaction with carbonyl compounds leads to alkenes, and this alkene synthesis is one example of the *Wittig reaction* (1953, 1956).

The mechanism of the Wittig reaction has been the subject of much discussion, but the evidence is now strongly in favour of the formation of an intermediate betaine, followed by ring closure and then fission.

$$Ph_3P{=}CHR^1 \quad \underset{\underset{O}{\|}}{CHR^2} \longrightarrow \begin{array}{l} Ph_3P{-}CHR^1 \\ \qquad\quad | \\ {}^-O{-}CHR^2 \end{array} \longrightarrow$$

$$\begin{array}{l} Ph_3P{-}CHR^1 \\ \;\;|\qquad\;\; | \\ \;O{-}CHR^2 \end{array} \longrightarrow \begin{array}{l} Ph_3P \quad CHR^1 \\ \;\;\| \;\;+\;\; \| \\ \;\;O \qquad CHR^2 \end{array}$$

Support for this mechanism comes from the observation that the entropy of activation is negative. This is in keeping with the formation of a cyclic intermediate. Furthermore, cyclic intermediates have been isolated in certain cases.

The reactivity of these ylids is associated with the carbanionic centre in the dipolar structure, and is decreased when the negative charge is delocalised. Delocalisation of charge may be brought about through resonance due to the presence of a suitable electron-attracting group (or groups) attached to the methylene carbon atom, *e.g.*,

$$Ph_3P{=}CH-CH{=}O \leftrightarrow Ph_3\overset{+}{P}-\overset{-}{C}H-CH{=}O \leftrightarrow Ph_3\overset{+}{P}-CH{=}CH-\overset{-}{O}$$

These ylids react readily with aldehydes, but react slowly (if at all) with ketones. In some cases, the ylid is so stable that it does not react even with aldehydes. Because of this variation in stability, the rate-determining step in the Wittig reaction may be either the formation of the betaine (for stable ylids) or betaine decomposition (for reactive ylids).

Since in the formation of the ylid only *one* a-hydrogen is lost, the structure of the ylid is definite and so is the structure of the alkene produced.

When the alkene product is capable of geometrical isomerism, it has been found that the *trans* isomer is favoured by non-polar solvents and the *cis* isomer by highly polar solvents. Thus, for the solvents in which the order of polarity is DMF > EtOH > THF > $Et_2O$ > PhH, the yield of *trans* isomer increases from right to left, and left to right for the *cis* isomer.

Alternatively, by carrying out the reaction in the presence of a Lewis base, *e.g.,* amines, $I^-$, there is a very large predominance of the *cis* isomer, *e.g.,* Shemyakin *et al.,* (1963) carried out the following stereospecific reactions.

$$Ph_3P{=}CHPh + EtCHO \longrightarrow \underset{\substack{96\% \\ (DMF + LiI)}}{(Ph)(H)C{=}C(Et)(H)} \qquad \underset{\substack{100\% \\ (PhH)}}{(Ph)(H)C{=}C(H)(Et)}$$

The explanation for these results is still the subject of discussion.

Many alkylidenephosphoranes are hydrolysed by water, whereas others require heating in alkaline solution, *e.g.*,

$$Ph_3P = CHMe \xrightarrow{H_2O} Ph_3\overset{+}{P}EtOH^- \longrightarrow Ph_3PO + PhH$$

These ylids are also oxidised by a small amount of oxygen (or air) to alkenes, but with excess of oxygen the carbonyl compound is obtained.

$$PhP{=}CR_2 \xrightarrow{O_2} Ph_3\underset{|}{P}{-}\underset{|}{C}R_2 \longrightarrow Ph_2PO + R_2CO$$
$$\qquad\qquad O{-}O$$

$$\xrightarrow{Ph_3P=CR_2} Ph_3PO + R_2C{=}CR_2$$

With *excess* of oxygen, all of the ylid is oxidised and consequently the second step cannot occur.

Ylids containing groups other than alkylidene may also be used in the Wittig reaction. Some examples of their preparation and uses are.

(i) Dichloromethylenetriphenylphosphorane.

(ii) Carbonyl compounds may be stepped up to aldehydes containing one more carbon atom.

$$Ph_3P + ClCH2OMe \longrightarrow [Ph_3\overset{+}{P}CH_2OMe]Cl^- \xrightarrow{EtONa}$$

$$Ph_3P = CHOMe \xrightarrow{PhCHO} Ph3PO + PhCH$$
$$= CHOMe \xrightarrow{H^-} [PhCH = CHOH] \longrightarrow PhCH_2CHO$$

(iii) Keto-ylids may be produced as follows.

$$Ph_3P = CH_2 \xrightarrow{RCOCl} [Ph_3\overset{+}{P}CH_2COR]Cl^-$$

$$\xrightarrow{Ph_3P=CH_2} Ph_3P = CHCOR + [Ph_3PMe]^+Cl^-$$

*Alkylphosphonous dihalides*, $RPX_2$, and *dialkylphosphinous halides*, *$R_2PX$.* Phosphonous dichlorides may be prepared as follows.

$$R_2Cd + 2PI_3 \longrightarrow 2RPCl_2 + CdCl_2$$
$$R_3Al + 3PCl_3 \longrightarrow 3RPCl_2 + AlCl_3$$

The phosphinous chlorides may be prepared by partial alkylation of phosphorus trichloride with lead tetra-alkyls. The chlorine in both types of halides is very reactive (these halides are acid chlorides), *e.g.*,

$$RPCl_2 + 2H_2O \longrightarrow RHPO(OH) + 2HCl$$

$$R_2PCl + H_2O \longrightarrow R_2HPO + HCl$$

The alkylphosphonous dihalides and dialkylphosphinous halides readily combine with, chlorine to form chlorophosphoranes, and oxygen also adds on to give the phosphonic dichloride or phosphinic chloride, *e.g.*,

$$RPCl_4 \xleftarrow{Cl_2} RPCl_2 \xrightarrow{O_2} RPOCl_2$$

*Alkylphosphonic dihalides,* $RPOX_2$, and *dialkylphosphinic halides,* $R_2POX$. These may be prepared as described above, but a better method is to treat the corresponding halogenophosphorane with sulphur dioxide, *e.g.*,

$$R_2PCl + Cl_2 \longrightarrow R_2PCl_3 \xrightarrow{SO_2} R_2POCl + SOCl_2$$

Alternatively, phosphonic and phosphinic acids or their esters give the corresponding chloride when treated with thionyl chloride or phosphorus pentachloride, *e.g.*,

$$R^1PO(OR^2)_2 + 2PCl_5 \longrightarrow R^1POCl_2 + 2POCl_3 + 2R^2Cl$$

$$R_2PO(OH) + SOCl_2 \longrightarrow R_2POCl + SO_2 + HCl$$

Both types of chlorides react with water *(i.e.,* behave as acid chlorides), *e.g.*,

$$RPOCl_2 + 2H_2O \longrightarrow RPO(OH)_2 + 2HCl$$

*Esters of phosphorous acid :* These are readily prepared by the reaction between an alcohol and phosphorus trichloride in the presence of a tertiary base.

$$3ROH + PCl_3 \xrightarrow{3PhNMe_2} (RO)_3P + 3PhNMe_2.HCl$$

In the *absence* of a base, the phosphite is attacked by the hydrogen chloride produced in the reaction to give alkyl chlorides.

Phosphites usually oxidise in the air and are readily oxidised by ozone to phosphates (see later). They are readily hydrolysed by dilute acids, but only slowly by dilute alkali.

Triphenyl phosphite reacts with alcohols and halogen to produce alkyl halide in good yield (Rydon *et al.,* 1954).

$$(PhO)_3P + ROH + X_2 \longrightarrow RX + XPO(OPh)_2 + PhOH$$

Instead of halogen, a reactive halide may be used, *e.g.,* methyl iodide, benzyl chloride or bromide. The reaction does not take place through a carbonium-ion intermediate, since no rearrangement occurs (see also the Arbusov reaction, below), *e.g.*, neopentyl alcohol is converted into neopentyl bromide.

$$(PhO)_3P + Me_3CCH_2OH + PhCH_2Br \longrightarrow Me_3CCH_2Br + PhCH_2PO(OPh)_2 + PhOH$$

An interesting point about phosphites is their structures. Free phosphorous acid is (I), which is

```
HO     O          HO
  \   //            \
    P                 P—OH
  /   \             /
HO     H          HO
   (I)               (II)
```

phosphonic acid. Mono- and di-alkyl phosphites have structures derived from (I), *i.e.,* they are phosphonic esters. It has been shown, however, that dialkylphosphonates, (Ia), are in equilibrium with the dialkyl phosphites, (IIa), and that the equilibrium lies almost entirely on the phosphonate side.

```
RO     H          RO
  \   /             \
    P                 P—OH
  /   \\            /
RO     O          RO
   (Ia)              (IIa)
```

A study of the ΔH values (obtained from bond energies) has shown that conversion of phosphonate to phosphite is an endothermic reaction, and so (Ia) $\longrightarrow$ (IIa) is thermodynamically unfavourable. Thus, the phosphoryl form (Ia) is the more stable one.

*Esters of phosphonous and phosphinous acids :* Esters of the 'alkyl-acid' may be prepared by reaction between an alkylphosphonous dichloride or dialkylphosphinous chloride and an alcohol in the presence of a tertiary base, *e.g.*,

$$R^1_2PCl + R_2OH + C_5H_5N \longrightarrow R^1_2POR_2 + C_5H_5N.HCl$$

*Esters of phosphoric acid :* As with phosphites, the preparation of alkyl phosphates from alcohol and phosphoryl chloride requires the presence of a tertiary base. Also, as with phosphites, phosphates are readily hydrolysed by dilute acids, but only very slowly (if at all) by dilute alkali.

$$3ROH + POCl_3 + 3C_5H_5N \longrightarrow (RO)_3PO + 3C_5H_5N.HCl.$$

Phosphate esters are very important biologically, and most naturally occurring phosphorus compounds contain a terminal unsubstituted–$PO(OH)_2$ group. Introduction of this group into a molecule is known as *phosphorylation.* One important method uses a protected phosphorylating agent, *e.g.*,

$$(PhO)_2POCl + ROH + C_5H_5N \longrightarrow C_5H_5N.HCl + (PhO)_2PO(OR)$$

$$\xrightarrow{H_2-Pd} (HO)_2PO(OR) + 2PhMe.$$

If the protected group is not used, then polymers containing the P–O–P linkage are obtained.

*Esters of phosphonic and phosphinic acids :* Alkylphosphonates and dialkylphosphinates may be prepared as follows.

$$R^1PO(Cl_2) + 2R^2ONa \longrightarrow R^1PO(OR^2)_2 + 2NaCl$$

$$R^1_2POCl + R^2ONa \longrightarrow R^1_2PO(OR^2) + NaCl$$

Their most useful method of preparation is the *Arbusov reaction* (1906). This is a good means of forming carbon-phosphorus bonds, and in its simplest form, is the reaction between a trialkyl phosphite and an alkyl halide to form a dialkyl alkylphosphonate.

$$P(OR^1)_3 + R^2X \longrightarrow R^1PO(OR^1)_2 + R^1X$$

The Arbusov reaction also includes the formation of alkyl dialkylphosphinates from dialkyl alkylphosphonites.

$$R^1P(OR^2)_2 + R^3X \longrightarrow R^1\ R^3PO(OR^2) + R^2X$$

The Arbusov reaction occurs in *two* stages, each by the $S_N2$ mechanism, and the reaction works best with primary alkyl halides, *e.g.*,

$$X\text{—}R_2\quad P(PR^1)_3 \longrightarrow R_2\text{—}P(OR^1)_2\text{—}O\text{—}R^1\quad X^- \longrightarrow$$

$$R_2PO(OR^1)_2 + R^1X$$

It should be noted that the product of the Arbusov reaction is a phosphonate *(i.e.,* the phosphoryl form) which, as we have seen above, is more stable than the phosphite form.

**Radical Reactions at Phosphorus**

The majority of reactions undergone by phosphorus compounds are heterolytic, but free-radical reactions may be initiated in the presence of, *e.g.*, acyl peroxides or by means of irradiation with ultraviolet light. Most free-radical reactions involve tervalent phosphorus compounds;

these give a seven-electron species, the *phosphino* free radicals. On the other hand, there is also the less common nine-electron species, the *phosphoranyl* free radicals. Examples of their formation are (where $R^2$. is generated from an acyl peroxide).

$$R^1_2PX + R^2 \longrightarrow R^1_2P. + R^2X$$

$$R^1_3P + R^2. \longrightarrow R^2R^1_3P$$

$$R_2P - PR_2 \xrightarrow{hv} 2R_2P.$$

*Oxidation :* t-Phosphines are oxidised by oxygen to phosphine oxide and phosphinate (together with small amounts of other products). According to Buckler (1962) the mechanism is.

$$R. + O_2 \longrightarrow RO_2. \xrightarrow{R_3P} R_3PO + RO. \xrightarrow{R_3P} R_3\dot{P}OR$$

$$\longrightarrow R_2POR + R. , etc.$$

Ttialkyl phosphites are usually oxidised by oxygen to phosphates very slowly, but in the presence of acyl peroxides or ultraviolet light, the oxidation is fast. Plumb *et al.*, (1963) have proposed the following mechanism.

$$R. + O_2 \longrightarrow RO_2. \xrightarrow{(RO)_3P} (RO)_3PO + RO. \xrightarrow{(RO)_3P}$$

$$(RO)4_P. \longrightarrow (RO)_3PO + R., etc$$

t-Phosphines and trialkyl phosphites are also oxidised by di-t-bytyl peroxide ($R^1$ = t-Bu), *e.g.*,

$$R^1 - O - O - R^1 \longrightarrow 2R^1O.$$

$$R^2_3P + R^1O. \longrightarrow R^2_3\dot{P}OR^1 \longrightarrow R^2_3PO + R^1., etc.$$

*Addition to alkenes :* Phosphines compounds containing a P – H bond add to alkenes (anti Markownikff) in the presence of acyl peroxides.

$$R^1_2PH + R_2 \longrightarrow R_2H + R^1_2P \xrightarrow{R^3CH=CH_2} R^1_2PCH_2CHR^3.$$

$$\xrightarrow{R^1_2PH} R^1_2PCH_2CH_2R^3 + R^1_2P., etc.$$

**Reactions with Thiyl Radicals**

Phosphines and trialkyl phosphites react with thiols in the presence of acyl peroxides (or on irradiation with ultraviolet light) to form phosphorus sulphides, *e.g.*,

$$R_2SH + R^1. \longrightarrow R^2S.R^1H$$

$$(R_3O)_3P \xrightarrow{R^2S.} (R^3O)_3\dot{P}SR^2 \longrightarrow (R^3O)_3PS + R^2., \text{ etc.}$$

**Reaction with Polyhalogenoalkanes**

t-Phosphines and trialkyl phosphites react with, *e.g.*, carbon tetrachloride in the presence of acyl peroxides.

$$CCl_4 + R^2. \longrightarrow R^2Cl + .Cl_3$$

$$(R^1O)_3P + .CCl_3 \longrightarrow (R^1O_3)\dot{P}CCl_3 \xrightarrow{CCl_4} .CCl_3 + (R^1O)_3$$

$$\overset{+}{C}Cl_3Cl^- \longrightarrow (R1O)_2PO(CCl_3) + R^1Cl$$

**Silicon Compounds**

Both carbon and silicon are Group IV elements, and their valence electrons are similar. C, $2s^2$, $2p^2$, Si, $3s^2$, $3p^2$. Hencer it can be anticipated that $sp^3$ hybridisation would operate in quadrivalent silicon compounds (as in quadrivalent carbon compounds), *e.g.*, X-rays analysis of $Me_4Si$ shows a tetrahedral arrangement. However, the chemical properties of these two elements differ in some respects. One reason is the difference in bond strengths for the same ligand, *e.g.*, some bond-strength orders are. Si – H < C – H, Si – O > C – O, so the free energies will be different. Silicon, unlike carbon, does not form multiple bonds such as occur in acetylene (sp), ethylene ($sp^2$), carbonyl compounds ($sp^2$), etc. The reason for this not certain, but as far as $p_\pi - p_\pi$ bonding is concerned, *e.g.*, in Si = O, the two orbitals have very different energies, and consequently there is very little overlap. Connected with this is the fact that silicon free radicals, *e.g.*, $Ph_3Si$, are much reactive than the corresponding carbon compounds, Ph3C. The latter are stabilised by resonance involving C = C bonds, there is no resonance stabilisation in the silicon compounds since the Si = C bond is absent. However, silicon can form multiple bonds, particularly with strongly electronegative ligands. In these compounds silicon has a covalency of five, and this is achieved by overlap of a lone pair ($2p^2$) of the donor atom with the empty 3d orbitals of silicon to form a $d_\pi - d_\pi$ bond. A consequence of the existence of five-co-ordinate silicon ($sp^3d$), for which there is no carbon analogue, is *e.g.*, the difference in acidity between $Me_3SiOH$ and $Me_3COH$. Carbon is more electronegative than silicon, and therefore the carbinol would be expected to be ore acidic than the silanol. In actual fact the reverse is true, and this may be explained on the basis that the (p → d) π-bond in the silanol reduces the negative charge on the oxygen atom, thereby increasing the acidity.

$$Me_3Si \leftleftharpoons OH \text{ or } Me_3Si—\ddot{O}H \longleftrightarrow Me\bar{S}i = \overset{+}{O}H$$

**Nomenclature of Organosilicon Compounds**

The following examples illustrate the classification and nomenclature of silicon compounds.

*Silanes* is the generic name of compounds of the general formula Sin $H_{2n+2}$, and prefixes di, tri, etc, indicate the number of silicon atoms present.

$$\underset{\text{silanes}}{Si_nH_{2n+2}} \quad \underset{\text{silane}}{SiH_4} \quad \underset{\text{disilane}}{H_3Si-SiH_3} \quad \underset{\text{trisilane}}{H_3SiSiH_2SiH_3}$$

Various substituted silanes are named by use of the appropriate prefixes or suffixes, *e.g.*,

$$\underset{\text{ethysilane}}{EtSiH_3} \quad \underset{\text{bromosilane}}{BrSiH_3} \quad \underset{\text{silana min e}}{H_3SiNH_2} \quad \underset{\text{silanedia min e}}{H_2Si(NH_2)_2}$$

$$\underset{\text{methylsilanol}}{MeSiH_2OH} \quad \underset{\text{disilane-1,2-diol}}{HOSiH_2SiH_2OH}$$

Compounds containing the Si – O– Si group are **siloxanes,** and those with the Si – N – Si group. *silazanes*, *e.g.*,

$$\underset{\text{methoxydisiloxane}}{MeOSiH_2-O-SiH_3} \quad \underset{\text{5-chlorotrisilazonal}}{\overset{5}{Cl}\overset{4}{Si}H_2\overset{3}{N}HSiH_2\overset{2}{N}H\overset{1}{Si}H_2OH}$$

*Alkylsilanes :* Many methods are available for the preparation of alkylsilanes, *e.g.*,

(i) $SiCl_4 + 4MeMgI \xrightarrow{\text{heat}} Me_4Si + 4MgClI$

(ii) $SiCl_4 + 2EtMgBr \longrightarrow Et_2SiCl_2 \xrightarrow{\text{LAH}} Et_2SiH_2$

(iii) $Cl_3SiH + 3EtLi \longrightarrow Et_3SiH + 3LiCl$

(iv) $2Et_3SiCl + 2Na \longrightarrow Et_3Si - SiEt_3 + 2NaCl$

The alkylsilanes are colourless oils and, expect for the lower monoalkylsilanes, are stable in air. The alkylsilicon hydrides are readily oxidised to silanols by oxygen in the presence of benzoyl peroxide. Tetra-alkysilanes are usually oxidised at the alkyl group, and are also readily attacked at the alkyl group by electrophilic reagents, *e.g.*,

$$C_2H_5)_4Si + Cl_2 \longrightarrow (C_2H_5)_3SiC_2H_4Cl + HCl$$

On the other hand, silicon hydrides give alkylsilicon chlorides, and form silanols with alkali (see below).

Chlorinated alkylisanes undergo β-elimination when warmed with alkali (cf, the E2 mechanism), *e.g.*,

$$HO^- \; SiEt_3—CH_2—CH_2—Cl \longrightarrow HOSiEt_3 + CH_2{=}CH_2 + Cl^-$$

The corresponding γ-chloro-compounds eliminate cyclopropane.

$$HO^- \; SiR—CH_2—CH_2—CH_2—Cl \longrightarrow HOSiR_3 + \underset{CH_2}{CH_2—CH_2} + Cl^-$$

*Silenes :* These correspond to carbenes, dimethylisilene, $Me_2Si$, has been generated by reaction between the vapours of dimethyldichlorosilane and sodium or potassium in an atmosphere of helium.

*Alkylsilicon chlorides :* These may be prepared, *e.g.*,

(i) by partial alkylation of silicon chloride with a Grignard reagent (see above).

(ii) Silicon hydrides directly halogenated at the silicon atom. The mechanism is uncertain, but since reaction occurs.

$$R_3SiH + Cl_2 \longrightarrow R_3SiCl + HCl$$

with retention of configuration, the mechanism cannot be $S_N2$. There is also evidence against a *four centre reaction*, *e.g.*, BrCl gives $R_3SiBr$ but, on the basis of such a reaction, with silicon atom being subject to nucleophilic attack (see below), $R_3SiCl$ would be expected to be the predominant product.

$$\begin{array}{ccc} \overset{\delta+}{R_3Si} \cdots\cdots & \overset{\delta-}{H} & \\ \vdots & \vdots & \longrightarrow R_3SiCl + HBr \\ \overset{\delta-}{Cl} \cdots\cdots & \overset{\delta+}{Br} & \end{array}$$

(iii) Methylsilicon chlorides are prepared commerically by passing methyl chloride over a mixture of silicon and copper at 350°C, the product is a mixture os $SiCl_4$, $MeSiHCl_2$, $MeSiHCl_2$, $Me_2SiCl$, and disilanes, of which the dimethyl dichloride dichloride predominates. This is a free-radical, and is believed to proceed via methylocopper (Me – Cu). Arylsilicons may also be prepared this way from aryl chlorides.

The alkylsilicon chlorides are very reactive, the chlorine being readily displaced by various nucleophilic reagents, *e.g.*, they react with alcohols to form alkoxysilanes (see also below).

$$MeOH \; SiEt_3{-}Cl \longrightarrow Cl^- + Me\overset{H}{\overset{|}{O^+}}{-}SiEt_3 \xrightarrow{-H^+} MeOSiEt_3$$

They also add to alkenes when heated under pressure, and the reaction is catalysed by platinum catalysts, the mechanism is probably ionic.

$$>C{=}C< \; + Cl_3\overset{+}{Si}H^- \xrightarrow{H_2PtCl_6} >CH{-}\overset{|}{\underset{|}{C}}{-}SiCl_3$$

This is an example of *hydrosilylation.* The the other hand, this addition can be effected in the presence of acyl peroxides or by ultraviolet light radiation, the mechanism is now free-radical.

$$Cl_3SiH \xrightarrow[\text{or } h\nu]{R\cdot} Cl_3Si\cdot \xrightarrow{>C=C<} Cl_3Si{-}\overset{|}{\underset{|}{C}}{-}\overset{|}{\underset{|}{C}}\cdot \xrightarrow{Cl_3SiH}$$

$$Cl_3Si{-}\overset{|}{\underset{|}{C}}{-}\overset{|}{\underset{|}{C}}H + Cl_3Si\cdot;\text{ etc.}$$

*Silanols :* These are usually prepared by the hydrolysis of the corresponding chlorides, the mechanism is $S_N2$, *e.g.*,

$$H_2O \; SiMe_2Cl{-}Cl \longrightarrow Cl^- + H_2\overset{+}{O}{-}SiMe_2Cl \xrightarrow{-H^+}$$

$$HO{-}SiMe_2Cl \xrightarrow{H_2O} HO{-}SiMe_2{-}OH$$

Because of the case with which these hydroxysilanes lose water to form siloxanes, hydrolyxis is best carried out in buffered solution to maintain a neutral solution.

Silanols are also formed by the action of alkali on silicon hydrides, hydrogen is liberated.

$$R_3SiH + OH^- \xrightarrow{H_2O} R_3SiOH + H_2$$

*Siloxanes :* Disiloxanes are formed when are heated alone or with acid.

$$MeSiOH \xrightarrow{H_3O^+} Me_3Si - OH_2^+$$

$$MeSi\ddot{O}H \; SiMe_3—OH_2^+ \longrightarrow H_2O + Me_3Si—OH^+—SiMe_3$$

$$\xrightarrow{-H^+} Me_3Si—O—SiMe_3$$

They are also formed when alkoxysilanes are hydrolysed by dilute acid.

$$2Me_3SiOEt \xrightarrow{H_3O^+} MeSi - O - SiMe_3 + 2EtOH$$

*Silicones :* Hydrolysis of methylsilicon dichlorides without special precautions (see above) results first in the formation of diols, which then combine with elimination of water to form polymers (poly-siloxanes) of general formula $(R_2SiO)_n$. These are oils and are called *silicones,* and normally consist of mixtures of linear and cyclic polymers, *e.g.*,.

$$HOSiMe_2(OSiMe_2)_nOH \qquad Me_2Si(O—SiMe_2)_2O$$

By using a mixture of chlorides, $Me_2SiCl_2$ and $MeSiCl_3$, cross-linear) polymers are obtained. The main groups in silicones are methyl and phenyl, and according to the degree of polymerisation and cross-linking, silicone rubbers, greases, resins, etc, are obtained. These are used for various purpose, *e.g.*, water-repellents, lubricants, electrical insulators.

**Silanamines (silylamines).** These may be prepared by reaction between a monochloride and an amine, *e.g.*, N-methyltriethylsilanamine.

$$Et_3SiCl + Me_2NH \longrightarrow Et_3SiNMe_2 + Me_2NH.HCl.$$

If a primary amine is used, a silazane is also formed (see below).

**Silazanes.** Disilazanes may be prepared as follows.

$$Me_3SiNHEt + Me_3SiCl \longrightarrow Me_3SiNEtSiMe_3 + HCl$$

Silicon dichlorides react with ammonia to give linear and cyclic silazanes (cf. silanediols), *e.g.*,

$$Me_2SiCl_2 + 2NH_3 \longrightarrow [Me_2Si(NH_2)_2]$$

$$\xrightarrow{3\text{ mol.}} Me_2Si(NHSiMe_2)_2NH$$

## BORON COMPOUNDS

The electronic configuration of boron is $1s^2$, $2s^2$, 2p, and most boron compounds are tricovalent ($sp^2$) and contain only three electron pairs (open sextet), *e.g.*, $BF_3$, $R_3B$, etc. These have a planar structure and the Z–B–Z angle is 120°. In six-membered rings, the third 2p orbital may be used to form a π-bond which becomes part of the delocalised. On the other hand, because of the open sextet, boron compounds readily accept a lone pair of electrons form a donor atom, thereby completing their octet, *e.g.*, $Na^+ BF_4^-$, $F_3\overset{-}{B}-\overset{+}{N}H_3$. In these compounds, the boron becomes tetrahedrally hybrised ($sp^3$).

A difficult problem associated with boron chemistry is the structure of boron hydrides. The simplest hydride is diborane (often called borane), $B_2H_6$, and is a dimer of borane, $BH_3$, which has no free existence .The structure of diborane is accepted as having a bridged configuration, which may be as shown. The molecule contains 12 valency electrons. Each H – B bond is covalent (*i.e.*, contains an electron-pair ) and each B...H...B bridge is considered to be covered by a three-centre M.O. containing *one pair of electrons.* Thus, in effect, each B... H bond may be regarded as a 'half-bond'. This molecule does not have sufficient number of electrons for all ligands to be bound covalently by electron-pairs. Thus, diborane is an examples of an *electron-deficient* molecule. Monomers, *e.g.*, $R_3B$, and carbonium ions, $R_3C^+$, *i.e.*, compounds containing an open sextet, have also been referred to as electron-deficient molecules, but it is clear that these and diborane differ in that in the former all linked atoms are covalently bound by electron-pairs.

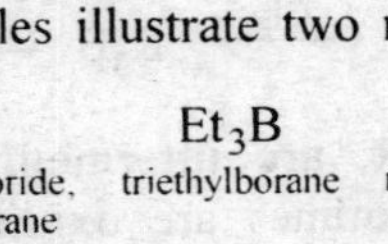

```
H     H     H
 \  /  \  /
  B      B
 /  \  /  \
H     H     H
```

### Nomenclature of Boron Compounds

The following examples illustrate two methods of nomenclature.

| $BH_3$ | $BCl_3$ | $Et_3B$ | $MeBCl_2$ |
|---|---|---|---|
| borane (borine) | boron trichloride, trichloroborane | triethylborane | methylboron dichloride, methyldichloroborane |

| $MeB(OH)_3$ | $Et_2BOH$ | $B(OH)_3$ |
|---|---|---|
| methylbrononic acid methyldihydroxyborane | diethylborinie acid; diethylhydroxyborane | orthoboric acid |

The nomenclature of some heterocyclic boron compounds is described in the text.

Other ways of naming have also been used, but recent literate uses the two described here

*Trialkylboranes* : Some convenient preparations are.

(i) Hydroboration

(ii) Reaction between a Grignard reagent and boron trifluoride or its etherate.

$$3R,gX + BF_3.(Et_2O)_2 \longrightarrow R_3B + 3MgFX + 2Et_2O$$

(iii) $B(OMe)_3 + 3RMgX \longrightarrow R_3B + 3MgXOMe$

(iv) $B(OR_1)3 + R_3^2Al \longrightarrow R_3^2B + Al(OR_1)_3$

Trialkylbvoranes are gases or liquids, and exist as monomers, and this monomeric form is attributed to steric effects preventing dimerisation. They readily form co-ordination complexes. *e.g.*, (Z = N. P etc.).

$$R_3^1B + ZR_3^2 \longrightarrow R_3^1\overset{-}{B} - \overset{+}{Z}R_3^2$$

The stability of these complexes depends on several factors. I effect of the R groups, electro-negativity of Z, steric effects of the R groups, etc. Trimethylamine complexes, in particular, have been used to store volatile boranes, which are regenerated with hydrogen chloride.

Trialkylboranes react with halogens (the lower members inflame), hydrogen halides, and boron halides on heating to give mono-and dialkylboron halides, *e.g.*,

$$R_2BCl + RCl \xleftarrow{Cl_2} R_3B \xrightarrow{HBr} R_2BBr + RH.$$

All volatile trialkylboranes are spontaneously inflammable, but controlled oxidation with dry oxygen gives a dialkyl alkylboranate or an alkyl dialkylborinate with wet oxygen.

$$RB)OR)_2 \xleftarrow{Cl_2} R_3B \xrightarrow[H_2O]{O_2} R_2BOR$$

Peroxides, $R_2BOOR$, are intermediates, and undergo rapid rearrangement. Trialkylboranes are oxidised by alkaline hydrogen peroxide to alcohols, and they readily react with carbon monoxide, and these carbonylated products, under suitable conditions, give carbonyl compounds, alcohols, etc. *e.g.*,

$$R_3B + CO \rightleftharpoons R_3\overset{-}{B}-\overset{+}{C}O \longrightarrow R_2BCOR \xrightarrow{LiBH_4} R_2BCHRO^- \xrightarrow{KOH} RCH_2OH.$$

Triarylboranes are prepared by method (ii). They are solids, relatively resistant to oxidation, and react with sodium or potassium to form highly coloured anion-radicals. *e.g.*,

$$Ph_3B + Na \longrightarrow Ph_3B.^- + Na^+.$$

*Alkylboron hydrides :* The bridging hydrogens of diborane cannot be replaced by alkyl groups. Thus, *five* derivatives are theoretically possible, all five methyl compounds have been prepared. One method of preparing alkyldiboranes is by equilibrium between volatile trialkylboranes and diborane.

$$2MeB + B_2H_6 \rightleftharpoons 2Me_2BHBH_2Me$$

Alkyldiboranes are readily attacked by water, oxygen, etc., and may be prepared in monomeric form as co-ordination compounds, *e.g.*., $R\bar{B}H_2-\bar{N}Me_3$.

*Alkylboron chlorides :* These may be prepared in various ways, *e.g.*,

(i) By reaction between trialkylboranes and $Cl_2$, HCl, or $BCl_3$ (above).

(ii) By reaction between boron trichloride and the appropriate amount of Grignard reagent (above).

(iii) $R_4Sn + 2BCl_3 \longrightarrow 2RBCl_2 + R_2SnCl_2$

*Orthoborates* may be prepared by reaction between boric acid or boron trichloride and an alcohol, or by transesterification (alcoholysis) with the appropriate alcohol.

$$3R^1OH + B(OH)_3 \longrightarrow (R^1O)_3B + 3H_2O$$

$$(R^1O)_3B + 3R^2OH \rightleftharpoons (R^2O)_3B + 3R^1OH.$$

Orthoborates are generally readıly hydrolysed by water.

$$(RO)_3B + 3H_2O \longrightarrow 3ROH + B(OH)_3.$$

It appears to be uncertain whetheı this reaction is stepwise or not.

*Alkylboronic acids,* $RB(OH)_2$, and *dialkylborinic acids,* $R_2BOH$. This are obtained on hydrolysis of the corresponding alkylboron chloride, and their esters are produced if an alcohol is used.

$$R^1_2BOH \xleftarrow{H_2O} R^1_2BCl \xrightarrow{R^2OH} R^1_2BOR^2$$

The acids and their esters may also be prepared by reaction between an orthoborate and the appropriate amount of Grignard reagent, followed by hydrolysis (to give the acid).

$$(R^1OR)_3B \xrightarrow{R^2MgX} R^2B(OR^1)_2 \xrightarrow{H_2O} R^2B(OH)_2.$$

The boronic and boronic acids are weak acids, and are less sensitive to oxidation than then trialkyl-boranes, and the former acids are less sensitive than the latter. Both acid are readily dehydrated on heating to *boroxines (boroxoles),* $(RBO)_3$, and *boronic anhydrides* $(R_2B)_2O$, respectively. Boroxines are cyclic anhydrides, the alkylboroxines are stable, but the parent compound (R = H) is not. Although this structure could exhibit aromatic character, calculations have shown that it is very small in boroxines (cf. borazines, below).

O—BR
RB O
O—BR

*Organoboron-nitrogen compounds* : It has already been mentioned that alkyl-dialkylboron chlorides react with secondary amines to form complex which. On the other hand, $Et_3N$, eliminate a molecule of HCl to form a *borazene,* $R_2BNR_2$. When they exist as monomers (aminoboranes), their structure is (I) but, unless the R groups are large (steric effect), the

$R_2B \leftleftharpoons NR_2$ (I)

$R_2B \leftarrow NR_2$
$R_2N$ $BR_2$
$R_2B—NR_2$
(II)

RH—NH
$R_2N$ BH
BH⇐NH
(IIIa)

BH—NH
HN BH
BH—NH
(IIIb)

borazenes usually exists as cyclic trimers (II. this is of the cyclohexane type), and are called *cyclohexane.*

*Borazines (borazoles)* : These are six-membered rings, the parent compound being (IIIa or IIIb), which may be prepared by heating diborane with ammonia. If borazine has substituents on the N atoms, this is indicated by prefixing the substituent with N-if on the B atoms, by B-B-Trichloro-N trialkylborazines are conveniently prepared by heating boron trichloride with amines, *e.g.*,

$$3BCl_3 + 3RNH_2 \longrightarrow B_3Cl_3N_3R_3 + 6HCl$$

Borazines are of special interest in that they exhibit aromatic character, they contain a closed shell of six electrons. Although borazines are 'aromatic', this aromaticity is not very pronounced, *e.g.*, they are far more reactive than benzene, they readily add on three molecules of HX, H2O, etc., the negative part of the addendum becoming attached to the boron atoms.

*Complex anions of boron :* These are of the type $M+ BR_4^-$. Sodium and lithium borohydrides (R = H), also called hydroborates, are useful reducing agents in organic chemistry. Tetra-alkyl and -arylborates (R = R or Ar) may be prepared in several ways, *e.g.*,

$$B_3^1B + R_2Li \longrightarrow Li\,BR_3^1R^2$$

Sodium tetraphenlyborate is used as a regent for estimating potassium, nitrogen bases, etc. It may be prepared as follows.

$$BF_3 + 4PhMgBr \longrightarrow 3MgClF + Ph_4B^-MgCl^+$$

$$\xrightarrow{\text{aq.NaCl}} Na + Bh_4^-$$

## EXERCISES

1. The reaction $RX + RS^-Na^+ \longrightarrow R_2S + NaX$ is faster than the reaction $RX + RO^-Na^+ \longrightarrow R_2O + NaX$. Explain.
2. Complete the following equations.

   (i) $CH_2 = CHCMe = CH_2 + SO_2 \longrightarrow ?$

   (ii) $MeNH_2 + CS_2 + NaOH \xrightarrow[\text{heat}]{H_2O} ?$

   (iii) $MeCONH_2 + P_2S_5 \xrightarrow{C_6H_6} ?$

   (iv) $CH_2CHCN + HOH_2CH_2SH \xrightarrow{\text{heat}} ?$

   (v) $\underset{\diagdown\ O\ \diagup}{CH_2{-}CH_2} + H_2S \longrightarrow$

(vi) MeCH = CHCH2SCN $\xrightarrow{\text{heat}}$ – CH(NCS) – (part of product) ?

(vii) $CH_2 = CH_2 + (SCN)_2 \longrightarrow$ ?

(viii) $Me_2C(SEt)_2 \xrightarrow{KMnO_4}$ ?

(ix) $PHSO_2Me \xrightarrow{EtONa}$

(x) MeCOCl + EtSH $\longrightarrow$ ?

(xi) $(RS)_2Hg \xrightarrow{\text{heat}}$ ?

(xii) $Et_2S + H_2SO_4 \longrightarrow\ ?\ \xrightarrow{H_2O}$ ?

(xiii) $Et_2S$ + BrCN $\longrightarrow$ ?

3. Name the following compounds.

| | |
|---|---|
| (i) $PhPO(OH)_2$, | (ii) $HOP(OEt)_2$, |
| (iii) $EtPCl_2$, | (iv) $MePOCl_2$ |
| (v) $H_2PCH_2CH_2Cl$, | (vi) $H_3P = NH$, |
| (vii) $(HO)POCl_2$, | (viii) $POCl_2(NH_2)$, |
| (ix) HPOCl(OH), | (x) $H_2POCl$, |
| (xi) $MeHPO(NH_2)$, | (xi) $Et_2PCl$. |

4. Suggest a mechanism for the removal of oxygen from epoxides by $Ph_3P$ to form alkenes (use ethylene oxide as your example).

5. Suggest a preparation for each of the following.

(i) $Ph_3P$,

(ii) MeEtPH,

(iii) $Et_3PO$,

(iv) diethylphosphinous chloride,

(v) diethyl methylphosphonate.

(vi) dimethylphosphonic chloride,

(viii) ethylphosphonic diamide.

6. Complete the following equations

(i) $PhCH = CHCH_2PPh_3{}^+CR^-$ + PhCHO $\xrightarrow{EtOLi}$ –?

(ii) 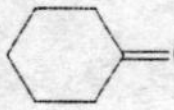 Cyclohexanone $=O + Ph_3P = CH_2 \longrightarrow ?$

(iii) $Ph_3P = CCl_2 + p\text{-}Me_2NC_6H_4CHO \longrightarrow ?$

(iv) $Ph_3P + CHCl_3 \xrightarrow{t-BuOK} ?$

(v) $Ph_3P + ClCH_2COMe \longrightarrow ? \xrightarrow{NaCO_3} ? \xrightarrow{PhCHO} ?$

(vi) $PhCO(CH_2)_4Br \xrightarrow{Ph_3P} ? \xrightarrow[EtOH]{EtONa}$

$PhCO(CH_2)_3CH = PPh_3 \longrightarrow$

(vii) $PhNO + Ph_3P = CHPh \longrightarrow ?$

(viii) $Et_2PCl + Et_2PNa \longrightarrow ?$

(ix) $Ph_3P + Et_3\overset{+}{N}-\overset{-}{O} \longrightarrow ?$

(x) $R^1_3P + R^2_2SO \longrightarrow ?$

(xi) $Ph_3P = CH_2 + MeCH = CHCOMe \longrightarrow ?$

(xii) $(EtO)_3P + Cl_2 \longrightarrow ?$

(xiii) $(EtO)_3P + Me_3\overset{+}{N}-\overset{-}{O} \longrightarrow ?$

(xiv) $Me_2SO + Ph_3P \longrightarrow ?$

(xv) $Ph_4P^+I^- + PhLi \longrightarrow ?$

7. Name the following compounds.

(i) $H_3Si(SiH_2)_2SiH_3$;

(ii) $MeHSi(NH_2)_2$,

(iii) $EtSi(OH)_3$,

(iv) $Me_2SiCl_2$,

(v) $Et_2SiHOOCMe$,

(vi) $EtOSiH_2(OSiH_2)_3SiH_2NH_2$,

(vii) $SiH_3NMeSiH(OEt)NHSiH_3$.

8. Suggest a synthesis for each of the following.

(i) $n\text{-}Pr_4Si$, (ii) $MeSiH_3$, (iii) $Et_3SiSiEt_3$; (iv) $Me_3SiCH_2Cl$,

(v) $Me_3SiNEt_2$, (vi) $Me_3SiNMeSiMe_3$, (vii) $Me_3SiEt$.

9. Complete the following equations :

(i) $MeCH = CHMe + Me2SiHCl \xrightarrow{cat.} ?$

(ii) $2CH_2 = CHCH_2Cl + Si \xrightarrow{cu} ?$

(iii) $Me_3SiCl + BrCl \longrightarrow ?$

(iv) $Me_3SiCl \xrightarrow{H_2O} ?$

(v) $Et_3SiCl + Me_3SiOH \longrightarrow ?$

(vii) $Me_3SiCH_2CHClCH_3^- \longrightarrow ?$

10. Write out the structures of :

(i) B-trichloroborazine, (ii) dimethoxyborane, (iii) methyl dimethylborinate, (iv) diethyl methyl-boronate, (v) trimethyl borate, (vi) ethoxy (methoxy) borane, (vii) acetoxyborane.

11. Write out the structures of the five possible methyldiboranes (mono, di, etc.). Comment.

12. Complete the following equations.

(i) $B(OMe)_3 + 3EtMgBr \longrightarrow ?$

(ii) $3PhMgBr + BF_3 \longrightarrow ?$

(iii) $Et_3B + Br_2 \longrightarrow ?$

(iv) $2(MeO)_3B + NaH \longrightarrow ?$

(v) $Me_3B + HCl \longrightarrow ?$

(vi) $EtBCl_2 + 2MeOH \longrightarrow ?$

(vii) $B(OMe)_3 + 2EtMgBr \longrightarrow ? \xrightarrow{H_2O} ?$

(viii) $MeB(OH)_2 \xrightarrow{heat} ?$

13. Suggest mechanisms for the following rearrangements.

(i) $EtSCHMeCH_2OH \xrightarrow{HCl} EtSCH_2CHClMe$

(ii) $Ph_3P + ROOR \longrightarrow Ph_3PO + R_2O$ [Not free radical]

14. Convert.

(i) $ClCH_2-C_6H_4-CH_2Cl$ into $PhCH{=}CH-C_6H_4-CH{=}CHPh$;

(ii) EtMeCHBr into EtMeC = $CHPr^n$.

15. Complete the following equations.

(i) $Ph_2CO + Ph_3P = CHOMe \longrightarrow ? \xrightarrow{H^+}$

(ii) $Ph_2CN_2 + SO_2 \longrightarrow ? \xrightarrow{ROH} ?$

(iii) $MeSC \equiv CH + HCl \longrightarrow ?$

(iv) $(R^1O)_3P + R^2COCl \longrightarrow ?$

(v) $Et_3B + Et_3N \longrightarrow ?$

(vi) $Et_3B + O_2(H_2O) \longrightarrow ?$

(vii) $B(OMe)_3 + 2EtMgBr \longrightarrow ? \xrightarrow{H_2O} ?$

(viii) $EtB(OH)_2 \xrightarrow{heat} ?$

(ix) $? \xleftarrow{EtOH} CH_2\text{—}CH_2 \xrightarrow{Me_2NH} ?$ (with $CH_2$–$CH_2$ bridged below by O, epoxide ring)

(x) $(CH_2SH)_2 + R_2CO \xrightarrow[BF_3.Et_2O]{AcOH} ? \xrightarrow[H_2]{Ni} ?$

(xi) $MeP(OEt)_2 + EtI \longrightarrow ?$

(xii) $Et_2PNEt_2 + MeI \longrightarrow ?$

(xiii) $R^1SSR^2 \xrightarrow[Ni]{Raney} ?$

(xiv) $3BCl_3 + 3EtNH_2 \xrightarrow{heat} ?$

16. **Each of the following compounds showed only one singlet line in its NMR spectrum. (i) $C_2H_3F_3O_2S$,**

**(ii) $C_2H_4O_4S$. Suggest a structure for each compound.**

# 2

# NITROGEN COMPOUNDS

## INTRODUCTION

The cyan compounds are characterised by the presence of a cyan group (CN). They are considered to be derivatives of hydrogen cyanide (HCN). Since the cyanide ion (: C ≡ M:) has an unshard electon pair onboth the carbon and nitrogen atoms, the proton ($H^+$) can attach itself to either of these atoms. It is now proposed to deal with many other nitrogen compounds, most of which may be regarded as alkyl nitrogen compounds.

*Hydrocyanic acid, hydrogen cyanide* (*prussic acid*) b.p. 26°C. was discovered by Scheele (1782) who obtained it from bitter almonds which contain the glycoside *amygdalin.* Amygdalin, on hydrolysis with dilute acid, yields hydrogen cyanide, benzaldehyde and glucose.

$$C_{10}H_{27}O_{11}N + 2H_2O \xrightarrow{\text{acid}} HCN + C_6H_5CHO + 2C_6H_{12}O_6 .$$

Hydrogen cyanide also occurs in the leaves of certain plants, *e.g.*, laurel. It may be prepared by heating sodium cyanide with concentrated sulphuric acid, the gas being dried over calcium chloride

$$NcCN\ \ H_2SO_4 \longrightarrow HCN + NaHSO_4\ \ (93\text{-}97\%).$$

Hydrogen cyanide is prepared industrially in several ways, *e.g.*, by passing a mixture of ammonia, at and methane over a platinum rhodium gauze catalyst at 1000°C.

$$2NH_3 + 3O_2 + 2CH_4 \longrightarrow 2HCN + 6H_2O .$$

Hydrogen cyanide is a very weak acid, and hydrolyses slowly in aqueous solution, and more rapidly in the presence of inorganic acids to form first formamide, and then ammonium formate.

$$H-C\equiv N \xrightarrow{H_2O} HCONH_2 \xrightarrow{H_2O} HCO_2NH_4.$$

Two types of organic derivatives of hydrocyanic acid are known, alkyl cyanides, RCN, and alkyl isocyanides, RNC. Their formation from cyanides of the alkali metals may be explained as follows. Alkali cyanides are ionic $(:C\equiv N^-:)$ and the cyanide ion is ambident and so can form a covalent bond either from the carbon or the nitrogen. Silver cyanide. $AgC\equiv N$, is essentially covalent, and hence the lone pair on the nitrogen is mainly available for covalent bond formation, resulting in the predominant formation of isocyanides.

## ALKYL CYANIDES

These are also knows as *nitriles* or *carbonitriles.*

### Nomencluature

This group of compounds is usually named either as the *alkyl cyanides* (*i.e.*, the alkyl derivatives of hydrogen cyanide), or as the *nitrile* of the acid which is poduced on hydrrolysis, the suffixic of thhe trivial name being repladed by *–onitrile, e.g.*,

| | |
|---|---|
| HHCN | hydrogen cyanide or formonitrile |
| $CH_3CN$ | methyl cyanide or accetonitrile |
| $(CH_3)_2CCHCN$ | isopropyl cuamode pr ospnbutyronitrile |

### General Methods of Preparation

1. By dehydration of acid amides with phosphorus pentoxide. High molecular a eight acid amides are dehydrated to the corresponding cyanide by heat alone.

Cyanides are prepared industrially by passing a mixture of carboxylic acid and ammonia over alumina at 500°C. Thus reaction probably occurs as follows.

$$RCO_2H + NH_3 \longrightarrow RCO_2NH_4$$

$$\xrightarrow{Al_2O_3} H_2O + RCONH_2 \xrightarrow{Al_2O_3} H_2O + RCN$$

Another industrial method is the *ammoxidation* of alkanes.

$$2RCH_3 + 2NH_3 + 3O_2 \xrightarrow[500-600°C]{cal.} 2RCN + 6H_2O$$

This term describes the process involving oxidation in the presence of ammonia.

2. By the dehydration of aldoximes with phosphorus pentoxide, or better, with acetic anhydride.

$$RCH = NOH \xrightarrow[-H_2O]{(CH_3CO)_2O} RCN \quad (g).$$

Aldehydes may be converted directly to cyanides by refluxing a solution of an aldehyde in formic acid with hydroxylamine hydrochloride and sodium acetate (30–42 per cent, van Es. 1965).

3. The most convenient method is to heat an alkyl halide with potassium cyanide in aqueous ethanolic solution, a small amount of isocyanide is also obtained.

$$RX + KCN \longrightarrow RCN + KX \text{ (g)}.$$

This method is satisfactory only if R is a primary or secondary alkyl group. If it is a tertiary group try little cyanide is obtained, the tertiary halide being converted into the corresponding alkene. Tertiary alkyl cyanides are best prepared by method 4 below. Cyanide yield form primary chlorides has been much improved by using dimethyl sulphoxide or dimethylformamide as solvent (Smiley *et al.,* 1980. Friedman *et al.,* 1960)

Many cyanides, particularly the lower members, may be readily prepared by warming the potassium alkyl sulphate with potassium cyanide (some isocyanide is also obtained).

$$ROSO_2OK + KCN \longrightarrow RCN + K_2SO_4 \text{ (g –v.g)}$$

Tosylates, *i.e.*, p-toluenesulphonates, are also useful for cyanide preparation.

$$p\text{–}CH_3C_6H_4SO_3R + KCN \longrightarrow RCN + p\text{-}CH_3C_6H_4SO_3K.$$

4. By the interaction of a Grignard reagent and cyanogen chloride.

$$MRgCl + ClCN \longrightarrow RCN + MgCl_2$$

This is the best method of preparing tertiary alkyl cyanides.

**General Properties**

The alkyl cyanides are stable neutral substances with fairly pleasant smells, and are not as poisonous as hydrogen cyanide. The lower members are liquids which are soluble in water (with which they can form hydrogen bonds), but the solubility diminishes as the molecular weight increases. Since they cannot form intermolecular hydrogen bonds, alkyl cyanides are much more volatile than their corresponding acids.

The infra-red absorption band of the cyano group (C–N stretch) in alkyl cyanides is 2260–2240 $cm^{-1}$(w. m). Fig. 2.1 is the infra-red spectrum of methyl cyanide (as a film).

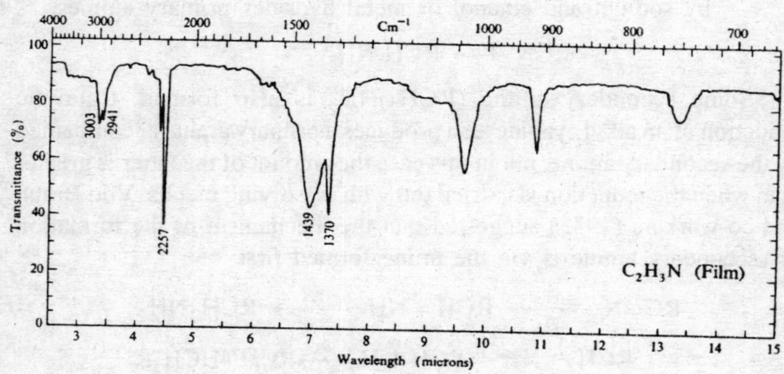


**Fig. 2.1**

2257$cm^{-1}$ $C \equiv N$ (str.) in saturated alkyl cyanides. 3003 and 2941 $cm^{-1}$. C – H (str.) in Me (and or $CH_2$) 1439 $cm^{-1}$. C – H (def.) in Me (and or $CH_2$). 1370 $cm^{-1}$. C – H (def.) in Me.

**Reactions**

1. The alkyl cyanides are hydrolysed by acids or alkalis to the corresponding acid via the intermediate formation of

$$RC \equiv N \xrightarrow{H_2O} \left[ R - C \begin{matrix} \diagup OH \\ \diagdown\!\!\diagdown NH \end{matrix} \right] \longrightarrow RCONH_2 \xrightarrow{H_2O} RCO_2R^2 + NH_3$$

When a solution of alkyl cyanide in an alcohol is heated with concentrated sulphuric acid or hydrochloric acid, the *ester* is obtained.

$$R_1CN + R_2OH + H_2O \longrightarrow R_1CO_2R_2 + NH_3.$$

It dry hydrogen chloride is passed into the solution of an alkyl cyanide in anhydrous alcohol, the *imidic ester hydrochloride* is formed and when treated with *dry* hydrogen chloride, alkyl cyanides form the imidochloride.

$$RC\equiv N + 2HCl \longrightarrow RCCl = NH_2^+ Cl^-.$$

Alkyl cyanides combine with dry ammonia to form *amines* and with acid anhydrides on heating to form *tertiary acid amides.*

$$RC(=NH)NH_2 \xrightarrow{NH_3} RCN \xrightarrow{RCOH_2O} (RCO)_3N$$

2. Alkyl cyanides may be converted into aldehydes and are reduced by sodium and ethanol or metal hydrides primary amines.

$$RCN \longrightarrow RCH_2NH_2$$

Some secondary amine $(RCH_2)NH_2$ is also formed. Catalytic reduction of an alkyl cyanide also produces a primary amine accompanied by the secondary amine, but in this case the amount of the latter is greater than when the reduction is carried out with dissolving metals. Von Braun and co-workers (1933) suggested that the mechanism of the formation of secondary amine is via the imine formed first.

$$RC\equiv N \xrightarrow[N_2]{H_2} RCH = NH \xrightarrow{H_2} RCH_2NH_2$$

$$RCH = NH + RCH_2NH_2 \longrightarrow \underset{\substack{|\\ NH_2}}{RCHNHCH_2R}$$

$$\longrightarrow NH_3 + RCH = NCH_2R \xrightarrow{H_2} RCH_2NHCH_2R$$

Support for this mechanism is to be found in the fact that the formation of secondary amine is prevented if the reduction is carried out with a Raney nickel catalyst *in the presence of ammonia.*

Alternatively, the formation of secondary amine may be prevented by carrying out the hydrogenation in acetic anhydride solution containing sodium acetate (Gould *et al.*, 1960). Using these conditions, the primary amine formed is acetylated, this acetyl derivative cannot react with the imine.

3. Alkyl cyanides react with Grignard reagents to form ketones.

$$R^1C\equiv N + R^2MgX \longrightarrow P^lR^2C = NMgX \xrightarrow{H^+} R^1COR^2$$

4. Alkyl cyanides undergo condensation reactions in the presence of sodium, only the a-hydrogen atoms being involved. These condensations are to be expected in view of the fact that the cyano group is strongly electron attracting. When the reaction is carried out in ether, two molecules of cyanide condense.

$$CH_3CH_2C{\equiv}N + CH_3CH_2CN \xrightarrow[\text{ether}]{\text{Na}} CH_3CH_2\overset{\overset{\large NH}{\|}}{C}-\overset{\overset{\large CH_3}{|}}{CH}-CN$$

This is known as the *Thorpe nitrile condensation* (1904).

On the other hand, alkyl cyanides with a-hydrogen atoms can condense with esters if the reaction is carried out in ether solution in the presence of sodamide (Levine and Hauser, 1946), *e.g.*, methyl cyanide condenses with ethyl propionate to form propionylmethyl cyanide.

$$CH_3CN + NaNH_2 \longrightarrow [:\bar{C}H_2CN]Na^+ + NH_3 \xrightarrow{CH_3CH_2CO_2C_2H_5} CH_3CH_2COCH_2CN + C_2H_5OH \quad (40\%)$$

## ALKYL ISOCYANIDES

These are also known as *isonitriles* or *carbylamines.*

### General Methods of Preparation

1. By heating an alkyl iodide with silver cyanide in aqueous ethanolic solution, a small amount of cyanide is also formed.

$$RI + AgCN \longrightarrow RNC + AgI$$

2. By heating a mixture of a primary amine and chloroform with ethanolic potassium hydroxide.

$$RNH_2 + CHCl_3 + 3KOH \longrightarrow RNC + 3KCl + 3H_2O.$$

The mechanism of the isocyanide reaction is not certain. Robinson (1961) has suggested that the reaction proceeds via the intermediate formation of dichloromethylene, which is produced from the chloroform in alkaline solution.

$$RNH_2\ \ddot{C}Cl_2 \longrightarrow R\overset{+}{N}H_2-\overset{-}{C}Cl_2 \xrightarrow{-H^+} RNH-CCl_2 \xrightarrow[-H^+]{-Cl^-} RN=\bar{C}Cl \xrightarrow{-Cl^-} R\overset{+}{N}\equiv\bar{C}$$

3. Dehydration of N-substituted formamides with phosphoryl chloride in pyridine in pyridine produces isocyanides (Ugi *et al.*, 1960).

$$RNHCHO \xrightarrow{POCl_3} RNC\ (g)$$

## General Properties

The alkyl isocyanides are poisonous, most unpleasant-smelling liquids, with lower boiling points the isomeric cyanides. They are not very soluble in water, the nitrogen atom not having a lone part of electrons available for hydrogen bonding. Alkyl isocyanides absorb in the region 2185-2120 $cm^{-1}$(s)

## Reactions

1. Alkyl isocyanides are hydrolysed to an amine and formic acid by dilute acids, *but are not hydrolysed by alkalis.*

$$RNC + 2H_2O \xrightarrow{\text{acid}} RNH_2 + HCO_2H$$

2. Alkyl isocyanides are reduced to secondary amines either by dissolving metals or by catalytic reduction.

$$RNC \longrightarrow RNHCH_3$$

3. Alkyl isocyanides add on halogen to form *alkyliminocarbonyl halides,* sulphur to form *alkyl isocyanates,,* and are readily oxidised by mercuric oxide to *alkyl isocyanates.*

$$RNC + X_2 \longrightarrow RNCX_2$$

$$RNC + S \longrightarrow RNCS$$

$$RNC + HgO \longrightarrow RNCO + Hg$$

4. When alkyl isocyanides are heated for a long time, they rearrange to the cyanide.

$$RNC \longrightarrow RCN$$

*Cyanogen (ethanedinitrile)* may be prepared by heating the cyanides of mercury or silver.

$$Hg(CN)_2 \longrightarrow (CN)_2 + Hg$$

Cyanogen is a poisonous colourless gas. b.p – 21°C. and is stable is spite of the fact that its beat of formation is + 292.9 kJ $mol^{-1}$. It is hydrolysed by water, the main products being hydrocyanic acid and cyanic acid, and minor products being oxamide and ammonium formate.

$$(CN)_2 \xrightarrow{H_2O} HCN + HNCO + (CONH_2)_2 + HCO_2NH_4 .$$

When cyanogen is hearted at 400°C, it polymerises to *paracyanogen,* which, at 800°C, regenerates cyanogen. The structure of cyanogen is probably best represented as a resonance hybrid (I) being the most important.

$$:N{\equiv}C{-}C{\equiv}N: \longleftrightarrow :\overset{+}{N}{=}C{=}C{=}\ddot{\overset{..}{N}}: \longleftrightarrow :\ddot{\overset{..}{N}}{=}C{=}C{=}\overset{+}{N}:$$
(I) (II) (III)

*Cyanogen halides :* XCN (X = Cl) (b.p. 12 °C). Br (m or l m.p. 146.5°C) are readily prepared by the action of halogen on sodium cyanide.

$$NaCN + X_2 \longrightarrow XCN + NaX.$$

When pure, cyanogen halides are stable, but in the presence of impurities they readily trimerise in cyanuric halides $X_3C_3N_3$ (cyclic compounds). They are stable to water, but are hydrolysed by alkali

$$XCN + 2KOH \longrightarrow KNCO + KX + H_2O$$

They readily react with amines to form alkyl cyanamides *e.g.*,

$$2RNH_2 + XCN \longrightarrow RNHCN + RNH_3^+ X^-$$

*Cyanic acid :* HNCO. Urea, on dry distillation, gives *cyanuric acid.*

$$3CO(NH_2)_2 \longrightarrow H_3N_3C3O_3 + 3NH_3.$$

Cyanuric acid is colourless, crystalline solid not very soluble in water, and strongly acid reacting as a mono-. dt⁻ and tribasic acid. It has a cyclic structure (a *triazine* derivative) and X-rays analysis indicates that the acid is best represented as a resonance hybrid.

[Structures: cyanuric acid triketo form (HN, NH, O, N, O, H) ⟷ charged form with $O^-$, $\overset{+}{H}N$, $\overset{+}{N}H$, $^-O$, $\overset{+}{N}$, $O^-$, H ⟷ ...]

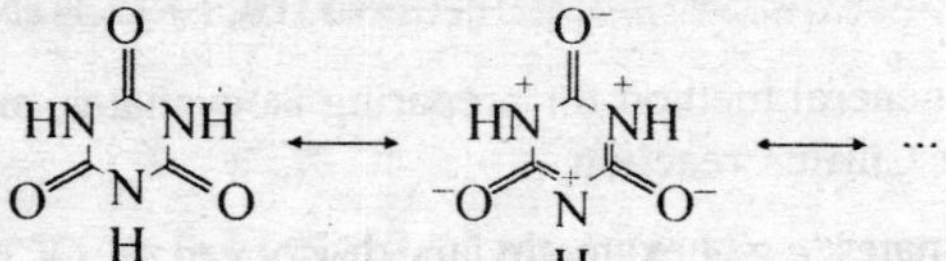

When cyanuric acid is heater, it does not melt but decomposes into cyanic acid vapour which when condensed below 0°C, gives a colourless condensate.

$$H_3N_3C3O_3 \longrightarrow 3HNCO$$

Cyanic acid is a colourless volatile, strongly acid liquid which, above 0°C, readily polymerises to cyanuric acid and *cyamelide,* (HNCO). Aqueous solutions of cyanic acid rapidly hydrolyse to give carbon dioxide and ammonia.

$$NHCO + H_2O \longrightarrow CO_2 + NH_3.$$

There are two possible structures for cyanic acid. H—O—C≡N and H–N=C=0. There is very little evidence for the existence of the format cyanic acid). *e.g.*, infra-red absorption spectra indicate the latter (isocyanic acid).

When a solution of potassium isocyanate (usually called potassium cyanate) and ammonium sulphate is evaporated to dryness, urea and potassium sulphate are obtained, the urea being formed by the molecular rearrangement of the intermediate product ammonium isocyanate.

$$NH_4NCO \longrightarrow CO(NH_2)_2$$

Cyanic acid reacts with alcohol to form *urethans,* excess of cyanic acid converts urethans into *allophanates* which are extras of the unknown acid, *allophanic acid.* $NH_2ONHCO_2H$.

$$ROH + HNCO \longrightarrow NH_2CO_2R \xrightarrow{HNCO} NH_2CONHCO_2R$$

N-substituted urethans are formed by the reaction between alkyl or aryl isocyanates and an alcohol. Phenyl isocyanate is also used to substituted primary and secondary amines, with which it forms substituted ureas.

Phenyl isocyanate is usually prepared by the action of carbonyl chloride on aniline.

$$C_6H_5NH_2 + COCl_2 \longrightarrow HCl + C_2H_5NHCOCl$$

$$\xrightarrow{Heat} HCl + C_6H_5NCO\,.$$

This is a general method for preparing isocyanates, another general method is the Curtius reaction

Di-isocyanates, *e.g.*, hexamethylane di-isocyanate. $OCN(CH_2)_6NCO$, are used to prepare poly-urethan plastics by reaction with di-and polyhydroxyalcohols *e.g.*, tetramethylene glycol.

$$HO(CH_2)_4OH + OCN(CH_2)_6NCO + HO(CH_2)4OH + \ldots \longrightarrow$$

$$HO(CH_2)_4OCONH(CH_2)_6NHCOO(CH_2)_4O \ldots$$

*Cyanamide :* m. C. may be prepared by the action of ammonia on cyanogen chloride.

$$ClCN + NH_3 \longrightarrow NH_2CN + HCl.$$

It is also readily prepared by the action of water and carbon dioxide on calcium cyanamide.

Cyanamide is converted into *guanidine (q.x) by ammonia, and into* thiourea (q.v). by hydrogen sulphide. When cyanamide is melted it forms the dimer *dicyandiamide.* $(NH_2)_2C{=}NCN$. and the trimer *melamine.* Which is a cyclic compound. X-rays analysis has shown that all the C–N bond length are the same (Hughes. (1994). This can be explained on the assumption that melamine is a reaction hybrid of the amino-form (ef. cyanuric acid).

Melamine is manufactured by heating urea and passing the gaseous products over a heated calcium

$$CO(NH_2) \xrightarrow{-NH_3} HNCO \xrightarrow{6\ mol.} C_3H_6N_6 + 3CO_2\,.$$

Melamine is used for making melamine-formaldehyde plastics.

Cyanamide itself is a tautomeric compound (amidine system), and Raman spectra investing have shown that two forms are present in equilibrium in the solid or fused state, on in solution

$$H_2N-C{\equiv}N \rightleftharpoons HN{=}C{=}NH$$

Cyanamide forms salts, the most important of which is the calcium salt. This is manufactured heating calcium carbide mixed with 10 per cent its weight of calcium chloride in a stream of nitrogen 800 C.

$$CaC_2 + N_2 \longrightarrow CaNCN + C$$

The calcium cyanamide carbon mixture is used as a fertiliser, it is hydrolysed in the soil to cyanamide which is then hydrolysed to urea, which in turn, is converted into ammonium carbonate by the in the soil.

$$CaNCN \xrightarrow{H_2O} Ca(OH)_2 + NH_2CN \xrightarrow{H_2O}$$

$$CO(NH_2)_2 \xrightarrow{H_2O} (NH_4)_2CO_3$$

Calcium cyanamide is also used to prepare urea industrially.

*Carbodi-imide :* This are most conveniently prepared by heating disubstituted thioureas yellow mercuric oxide.

$$RNHCSNHR + HgO \longrightarrow RN = C = NR + HgS + H_2O$$

Carbodi-imides are very powerful dehydrating agents and are used for removińg the elements of in various reactions, the most common one used for this purpose being *dicyelohexylcarbodi* 20°C. *e.g.,*

$$2ROCO_2H + C_6H_{11}N = C = NC_6H_{11} \longrightarrow (RCO)_2O + (C_6H_{11}NH)_2CO$$

## NITROALKANES

### Nomenclature

The nitroalkanes are named as the nitro-derivative of the corresponding alkane, the of the nitro-groups being indicated by numbers. *e.g.,*

$CH_3NO_2$ nitromethane

$CH_3CH(NO_2)CH_2CH(NO_2)CH_2CH_3$ 2,4-dinitrohexane

The nitroalkanes the structure os which is $RNO_2$, are isomeric with the alkyl nitrites. RONO. The evidence that may be adduced for these respective formula is to be found in the study of the reactions of these two groups of compounds. The reactions of the nitrites have already been described, and those of the nitroalkanes are described below. It is however, worth while at this stage to mention the reaction which most clearly indicates their respective structures viz, reduction. When alkyl nitrites are reduced an alcohol and ammonia or hydroxylamine are formed. This shows that the alkyl group in nitrites is attached to an oxygen atom.

$$R-ONO \xrightarrow{e:H^+} ROH + NH_3 + H_2O.$$

On the other hand, when nitroalkanes are reduced, a primary amine is formed. This shows that the alkyl group is attached to the nitrogen atom since the structure of a primary amine is known to be $R-NH_2$ from its method of preparation.

$$RNO_2 \xrightarrow{e:H^+} RNH_2 + 2H_2O.$$

This structure of the nitro-compounds is best represent as a resonance hybrid.

$$R-N(=O)\rightarrow O \longleftrightarrow R-N(\rightarrow O)=O \quad \text{or} \quad R-\overset{+}{N}(=O)-\overset{-}{O} \longleftrightarrow R-\overset{+}{N}(-\overset{-}{O})=O$$

For most purpose, however, the non-committal formula $RNO_2$ is satisfactory.

From the M.O point of view, the nitro-group is conjugated (Fig. 2.2) and delocalisation of bonds increases its stability (the two oxygen atoms are equivalent; *cf.* the carboxylate ion).

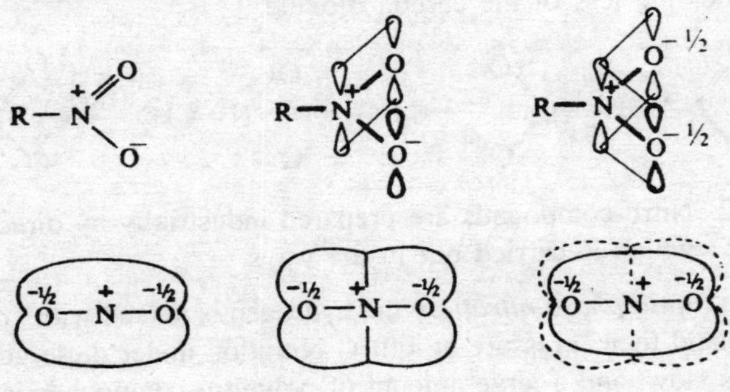

**Fig. 2.2**

## General Methods of Preparation

1. By heating an alkyl halide with silver nitrite in aqueous ethanolic solution.

$$RX + AgONO \longrightarrow RNO_2 + RONO + AgX$$

This method is only useful for the preparation of primary nitroalkanes. With s-halides the said is about 15 per cent, and with t-halides 0–5 per cent, the yield of nitrite increases from primary to tertiary halides (Kornblum *et al.,* 1955).

Kornblum *et al.,* (1956) have shown the sodium nitrite reacts with alkyl halides to give good yields of nitroalkane (55–62 per cent) their with alkyl nitrites (25–33 per cent). The success of the reaction depends on the use of the formamide as solvent, and the addition of urea increases the solubility of sodium nitrite. The bromides and iodides are most satisfactory, the chlorides react too slowly to be useful. On the other hand, if dimethyl sulphoxide is used as solvent instead of D.M,F., the addition of urea is unnecessary (Smiley *et al.,* 1960. Friedman *et al.,* 1960). It should be noted that the nitrite is ambident.

Nitromethane may be prepared by boiling an aqueous solution of sodium nitrite with a halogen-acetic acid, *e.g.,*

$$ClCH_2CO_2Na + NaNO_2 \longrightarrow NaCl + [CH_2(NO_2)CO_2H]$$
$$\longrightarrow CH_3NO_2 + CO_2 \quad (35\text{-}38\%)$$

The ready decarboxylation of the intermediate nitroacetic acid may be explained on the assumption that the electron attracting nitro-group facilities the less of the carbon dioxide.

$$O=\overset{+}{N}(O^-)-CH_2-C(=O)O^- \longrightarrow CO_2 + {}^-O-\overset{+}{N}(O^-)=CH_2 \xrightarrow{H_2O} O=\overset{+}{N}(O^-)-CH_3$$

2. Nitro-compounds are prepared industrially by direct nitration, which is carried out in the ways.

In *liquid phase nitration,* the hydrogen is heated with concentrated nitric acid then pressure at 140°C Nitration under these conditions is always slow, and a large amount of polynitro-compounds is produced.

In *vapour phase nitration,* the hydrocarbon is heated with nitric acid (or with oxides of nitrogen) at 150 – 475°C), each hydrocarbon has its optimum temperature, *e.g.*,

$$CH_3CH_2CH_3 \xrightarrow[400°C]{HNO_3} CH_3CH_2CH_2NO_2 + CH_3CH(NO_2)$$
$$CH_3 + CH_3CH_2NO_2 + CH_3NO_2$$

Vapour-phase nitration is more satisfactory than liquid phase nitration.

Under the above conditions (*i.e.*, at 400°C), the ease of substitution by the nitro group r-H > s-H > prim H. Also any alkyl group present in the alkane can be replaced by a alkyl group, *i.e.*, chain fission takes place, *e.g.*, isopentane yields nine nitroalkanes.

Vapour-phase nitration is accepted as proceeding by a free-radical mechanism, but the details or uncertain

3. Kornblum et al (1956) have prepared t-nitro-compounds by the oxidation of t-carbinamines with potassium permanganate.

$$R_3CNH_2 \xrightarrow{KMnO_4} R_3CNO_2 \qquad (70\text{-}80\%)$$

4.. The addition of nitrous fumes to alkenes is complicated, and much of the earlier work is of doubtful nature. Stevens (1960) has carried a detailed investigation of the addition of pure dioxide tetroxide in ether to l-alkenes in the presence of oxygen, and showed that the major products and dinitro-compound (I), nitro-nitrite (II), nitro-nitrate (III) and nitro-ketone (IV).

$$RCH=CH_2 \xrightarrow[O_2]{N_2O_4} \underset{\displaystyle X}{RCHCH_2NO_2} + \underset{(IV)}{RCOCH_2NO_2}$$

(I) X = $NO_2$, (II) X = ONO, (III), X = $ONO_2$. It is the oxygen which is responsible for the formation of (III) and (IV) both being formed at the expense of (I) expense of (I). According to Schechter *et al.,* (1952). Stevens *et al.,* (1958), (1) and (II) are formed by the addition of the nitronium free-radical:

$$N_2O_4 \rightleftharpoons 2.NO_2$$

$$RCH=CH_2 \xrightarrow{.NO_2} R\dot{C}H-CH_2 \xrightarrow{.NO_2} \underset{\displaystyle NO_2}{RCHCH_2NO_2} + \underset{\displaystyle ONO}{RCHCH_2NO_2}$$

## General Properties

The nitroalkanes are colourless (when pure), pleasant-smelling liquids which are sparingly while water. Most of them can be distilled at atmospheric pressure.

There are two infra-red absorption bands of the nitro-group ($NO_2$ vibrations) in alkyl primary secondary nitro-compounds, 1565-1545 (s) and 1385 – 1360 $cm^{-1}$ (s), ad for the tertiary compounds. 1545–1530 (s) and 1360 – 1340 $cm^{-1}$ (s). These regions clearly distinguish compounds from alkyl nitrites, in which the nitroso-group (N = O stretch, R—O—N=O) verbs in the region 1680-1610 $cm^{-1}$ (v.s)

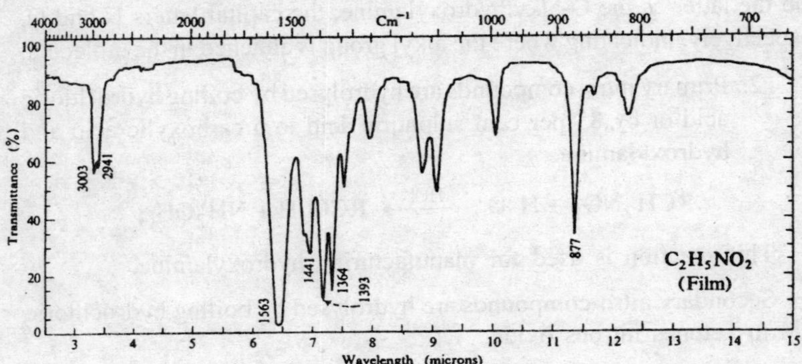


Fig. 2.3

Wintanoski *et al.*, (1964), from their NMR studies, have shown that it is possible to identify prim s-and t-nitroalkanes.

Infra-red spectrum of nitroethane (as a film) is shown in Fig. 2.3 and 1393 $cm^{-1}$, $NO_2$ vibrations in alkyl nitro-compounds. 877 $cm^{-1}$. C —N (str.) or $NO_2$ (def.). and 2941 $cm^{-1}$, C—H (str.) in Me and/or $CH_2$. 1441 $cm^{-1}$, C—H (def.) in Me and/or $CH_2$. $cm^{-1}$ C—H (def.) in Me.

**Reactions**

1. Nitroalkanes are reduced catalytically, by lithium aluminium hydride, or by metal in acid solution to primary amines.

$$RNO_2 \longrightarrow RNH_2 + 2H_2O$$

When the reduction with metal is carried out in *neutral* solution, *e.g.*, with zinc dust and ammonium chloride solution, nitro-compounds are converted into n N-alkylhydroxylamine.

$$RNO_2 \xrightarrow{Zn/NH_4Cl} RNHOH + H_2O$$

When stannous chloride and hydrochloric acid are used as the reducing agent, nitro-compounds are converted into a mixture of N-alkylhydroxylamine and oxime.

$$RCH_2NO_2 \xrightarrow{SnCl_2/HCl} RCH_2NHOH + RCH = NOH$$

Since hydroxylamine can give rise to two types of derivatives, *e.g.*, RNHOH and $NH_2OR$, it is necessary to distinguish one form the other. A common method is to name the former as the N-alkylhydroxylamine, and the latter as the O-alkylhydroxylamine, the capital letters N and O, respectively, indicating where the alkyl group is attached in the molecule.

2. Primary nitro-compounds are hydrolysed by boiling hydrochloric acid or by 85 per cent sulphuric acid to a carboxylic acid and hydroxylamine.

$$RCH_2NO_2 + H_2O \xrightarrow{HCl} RCO_2H + NH_2OH$$

This reaction is used for manufacturing hydroxylamine.

Secondary nitro-compounds are hydrolysed by boiling hydrochloric acid to ketones nitrous oxide.

$$2R_2CHNO_2 \xrightarrow{HCl} 2R_2CO + N_2O + H_2O$$

Tertiary nitro-compounds are generally unaffected by hydrochloric acid.

3. Primary and secondary nitro compounds. *i.e.*, those containing α-hydrogen atoms, tautomerism.

$$\underset{\text{(I)}}{RCH_2\overset{+}{N}(=O)O^-} \rightleftharpoons \underset{\text{(II)}}{RCH=N(OH)O^-}$$

The nitro-from (I) is often called the *pseudo-acid* from, (II), the isonitro-form, is known as *aci-form* or *nitronic acid.* This is an example of a triad system, the *nitro-acinitro* system the equilibrium is almost completely on the left, and this may be due to the nitro-form the stabilised by resonance (of the nitro-group).

These nitro-compounds dissolve in aqueous sodium hydroxide to form salts. *e.g.*,

$$\underset{\text{(I)}}{CH_3\overset{+}{N}(=O)O^-} + OH^- \longrightarrow H_2O + \underset{\text{(Ia)}}{{}^-CH_2-\overset{+}{N}(=O)O^-} \longleftrightarrow \underset{\text{(IIa)}}{CH_2=\overset{+}{N}(O^-)O^-}$$

The removal of the proton is made possible by the strong – 1 effect of the nitro-group *e.g.*, adjacent methylene group is an active methylene group is an active methylene group. At the same time, the extended jugation in the conjugate base stabilises it with respect to its acid.

When the sodium salt is acidified at low temperature, there is not always an increase separation of oily drops (of the nitro-form). On standing, however, the acidified solution slowly oily drops. This time factor can be explained by protonation (IIa) to product *aci*-form (soluble) which then slowly tautomerises to the insoluble nitro-form (see also.)

When the sodium salt solution of the nitronic acid is acidified with 50 per cent sulphuric at room temperature, in aldehyde (from a primary nitro-compound) and a ketone secondary nitro-compound) is obtained, *e.g.*, ($R_2$ is either an alkyl group or a hydrogen

$$2R^1R^2C=N\bar{O}_2\overset{+}{Na} + 2H_2SO_4 \longrightarrow 2R^1R^2CO + N_2O$$

$$2NaHSO_4 + H_2O$$

This reaction is known as the Nef carbonyl synthesis (1894).

When treated with stannous chloride and hydrochloric acid, the sodium salt of the acid is reduced to the aldoxime or ketoxime.

$$R^1R^2C{=}N\bar{O}_2N\overset{+}{a} \xrightarrow{SnCl_2/HCl} R^1R^2C{=}NOH$$

4. Primary and secondary nitro-compounds are readily halogenated in alkaline solution in the α–position only.

$$R_2C{=}N\bar{O}_2N\overset{+}{a} + Br_2 \xrightarrow{NaOH} R_2CBrNO_2 + NaBr$$

Primary nitro-compounds can form the mono- and dibromo-derivatives, whereas secondary from only the monobromo-derivative, nitromethane is exceptional in that is can form the bromo-derivative, *Chloropicrin* is manufactured by the reaction between nitroethane, chlorine and sodium hydroxide.

5. Nitro-compounds react with nitrous acid, the product depending on the nature of the alkyl group.

Primary nitro-compounds from *nitrolic acids*; these are crystalline substances which dissolve in sodium hydroxide to give red solutions:

$$RCH_2NO_2 + HONO \longrightarrow R{-}C({=}NOH)(NO_2) + H_2O$$

Secondary nitro-compounds form *pseudonitroles* (Ψ-nitroles), these are colourless crystalline substances which dissolve in sodium hydroxide to give blue solutions the blue colour is problem to the presence of the nitroso-group, see below.

$$R_2CHNO_2 + HONO \longrightarrow R_2C(NOH)(NO_2) + H_2O$$

Tertiary nitro-compounds do not react with nitrous acid since they have no α-hydrogen atom.

These reactions with nitrous acid are the basis of the red white and blue test for the nature of monohydric alcohols (Victor Meyer *et al.,* 1874)

6. Owing to the presence of active α-hydrogen atoms, primary and secondary nitro-compounds undergo condensation with

aldehydes, *e.g.*, nitromethane condenses with benzaldehyde in the source of ethanolic potassium hydroxide to form ω-nitrostyrene.

$$CH_3NO_2 + OH^- \rightleftharpoons H_2O + {}^-CH_2NO_2 \xrightarrow{PhCHO}$$

$$\underset{\displaystyle PhCHCH_2NO_2}{\overset{O^-}{|}} \xrightarrow{H_2O} \underset{\displaystyle PhCHCH_2NO_2}{\overset{OH}{|}} \xrightarrow{-H_2O} PhCH{=}CHNO_2$$

Nitroalkanes can act as addenda in the Michael condensation, *e.g.*, with mesityl oxide nitromethane forms 4,4-dimethyl-5-nitropentan-2-one

$$(CH_3)_2{=}CHCOCH_3 + CH_2{=}NO_2^-Na^+ \longrightarrow$$

$$\underset{\displaystyle\quad CH{=}NO_2^-Na^+}{(CH_3)_2\overset{|}{C}{-}CH_2COCH_3} \xrightarrow{HCl} \underset{\displaystyle\quad CH_2NO_2}{(CH_3)_2\overset{|}{C}{-}CH_2COCH_3}$$

Primary and secondary nitro-compounds also undergo the Michael reaction.

## NITROSOALKANES

The nitrosoalkanes contain nitroso-group. – N = O, directly attached to a carbon atom. They are named as the nitroso-derivatives of the corresponding alkane, *e.g.*,

$(CH_3)_3CNO$ 2-methyl-2-nitrosopropane

### General Methods of Preparation

1. By the addition of nitrosyl chloride or bromide to alkanes where by alkenes, nitrosohalides are formed

$$CH_2{=}CH_2 + NOCl \rightarrow ClCH_2CH_2NO \xrightarrow{2\text{–mol.}} (ClCH_2CH_2NO)_2$$

2. By the action of nitrous acid on certain types of compounds, *e.g.*, secondary nitroalkanes.
3. By the oxidation of primary amines containing a tertiary alkyl group with *e.g.*, Caro's acid (peroxy-mono) sulphuric acid).

$$R_3CNH_2 + 2[O] \xrightarrow{H_2SO_4} R_3CNO + H_2O$$

On the other hand. Emmons (1957) has prepared primary, secondary and tertiary nitroso-compounds by oxidation of amines with neutralised peracetic acid in methylene dichloride (yield. 33–80 per cent.

N-Substituted hydroxylamines may also be oxidised (aqueous potassium dichromate-sulphuric acid to nitroso compounds.

$$\text{RNHOH} \xrightarrow{[O]} \text{RNO}.$$

**General Properties**

Nitroso-compounds, which are usually blue or green liquids, tend to associate to give colourless which (smelling like camphor) which are dimers. These bimolecular solids regenerate the moment of fused or when dissolved in solution. Chilton *et al.,* (1935) and Gowenlock *et al.,*) have examination absorption spectrum of the two solid forms of nitrosomethane, and have concluded that they geometrical isomers.

$$\underset{cis}{\begin{matrix} \text{Me} & & \text{Me} \\ & \overset{+}{\text{N}}=\overset{+}{\text{N}} & \\ {}^{-}\text{O} & & \text{O}^{-} \end{matrix}} \qquad \underset{trans}{\begin{matrix} \text{Me} & & \text{O}^{-} \\ & \overset{+}{\text{N}}=\overset{+}{\text{N}} & \\ {}^{-}\text{O} & & \text{Me} \end{matrix}}$$

Nitroso-compounds exist as such only when the nitroso-groups is attached to a tertiary carbon atom.

If the nitroso-group is attached to a primary or secondary carbon atom, the nitroso-compounds generally unstable, tending to rearrange to the oxime.

$$\underset{\substack{\text{blue and}\\ \text{stable}}}{R_2CH-NO} \longrightarrow \underset{\substack{\text{colourless}\\ \text{and stable}}}{R_2C=NOH}$$

This is known as the *nitro-oximino* triad system. This system is *potentially* tautomeric, but and is known of the isomerisation of the oxime directly to the nitroso-compound.

It has also heat that the bimolecular nitroso-chlorides rearrange, if possible, to the oxime when treated with the hydroxide.

$$(ClCH_2CH_2NO)_2 \xrightarrow{OH^-} 2ClCH_2CH=NOH$$

The so-called isonitroso-compounds (oximino-compounds) are actually the oximes, *e.g.*, acetone, $CH_3COCH = NOH$, formed by the action of nitrous acid on acetone, is really the half of methylglyoxal (q.v.)

Nitroso-compounds may be oxidised to the nitro-compound by nitric acid, and reduced to primary amine by, *e.g.*, tin and hydrochloric acid.

$$R_3CNO_2 \xleftarrow{HNO_3} R_3CNO \xrightarrow{Sn/HCl} R_3CNH_2$$

**Monoamines**

Amines are derivatives of ammonia in which one or more hydrogen atoms have been replaced in alkyl groups. The amines are classified as primary, secondary or tertiary amines according one, two or three hydrogen atoms in the ammonia molecule have been replaced by alkyl groups.

Thus the general formula of primary, secondary and tertiary amines may be written $NH_2$, $R_2NH$ and $R_3N$, and each is characterised by the presence of the *amino-group* – $NH_2$, is amino-group $\rangle NH$, and the *tertiary nitrogen atom* $\rangle NH$—, respectively.

In addition to the amines the tetra-alkyl derivatives of ammonium hydroxide, $[R_4N]–OH^-$, are known, these are called the *quaternary ammonium hydroxides.*

**Nomenclature**

The suffix of the series if *amine*, and each member is named according to the alkyl groups attached to the nitrogen atom, *e.g.*,

| | |
|---|---|
| $CH_3NH_2$ | methylamine |
| $(CH_3)_2NH$ | dimethylamine |
| $(CH_3)_3N$ | trimethylamine |
| $(C_2H_5)_2NCH(CH_3)_2$ | diethylisopropylamine |

The amines are said to be simple when all the alkyl groups are the same, and mixed when the groups are different.

The method of naming a quaternary ammonium hydroxide is illustrated by the following examples.

| | |
|---|---|
| $(CH_3)_4N^+OH^-$ | tetramethylammonium hydroxide |
| $[(CH_3)_3\overset{+}{N}C_2H_5]OH$ | ethyltrimethylammonium hydroxide |

## General Methods of Preparation

Many methods are available for preparing amines, but it is instructive to consider them in the following groups.

### Methods for the Three Classes of Amines

1. *Hofmann's method* (1850). This is by means of *ammonolysis,* an alkyl halide and an ethanolic atom of ammonia being heated in a sealed tube at 100°C. A mixture of all three classes of atom in obtained, together with some quaternary ammonium compound.

$$RX + NH_3 \longrightarrow RNH_3{+}X{-}$$

$$RNH_3{}^+X^- + NH_3 \rightleftharpoons RNH_2 + NH_4X$$

$$RNH_2 + RX \longrightarrow R_2NH_2{}^+X^- \xrightleftharpoons{NH_3} NH_4X + R_2NH$$

$$\xrightarrow{RX} R_3NH^+X^- \xrightleftharpoons{NH_3} NH_4X + R_3N \xrightarrow{RX} R_4N^+X^-$$

The alkylation of amines works well with primary alkyl halides, fairly well with many s-alkyl halides but t-alkyl halides give alkenes, *e.g.*,

$$H_2\ddot{N}\ H{-}CH_2{-}CMe_2{-}Cl \longrightarrow CH_2{=}CMe_2 + NH_4Cl$$

The order of reactivity of halides is RI > RBr > RCl, and the mechanism of the reaction is $S_N2$

2. A mixture of the three types of amines may be prepared by the ammonolysis of alcohols the alcohol and ammonia are heated under pressure in the presence of a catalyst, *e.g.*, chromite or alumina.

$$ROH \xrightarrow{NH_3} H_2O + RNH_2 \xrightarrow{ROH} H_2O + R_2NH$$

$$\xrightarrow{ROH} H_2O + R_3N$$

### Separation of Amine Mixtures

When the mixture contains the three amine salts and the quaternary salt, it is distilled potassium hydroxide solution. The three amines distiil, leaving the quaternary salt unchanged solution, *e.g.*,

$$R_2NH_2{}^+X^- + KOH \longrightarrow R_2NH + KX + H_2O$$

The distillate of mixed amines may now be separated into the individual amines as follows.

*Fractional distillation.* This is the most satisfactory method and is now used industrially, success being due to the high efficiency of industrial fractionation apparatus.

*Hinsberg's method* (1890). The mixture of amines is treated with an aromatic sulphonyl chloride. Benzenesulphonyl chloride, $C_6H_5SO_2Cl$, was used originally, but has now been replaced by p-toluenesulphonyl chloride, $CH_3C_6H_4SO_2Cl$. After treatment with this acid chloride, the solution is made alkaline with potassium hydroxide. Primary amines form the N-alkyl-sulphonamide, *e.g.*, ethylamine forms N ethyl-p-toluenesulphonamide.

$$CH_3C_6H_4SO_2Cl + C_2H_5NH_2 \longrightarrow CH_3C_6H_4SO_2NHC_2H_5 + HCl$$

These sulphonamides are often soluble in potassium hydroxide solution due to the formatior of the potassium salt. The toluensulphonyl group is strongly electron-attracting, and there facilitates the release of the proton attached to the nitrogen atom.

$$CH_3C_6H_4SO_2NHC_2H_5 + KOH \longrightarrow [CH_3C_6H_4SO_2\bar{N}C_2H_5] \leftrightarrow$$

$$\left[\begin{array}{c} O^- \\ | \\ CH_3C_6H_4S{=}NC_2H_5 \\ \| \\ O \end{array}\right] K^+ + H_2O$$

Secondary amines form N, N-dialkyl-sulphonamides, which are insoluble in potassium hydroxide solution because there is no α-hydrogen atom, *e.g.*, diethylamine forms N,N-diethyle toluenesulphonamide.

$$CH_3C_6H_4SO_2Cl + (C_2H_5)_2NH \longrightarrow CH_3C_6H_4SO_2N(C_2H_5)_2 + HCl$$

Tertiary amines do not react with p-toluenesulphonyl chloride.

The alkaline solution is distilled, and the tertiary amine thereupon distils off. The residence liquid is filtered, the filtrate contains the primary amine derivative which, on acidification give the alkyl-sulphonamide the residual solid is the dialkyl -sulphonamide. The amines are to generated from the sulphonamides by refluxing with 70 oer cent sulphuric acid or 25 per cent hydrochloric acid.

Unfortunately, the Hinsberg separation is not always successful because many monoalky suphonamides are insoluble in alkali. This insolubility appears to be general for the highe primary amines. *e.g.*, the sulphonamide form n-heptylamine.

**Preparation of Primary Amines**

1. By the reduction of nitro-compounds with metal and acid, or with hydrogen and a nickel catalyst, or with LAH
2. (i) By the reduction of alkyl cynides.

   (ii) By the reduction of oximes or amides with sodium and ethanol, LAH, or catalytically

On the other hand, oximes are reduced by diborane to N-substituted hydroxylamines (Feuer *et al.,* 1965)

$$R_2C{=}NOH \xrightarrow{B_2H_4} R_2CHNHOH \quad (50\text{-}90\%)$$

3. By passing a mixture of aldehyde or ketone and a large excess of ammonia and hydrogen under pressure (20–150 atm) over Raney] nickel at 40–150°C. The reaction may also be carried out at I in the presence of excess of ammonium chloride and Adams' platinum catalyst. A possible mechanism is via an imine.

$$R_2C{=}O \;\; \ddot{N}H_2 \rightarrow R_2\overset{O^-}{\overset{|}{C}}{-}NH_3^+ \rightarrow R_2\overset{OH}{\overset{|}{C}}{-}NH_2 \xrightarrow{-H_2O}$$

$$R_2C = NH \xrightarrow[Ni]{H_2} R_2CHNH_2$$

   Small amounts of secondary and tertiary amines are obtained as by products. This process of producing alkyl groups into ammonia or primary or secondary amines by means of an aldehyde or ketone in the presence of a reducing agent is known as *reductive alkylation.*

4. Aldehydes and ketones react with ammonium formate or with formamide to give formyl activities of primary amines (*Leuckart reaction*. 1885).

$$\rangle C{=}O + 2HCO_2NH_4 \rightarrow \rangle CHNHCHO + 2H_2O + CO_2 + NH_3$$

$$\rangle C{=}O + 2HCONH_2 \rightarrow \rangle CHNHCHO + CO_2 + NH_3$$

These formyl derivatives are readily hydrolysed by acid to the primary amine.

The mechanism of the Leuckart reaction is not settled, but it appears that an imine is formed first and this is then reduced to an amine by hydride ion transfer from formic acid.

$$\rangle C{=}O + NH_2{-}CHO \longrightarrow \rangle C(O^-){-}NH_2CHO \longrightarrow \rangle C(OH){-}NHCHO \xrightarrow{-H_2O}$$

$$\rangle C{=}NCHO \longrightarrow \rangle CHNHCHO + CO_2$$

$$(H{-}COOH)$$

5. *Gabriel's phthalimide synthesis (1887)* : In this method phthalimide is converted by means ethanolic potassium hydroxide, into its salt which, on heating with an alkyl halide, gives the N-alkyl phthalimide. This is then hydrolysed to phthalic acid and a primary emine by heating with 20 per cent hydrochloric acid under pressure, or by refluxing with potassium hydrogen solution.

$$C_6H_4(CO)_2NH \xrightarrow{KOH} C_6H_4(CO)_2\overset{-}{N}\overset{+}{K} \xrightarrow{RI}$$

$$C_6H_4(CO)_2NR \xrightarrow{H_2O} C_6H_4(CO_2H)_2 + RNH_2$$

When hydrolysis is difficult, the alkyl phthalimide can be treated with hydrazine to give the acid.

$$C_6H_4(CO)_2NR + \begin{matrix} H_2N \\ | \\ H_2N \end{matrix} \longrightarrow C_6H_4\begin{matrix} CONH \\ | \\ CONH \end{matrix} + RNH_2$$

Gabriel's synthesis is a very useful method since it gives a pure primary amine, and has been improved by using dimethyl formamide as solvent.

6. By means of a Grignard reagent (or a trialkylbronane) and chloramine. The product is

$$RMgX + ClNH_2 \longrightarrow RNH_2 + MgXCl$$

Alternatively, a primary amine may be prepared by interaction of a Grignard reagent O-methylhydroxylamine.

7. A good method for preparing primary amines containing a t-alkyl group is by the reaction (1948). This reaction depends on the formation of intermediate carbonium ions, are hence is mainly used with t-alcohols (or alkenes that can product t-carbonium ions). The reactions occurs between alcohol (or alkene) and alkyl cyanide (or hydrocyanic acid) in the presence of sulphuric acid to produce an amide, *e.g.*,

$$R_3COH \xrightarrow{H^+} H_2O + R_3C^+ \xrightarrow{NaCN} R_3C\overset{+}{N}{\equiv}CH \xrightarrow{H_2O}$$

$$R_3CNHCHO \xrightarrow{OH^-} R_3CNH_2 + HCO_2$$

8. Other methods for preparing primary amines are. Hofmann's preparation of primary very convenient method), the Curitus reaction, Lossen reaction means not a good method), Schmidt reaction with acids or with ketones these preparations may be dangerous).

Wurtz (1849) discovered amines by boiling alkyl isocyanates with alkali.

$$RNCO + 2KOH \longrightarrow RNH_2 + K_2CO_3$$

This method is not very useful in practice owing to the inaccessibility of the isocyanates.

## Preparation of Secondary Amines

1. By heating a primary amine with the calculated quantity of alkyl halide.

$$R^1NH^2 + R_2X \longrightarrow R^1R^2NH_2^+X^-.$$

This reaction usually result in some over-alkylation, but this may be avoided as follows.

$$R^1NH_2 \xrightarrow[OH^-]{TsCl} Ts\bar{N}R \xrightarrow{R^2X} TsNR^1R^2$$

$$\xrightarrow{H^+} TsOH + R^1R^2NH$$

2. Aniline is heated with an alkyl halide, and the product, dialkylaniline, is treated with nitrous and, the p-nitroso-dialkylaniline so formed is then boiled with sodium hydroxide solution. A pure secondary amine and p-nitrosophenol are produced.

$$\underset{}{\overset{NH_2}{\bigcirc}} \xrightarrow{RX} \overset{NR_2}{\bigcirc} \xrightarrow{HNO_2} \underset{NO}{\overset{NR_2}{\bigcirc}} \xrightarrow{NaOH} \underset{NO}{\overset{OH}{\bigcirc}} + R_2NH$$

This is one of the best methods of preparing pure secondary amines.

3. Pure secondary amines may be prepared from calcium cyanamide as follows.

$$CaNCN \xrightarrow{NaOH} Na_2NCN \xrightarrow{2RX} R_2NCN$$

$$\xrightarrow[OH^-]{H^+ \text{ or}} R_2NH + CO_2 + NH_3$$

The catalytic reduction of imines may be used to prepare secondary amines, but this method is usually confined to amines containing a phenyl group (since the aldehyde is commonly benzaldehyde). This may be avoided as follows.

$$PhCHO + R^1NH_1 \longrightarrow PhCH{=}NR^1 \xrightarrow[Ni]{H_2} PhCH_2NHR^1$$

$$\downarrow R^2X$$

$$[PhCH{=}NR^1R^2]X^- \xrightarrow{H^+} PhCHO + R^1NHR^2$$

On the other hand, reductive alkylation of any carbonyl compound in the presence of a primary amine gives secondary amines.

**Preparation of Tertiary Amines**

1. The best method is to heat an ethanolic solution of ammonia with alkyl halide which is used in slight excess required by the equation.

$$3RX + NH_3 \longrightarrow R_3NH^+ X^- + 2HX$$

2. By the reductive alkylation of a carbonyl compound (in excess) in the presence of a secondary amine *e.g.*,

$$R_2^1NH + R^2CHO + H_2 \xrightarrow{Ni} R_2^1NCH_2R^2 \;\; H_2O$$

### Preparation of Quaternary Compounds

This is only one satisfactory method of preparing quaternary compounds, viz, by heating ammonia with a very large excess of alkyl halide, a primary, secondary or tertiary amine may be used instead of ammonia. The starting materials will depend on the nature of the desired quaternary compound, *e.g.*, ethyltrimethylammonium iodide may be prepared by heating trimethylamine with ethyl iodide.

$$(CH_3)_3N + C_2H_5I \longrightarrow [(CH_3)_3NC_2H_5]^+I^-$$

### General Properties of the Amines

The lower members are gases, soluble in water (with which they can form hydrogen bonds). These are followed by members which are liquids, and finally by members which are solids. The solubility in water decreases as their molecular weight increases. All the volatile members have a powerful fishy smell, and are combustible.

The reactions of an amine depend very largely on the class of that amine.

Primary and secondary amines both contain the NH group, both types (n–H stretch) absorb in the region 3 500–3 300cm$^{-1}$ (v), but the two are distinguished by the fact that primary amines show two bands in this region, where as secondary amines generally show only one. If there is hydrogen bonding, then the region is 3400-3 100 cm$^{-1}$ (s). The C–N (stretch) absorption region of aliphatic amines is 1220–1020 cm$^{-1}$. The intensity of absorption is weak to medium, and this region is different for aromatic amines, for which the bands are strong. There is also a band due to N–H (def.), primary amines, 1650–1590cm$^{-1}$(cm)., secondary amines, 1650–1550 cm$^{-1}$(s). Because of the overlap of these regions, it is not usually possible to distinguish the two types of amines in this way.

Fig,. 2.4 is the infra-red spectrum of cyclohexylamine (as a film). 3333 and 3257 cm$^{-1}$ : N–H (st.) in primary amines. 3175 cm$^{-1}$ : N–H (str.); association? 1600 cm$^{-1}$ : N–H (def.) in primary (or secondary) amines.1087 cm$^{-1}$ : C–N (str.) in primary (or secondary) amines. 2940 and 2870 cm$^{-1}$ : C–H (str.) in (Me and/or) $CH_2$. 1450 cm$^{-1}$ : C–H (def.) in (Me and/or) $CH_2$.

Protons attached to nitrogen usually give rise to a single absorption line in the NMR spectrum. The reason for this is the same as that for alcohols, *i.e.*, there is rapid chemical exchange of the amino hydrogens. The following τ-values are characteristic of the NH proton, aliphatic and cyclic amines, 8-9 p.m, aromatic amines, 5.3-7.4, amides, 2.0-5.0. NMR studies have also been used to detect, in particular, N-methyl groups (τ-value 7.8-8.0).

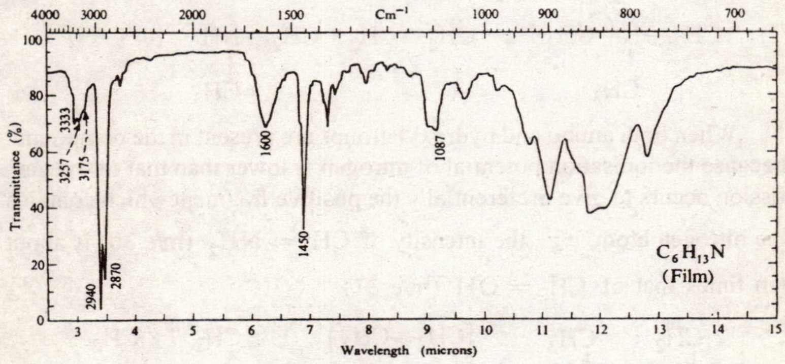


**Fig. 2.4**

A problem in connection with nitrogen compounds is that of restricted rotation *e.g.*,

$$\mathrm{Me_2\ddot{N}{-}C({=}O)R} \qquad \mathrm{Me_2\ddot{N}{-}\ddot{N}{=}O}$$

In both examples, the environments of the methyl groups are different and so these will give different signals.

When a compound contains an odd number of nitrogen atoms, its molecular weight is an odd number. The molecular on of amines is very weak or absent, and its most characteristic fragmentation pattern is viz β-cleavage, *e.g.*,

$$R{-}CH_2{-}\overset{+}{N}H_2 \longrightarrow R\cdot + \overset{+}{C}H_2{=}NH_2 \quad (m/e\ 30)$$

Thus, the base peak of primary amines is m/e 30 provided no branching occurs at the α-carbon atom. The presence of this peak, however, does not necessarily mean that the amine is primary, since s- and t-amines can also give rise to tis peak via. McLafferty rearrangements. The behaviour of these amines is similar to that of the ethers, β-cleavage occurs preferentially to eliminate the largest alkyl group, *e.g.*, ($R^1 > R^2$ and $R^3$).

$$R^1-CH_2-\overset{+}{N}R^2R^3 \longrightarrow R^1 + CH_2=\overset{+}{N}R^2R^3$$

If, *e.g.*, $R^2$ = Me and $R^3$ = Et, then a McLafferty rearrangement can now occur.

$$CH_2=\overset{+}{N}(CH_3)-CH_2-CH_2-H \longrightarrow CH_2=CH_2 + CH_2=\overset{+}{N}H(CH_3) \quad (m/e\ 44)$$

When both amino and hydroxyl groups are present in the compound, because the ionisation potential of nitrogen is lower than that of oxygen, fission occurs to give preferentially the positive fragment which contains the nitrogen atom, *e.g.*, the intensity of $CH_2=\overset{+}{N}H_2$ (m/e 30) is about ten times that of $CH_2=\overset{+}{O}H$ (m/e 31).

$$\underset{OH}{\overset{CH_2}{|}} + \underset{NH_2}{\overset{CH_2}{\|}} \longleftarrow \left[\underset{OH}{\overset{CH_2}{|}}-\underset{NH_2}{\overset{CH_2}{|}}\right]^{+} \longrightarrow \underset{^{+}OH}{\overset{CH_2}{\|}} + \underset{^{+}NH_2}{\overset{\cdot CH_2}{\|}}$$

The order of stability of fragments of the above type is.

$$CH_2=\overset{+}{N}H_2 > CH_2=\overset{+}{S}H > CH_2=\overset{+}{O}H$$

**Reactions Given by all Three Classes of Amines**

1. All the amines are basic, and they are stronger bases than ammonia.

$$RNH_2 + H_2O \rightleftharpoons RNH_3^{+} + OH^{-}$$

One factor that controls the basicity of amines will be the availability of the lone pair for protonation. At first sight, it would appear that as the number of alkyl groups increases, then because of their + 1 effect the lone pair will become more available, thereby increasing the basicity of the amine.

Inspection of the $pK_2$' in Table 13.1 shows that the introduction of one alkyl group into ammonia strengthens the base (remember . $pK_a = pK_w - pK_b$) a second alkyl group further strengthens the base, but the third alkyl group *decreases* the strength, *e.g.*,

$$NH_3 < Me_3N < MeNH_2 < Me_2NH$$

The reason for this order is not clear. It has been suggested that steric factor operates. Addition of the proton increase crowding and so sets up strain, which will be greatest in the tertiary amine, and consequently the stability of this molecule is decreased, *i.e.*, its basicity is reduced. However, when we consider that the three bonding pairs and the lone pair all occupy $sp^3$ orbitals, and that a lone pair causes crowding of bonding pairs, it would seem that protonation should relieve crowding by transforming the line pair into a bonding pair.

**Table 2.1**

| *Base* | *pKa* | *Base* | *pKa* |
|---|---|---|---|
| $NH_3$ | 9.27 | n-$PrNH_2$ | 10.58 |
| $MeNH_2$ | 10.62 | iso-$PrNH_2$ | 10.63 |
| $Me_2NH$ | 10.77 | n-$BuNH_2$ | 10.61 |
| $Me_3N$ | 9.80 | t-$BuNH_2$ | 10.45 |
| $EtNH_2$ | 10.63 | n-$aMNH_2$ | 10.63 |
| $Et_2NH$ | 10.93 | $NCCH_2NH_2$ | 5.34 |
| $Et_3N$ | 10.87 | | |

All the onium ions carry a positive charge and so are solvated. The greater the ion, the less will be the solvation and so the less stabilised is the ion. It seems possible that in going from $Me_2NH$ to $Me_3N$, although the availability of the lone pair is increased, the ion has now grown sufficiently in size at this stage to become less solvated and so the stabilisation lost due to the decrease in solvations is greater than the inductive effect making the lone pair available for protonation.

That the inductive effects are important is shown by the fact that, *e.g.*, in cyanomethylamine, in which the cyano group has a strong – I effect, the pKa value is must lower than for methylamine.

2. All the amines combine with acids to form salts, *e.g.*, methylamine combines with hydrochloric acid to form *methylyammonium chloride.*

$$CH_3NH_2 + HCl \longrightarrow [CH_3NH_3]^+Cl^-$$

The nitrogen is quadricovalent unielectrovalent in amine salts, but to show their relationship is the amines, their salts are often written as, *e.g.*, $CH_3NH_2HCl$, methylamine hydrochloride $[C_2H_5)_2NH]_2$, $H_2SO_4$, diethylamine sulphate.

3. All the amines combine with alkyl halides to form alkyl-substituted ammonium halides with more alkyl groups than the amine used.
4. When an amine salt is heated at high temperature, a molecule of alkyl halide is eliminated (*i.e.*, reverse of reaction 3, above).

$$R_3NH^+Cl^- \longrightarrow RCl + R_2NH \xrightarrow{HCl} RCl + RNH_2$$

$$\xrightarrow{HCl} RCl + NH_4Cl$$

If a t-alkyl group is present, this is eliminated as alkene (via pyrolysis of the alkyl halide). If a methyl group is present, this is generally eliminated preferentially.

When the methyl ether of a high boiling hydroxy compound is heated with concentration hydriodic acid at 100°C, methyl iodide is formed.

$$ArOCH_3 + HI \xrightarrow{100°C} ArOH + CH_3I$$

This is the basis of the *Zeisel method* for the quantitative estimation of methoxyl groups. When the temperature is raised to 150°C, N-methyl groups are converted into methyl iodide.

$$ArNHCH_3 + HI \xrightarrow{150°C} ArNH_2 + CH_3I$$

This is the basis of the *Herzing-Meyer method* for the quantitative estimation of methylimino groups. Thus methoxyl and methylimino groups can be estimated separately when both are present in the same compound.

**Reactions Given by Primary and Secondary Amines**

1. Primary and secondary amines react with acid chlorides or anhydrides to form the N-alkyl and aide, *e.g.*,

$$RNH_2 + (CH_3CO)_2O \longrightarrow CH_3CONHR + CH_3CO_2H$$

These monocycly derivatives are easily prepared and, since they are usually well-defined crystalline solids, are used to characterise primary amines.

It is difficult to prepare toe diacetyl derivative, excess of the acetylating agent and high temperature, however, often yield the diacetyl derivative.

$$RNHCOCH_3 + (CH_3CO)_2O \longrightarrow RN(COCH_3)_2 + CH_3CO_2H$$

Secondary amines form only the monoacyl d derivative.

$$R_2NH + CH_3COCl \longrightarrow CH_3CONR_2 + HCl$$

Salphonyl chlorides also react with primary and secondary amines to form N-alkyl-sulphonamides (see the Hinsberg separation,).

2. Primary and secondary amines form substituted ureas with pheynl isocyanate, and may be characterised by these derivatives.
3. Primary and secondary amines can participate in the *Mannich reaction* (1917). This is the condensation between formaldehyde (but other aldehydes may be used), ammonia or a primary or secondary amine (preferably as the hydrochloride) and a compound containing at least one give hydrogen atom, *e.g.*, ketones β-keto-esters, β-cyano-esters, nitroalkanes, alkynes (with $C\equiv H$, etc. The reaction is usually carried out in acid solution but may also be base-catalysed).

The mechanism of the Mannich reaction is uncertain. According to Masui *et al.,* (1970), the Mannich intermediate is formed as follows, *e.g.*, for diethylamine in acid solution.

$$2Et_2NH + CH_2O \rightleftharpoons H_2O + Et_2N - CH_2 - NEt_2 \overset{H^+}{\rightleftharpoons}$$

$$Et_2NH + Et_2\overset{+}{N} = CH_2 \longleftrightarrow Et_2N—CH_2$$

The subsequent steps are possibly, for acetophenone (the enol form appears to be the reactive species in all cases).

$$PhCOCH_3 \overset{H^+}{\rightleftharpoons} PhC(OH){=}CH_2 \quad \cdot CH_2{=}\overset{+}{N}Et_2 \longrightarrow Ph—C(={}^{+}OH)—CH_2CH_2NEt_2$$

$$\longrightarrow PhC(=O)—CH_2CH_2\overset{+}{N}HEt_2$$

The free base is obtained by basifying the solution at the end of the reaction

The products of the Mannich reaction are known as *Mannich bases,* and many are useful as intermediates in synthesis, *e.g.*,

(i) $PhCOCH_2CH_2NHR_2{}^{+}Cl^{-} \xrightarrow{heat} PhCOCH{=}CH_2$

$$\xrightarrow[\text{cal.}]{H_2} PhCOCH_2CH_3$$

(ii) $PhCOCH_2CH_2NR_2 \xrightarrow{CN^-} PhCOCH_2CH_2CN$.

4. Primary and secondary amines are methylated by heating with formaldehyde and an excess of formic at 100°C.

5. Primary and secondary amines form N-halogen-amines with halogen in weakly alkaline solution is.

$$RNH_2 \xrightarrow[\text{aq. } Na_2CO_3]{Br_2} RNHBr + RNBr_2$$

**Reactions Given by Primary Amines**

1. Primary amines form isocyanides when heated with chloroform and ethanolic potassium hydroxide

$$RNH_2 + CHCl_3 + 3KOH \longrightarrow RNC + 3KCl + 3H_2O$$

This reaction is used as a test for primary amines (or for chloroform)

2. When warmed with carbon disulphide, primary amines form a *dichiocarbamic acid,* which as decomposed by mercuric chloride to the *alkyl isothiocyanate.*

$$RNH_2 + CS_2 \rightarrow S{=}C\begin{matrix} \diagup NHR \\ \diagdown SH \end{matrix} \xrightarrow{HgCl_2} RNCS + HgS + 2HCl$$

This is known as the *Hofmann mustard oil reaction,* and may be used as a test for primary amines

3. Primary amines combine with aldehydes to form imines.

$$C_6H_5CHO + RNH_2 \longrightarrow C_6H_5CH = NR + H_2O$$

4. Primary amines react with nitrous acid with the evolution of nitrogen. The equation is usually written.

$$RNH_2 + HNO_2 \longrightarrow ROH + N_2 + H_2O$$

The reaction however, is far mor complicated than this equation indicates. When that nature of R permits it, the products are mixtures of isomeric alcohols, isomeric alkenes, and other compounds, *e.g.*, n-propylamine give n-propanol (7 per cent), isopropanol (32 per cent), propane (28 per cent), and propyl and isopropyl halides (when the nitrous acid is generated from $NaNO_2$ and HX). These products can be explained on the basis of the formation of an intermediate carbonium ion.

$$RNH_2 \xrightarrow{NaNO_2} R-\overset{+}{N}\equiv N + X^- \longrightarrow R^+ + N_2 + X^-$$

Thus, the reaction for the deamination of n-propylamine may be written.

$$CH_3CH=CH_2 \xleftarrow{-H^+} CH_3CH_2\overset{+}{C}H_2 \xrightarrow{H_2O} CH_3CH_2CH_2\overset{+}{O}H_2 \xrightarrow{-H^+} CH_3CH_2CH_2OH$$

$$CH_3CH_2\overset{+}{C}H_2 \xrightarrow{X^-} CH_3CH_2CH_2X$$

$$CH_3CH_2\overset{+}{C}H_2 \xrightarrow{H^-\ shift} CH_3\overset{+}{C}HCH_3 \xrightarrow[(ii)\ -H^+]{(i)\ H_2O} (CH_3)_2CHOH$$

$$CH_3\overset{+}{C}HCH_3 \xrightarrow{X^-} CH_3CHXCH_3$$

Whatever are the products formed by the action of nitrous acid on a primary amine nitrogen is always evolved. Thus this reaction may be used as a test for primary amines, since none of the other classes of amines liberate nitrogen.

The mechanism given above does not explain may other experimental results *e.g.*, n-proplyamine gives some cyclopropane (Silver. 1961. Skell *et al.*, 1962), s-butylamine gives a small amount of methylcyclopropane (Lee et at. 1965), and if the starting amine is an optically active form, then the corresponding alcohol produced is largely inverted.

$$\text{D-MeCH}_2\underset{\underset{NH_2}{|}}{\text{CHMe}} \xrightarrow{HNO_2} \triangleright\!\!-\text{Me} + \text{D-MeCH}_2\text{CHOHMe} + \text{L-MeCH}_2\text{CHOHMe}$$

The formation of inverted alcohol *without* rearrangement suggests that the diazonium ion is attacked directly.

$$H_2O\ R-N_2^+ \longrightarrow N_2 + H_2\overset{+}{O}-R \xrightarrow{-H^+} HO-R$$

On the other hand, some amines lead to a large amount of retention, *e.g.*, α-phenyl-s-butylamine (Streitwieser *et al.*, 1959). It must therefore be concluded that a number of paths are followed during deamination. One suggestion (Streitwieser, 1957) is that all acyclic products can be explained in terms of both the carbonium ion and the diazonium cation undergoing reaction.

Also the formation of cyclic products can be explained by internal electrophilic attack in a carbonium ion, *e.g.*,

$$H-CH_2-CH_2-\overset{+}{C}H_2 \longrightarrow H^+ + \underset{CH_2}{CH_2-CH_2}$$

5. *Oxidation of primary amines.* The products obtained depend on the oxidising agent used, and on the nature of the alkyl group *e.g.*,

(i) with potassium permanganate.

$$RCH_2NH_2 \xrightarrow{[O]} \underset{\text{aldimine}}{RCH=NH} \xrightarrow{H^+} RCHO + NH_3$$

$$R_2CHNH_2 \xrightarrow{[O]} \underset{\text{ketimine}}{R_2C=NH} \xrightarrow{H^+} R_2CO + NH_3$$

$$R_3CNH_2 \xrightarrow{[O]} R_3CN_2$$

(ii) with Caro's ($H_2SO_5$)

$$RCH_2NH_2 \xrightarrow{[O]} \underset{\text{N-alkyl-hydroxylamine}}{RCH_2-NHOH} + \underset{\text{aldoxime}}{RCH=NOH} + \underset{\text{hydroxamie acid}}{RC(OH)=NOH}$$

$$R_2CHNH_2 \xrightarrow{[O]} R_2C=NOH$$

$$R_3CNH_2 \xrightarrow{[O]} R_3CNO$$

**Reaction Given by Secondary Amines**

3. Secondary amines react with nitrous acid to form insoluble only *nitrosamines, nitrogen is not evolved.*

$$HO-NO + H^+ \rightleftharpoons H_2\overset{+}{O}-NO \rightleftharpoons H_2O + NO^+$$

$$R_2NH + NO^+ \rightleftharpoons R_2\overset{+}{N}NHNO \longrightarrow R_2NNO + H^+$$

Nitrosamines are yellow neutral oils which are steam-volatile. When warmed with a crystal of phenol a few drops of concentrated sulphuric acid, nitrosamines form green solution which when made alkaline with aqueous sodium hydroxide, turns deep blew. This procedure may be used a test for secondary amines, it is known as *Liebermanns's nitroso reaction.*

Nitrosamines are readily hydrolysed to the amine by boiling with dilute hydrochloric acid.

$$R_2NNO + H_2O \xrightarrow{HCl} R_2NH + HNO_2$$

Peroxytrifluoroacetic acid converts secondary nitrosamines into nitrosamines (Emmons, 1954).

$$R_2NNO \xrightarrow{CF_3CO_3H} R_2NNO_2 \qquad (73\text{-}95\%)$$

2. When warmed with carbon disulphide, secondary amines form a *dithiocarbamic acid,* which is not decomposed by mercuric chloride to the alkyl isothiocyanate (cf. primary amines).

$$S{=}C{=}S + R_2NH \longrightarrow S{=}C\begin{matrix} \diagup NR_2 \\ \diagdown SH \end{matrix}$$

3. Secondary amines readily form *enamines* with aldehydes containing α-hydrogen.

4. Secondary amines may be oxidised, the product depending on the oxidising agent used, *e.g.,*

$$R_2NH \xrightarrow{KMnO_4} \underset{\text{tetra-alkylhydrazine}}{R_2N - NR_2} ;$$

$$R_2NH \xrightarrow[\text{or } H_2O_2]{H_2SO_5} R_2NOH$$

**Reaction Given by Tertiary Amines**

1. Tertiary amines dissolve in *cold* nitrous acid to form the nitride salt, $R_3N.HNO_3$ or $[R_3NH]^+NO_2^-$,. If the amine is of the type $[RCH_2)_3N$ or $(R_2CH)_3N$, then, when the solution is warmed, the former produces an aldehyde, the latter a ketone, and in both cases a nitrosamine and nitrous oxide.
2. Tertiary amines are not affected by potassium paramagnet, but are oxidised to the *amines oxide* by Caro's acid, ozone or by hydrogen peroxide.

$$R_3N + [O] \longrightarrow R_3\overset{+}{N} - \overset{-}{O}$$

These amine oxides are basic and exist as $[R_3NOH]^+OH^-$ in solution, Amine oxides form addition compounds with gaseous hydrogen halide, *e.g.*, $[R_3NOH]^+Cl^-$, and with alkyl halides, *e.g.*, $[R_3NOR]^+Cl^+$.

Tertiary amines are oxidised at the a-hydrogen atoms by potassium permanganate, manganese dioxide, or mercuric acetate, the product is an enamine or its hydrolytic product (which may be further oxidised), *e.g.*,

$$R^1CH_2CH_2NR_2^2 \xrightarrow{Hg(OAC)_2} R^1CH{=}CHNR_2^2 \xrightarrow{H_2O}$$

$$R^1CH_2CHO + R_2^2NH$$

3. Tertiary amines react with cyanogen bromide to form a *dialkycyanamide* which, on hydrolysis, gives a secondary amine.

$$R_3N + BrCN \longrightarrow [R_3NCN]^+Br^- \longrightarrow RBr + R_2NCN$$

$$\xrightarrow[OH^-]{H^+ \text{ OR}} R_2NH$$

This method therefore offers a means of converting a tertiary amine in to a secondary, but is mainly used to open the ring of a cyclic amine containing a tertiary nitrogen atom.

## QUATERNARY AMMONIUM COMPOUNDS

The quaternary ammonium salts are white crystalline solids, soluble in water, and completely dissociated in solution (the nitrogen is quadricovalent unielectrovalent). When a quaternary ammonium halide is heated *in vacuo,* it gives the tertiary amine.

$$[R_4N]^+X^+ \longrightarrow R_3N + RX \text{ (g.v.g.,)}$$

Quaternary ammonium halides are converted into their corresponding hydroxides by moist silver oxide, methanolic potassium hydroxide, or by passing the halide over a basic ion exchange resin *e.g.*,

$$[R_4N]^+ X^- + \text{'AgOH'} \longrightarrow [R_4N]^+ OH^- + AgX$$

The quaternary ammonium hydroxides are white deliquescent crystalline solids which are is strongly basic as sodium and potassium hydroxides (all exist as ions in the solid state).

Their solutions absorb carbon dioxide, and will liberate ammonia from its salts.

Quaternary ammonium salts containing a long-chain alkyl group have detergent properties, and are known as *invert soaps* because their surface activity is due to the presence of the *positive* and not to a negative ion as in alkyl hydrogen sulphate detergents.

**Elimination Reactions**

Quaternary ammonium hydroxides (and their salts in the presence of alkoxide ions) undergo thermal decomposition to form alkenes, and do so by the E2 mechanism (the reader should revise the section on elimination reactions undergone by alkyl halide, before proceeding further).

The early work was done by Hofmann (1881), and the believed that only tetra-methylammonium hydroxide gave an alcohol (methanol) as one product, but Musker (1964) has shown that the major decomposition products are trimethylamine and dimethyl ether, the nor products are methanol and some other compounds.

$$2(CH_3)_4N + OH^- \longrightarrow 2(CH_3)_3N + (CH_3)_2O + CH_3OH$$

Musker worked with the solid substrate, but Tanaka *et al.,* (1966) showed that when this hydroxide is heated in aqueous solution under pressure, the decomposition proceeds according to the equation.

$$(CH_3)_4N^+ \; OH^- \longrightarrow (CH_3)_3N + CH_3OH$$

As outcome of his work, Hofmann proposed an impartial rule (the Hofmann rule) for the decomposition of 'onium hydroxides. *If an ethyl group is present, the predominant olefinic product is ethylene.* However, a detailed study of all cases where ' onium hydroxides give alkenes on heating, has led to a general statement of Hofmann's rule. *Elimination of hydrogen occurs referentially from that b-carbon atom which is joined to the largest number of hydrogen atoms, e.g.,*

$$\underset{\underset{\beta}{CH_2CH_3}}{\overset{Me}{Me\text{—}\overset{|}{\underset{|}{N^+}}}}\text{—}CH_2\overset{\beta}{C}H_2CH_3 \xrightarrow[\text{heat}]{OH^-} \underset{(98\%)}{C_2H_4} + \underset{(2\%)}{MeCH{=}CH_2}$$

Sulphonium salts also behave in a similar manner, *e.g.*,

$$MeCH_2\underset{\underset{\beta CH_3}{|}}{CH}SMe_2 \xrightarrow[\text{heat}]{EtO} \underset{(74\%)}{MeCH_2CH{=}CH_2} + \underset{(2\%)}{MeCH{=}CHMe} + Me_2S$$

From these examples it can be seen that the Hofmann generalised rule may be stated in an alternative way. *The predominant alkene is the least substituted ethylene, i.e., the one carrying the smallest number of alkyl groups.*

*Summary* : For alkene-forming reactions of ' hydroxides, the mechanism has been those to be E2 (see below). Some the most substituted alkene is generally the most stable isomer. Saytzeff's rule leads to the predominance of the most stable product and the Hofmann rule in the least stable one. Also, if the alkene produced is capable of geometrical isomerism, the term isomer usually predominates over the cis isomer the former is generally more stable than the latter.

*Mechanism of elimination in 'onium ions* : Kinetic studies have shown that the elimination is a second-order reaction, it is first-order in ammonium ion and in hydroxide ion. Thus, the general equation for the E2 mechanism may be written as follows.

$$HO^{-} \; H—CR_2—CR_2—\overset{+}{N}R_3 \longrightarrow H_2O + CR_2{=}CR_2 + R_3N$$

Since elimination proceeds by removal of a β-hydrogen, any factor that affects its case of removal will affect the rate of reaction. Hence, when alternative β-hydrogen are present, the one most readily removed will decide the structure of the predominating alkene product. One factor affecting the ease of removed is the acidity of the b-hydrogen, and the more acidic it is the easier it is to remove, it. Ingold *et al.,* (1927) showed that the inductive effect due to the positive charge on the nitrogen atom is an important factor in deciding the strength of this acidity, and so it follows that the Hofmann generalised rule is a consequence of the – I effect operating in 'onium ions (cf. the hyperoconjugative effect for the Saytzeff rule). The C–H bonds are weakened sufficiently for a β-proton to the eliminated by the E2 mechanism.

$$CH_3 \rightarrow \underset{}{\overset{H}{\overset{|}{C}H}} \rightarrow CH_2 \rightarrow \overset{+}{\underset{CH_3\ \ CH_3}{N}} \leftarrow CH_2 \leftarrow CH_2—H \xrightarrow{HO^-}$$

$$CH_3CH_2CH_2N(CH_3)_2 + CH_2 = CH_2 + H_2O \quad (90\%)$$

Since the terminal methyl group of the n-propyl group is electron-releasing, the positive charge induced on the β-carbon atom (of the propyl group) is partially neutralised. This 'tightens' the bonding of these β-hydrogen atoms, and so a less tightly bound β-hydrogen atom in the ethyl group is eliminated preferentially with the formation of ethylene.

In some case, the Hofmann generalised rule is not followed, *e.g.*, styrene and not ethylene is formed in the following elimination.

$$PhCH_2CH_2\overset{+}{N}Me_2Et \xrightarrow[\text{heat}]{OH^-} PhCH{=}CH_2 + Me_2NEt + H_2O$$

This may be explained on the basis that the b-hydrogen adjacent to the benzene ring (*benzylic* hydrogen) is more acidic than the b-hydrogen of the ethyl group because of stabilisation of the transition state by conjugation (which is absent for the ethyl group).

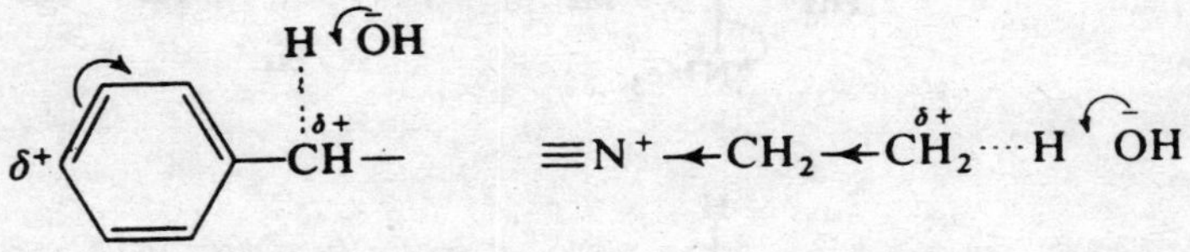

$$\delta^+ C_6H_5\text{—}\overset{\delta+}{C}H\text{—} \cdots H \; OH \qquad \equiv N^+ \leftarrow CH_2 \leftarrow \overset{\delta+}{C}H_2 \cdots H \; OH$$

It might also be noted that alcohols may be obtained in addition to alkenes, *e.g.*,

$$RCH_2CH_2\overset{+}{N}Me_2OH^- \longrightarrow RCH{=}CH_2 + RCH_2CH_2OH + Me_3N + H_2O$$

In some cases, no alkene is obtained at all, *e.g.*, quaternised tetrahydroquinoline.

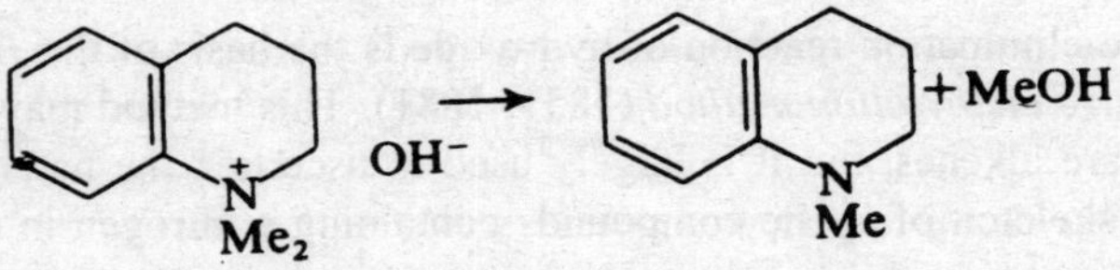

$$\text{(tetrahydroquinolinium, } \overset{+}{N}Me_2) \xrightarrow{OH^-} \text{(N-methyltetrahydroquinoline, } N\text{Me)} + MeOH$$

*Stereochemistry of the elimination reaction :* Just as it was shown that *trans* elimination occurs more readily than cis for alkyl halides, so experimental work has shown the same rule applies to' onium salts, *e.g.*, Cram *et al.,* (1956) examined the following reaction.

$$PhCHMeCHPh\overset{+}{N}Me_3I^- + EtO^- \longrightarrow PhMeC{=}CHPh + Me_3N + EtOH$$

The reaction was shown to be second-order (hence the bimolecular mechanism), and it was also shown that the *erythro* from of the ion gave the *cis* alkene, whereas the *threo* from gave the *trans alkene.* These results are in keeping with *trans* elimination.

The rates of reaction were very different, the *threo* diastereoisomer undergoing elimination about fifty times as fast as the *erythro.* In the *cis* product, the phenyl groups become eclipsed and consequently, due to the large steric repulsion, the energy of activation for the formation of the *cis* product is larger than that for the *trans*.

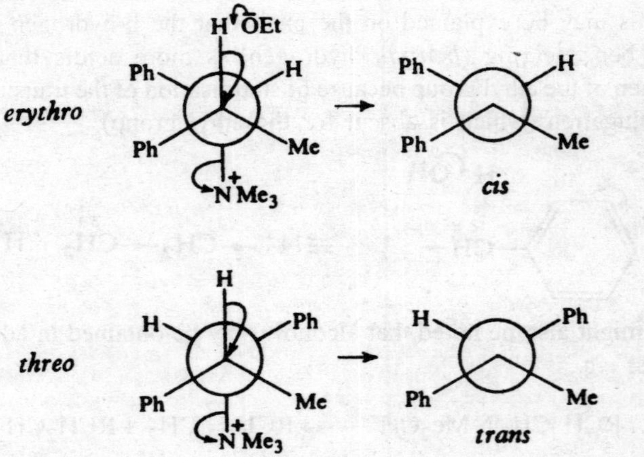

Fig. 2.5

*cis* Eliminations have been shown to occur with various' onium hydroxides and are far more common than was orginally believed.

The elimination reaction of hydroxide is the basis of the *Harmann exhaustive methylation method* (1851, 1881). This method may be used to prepare alkenes, but it is largely used to ascertain the nature of the carbon skeleton of cyclic compounds containing a nitrogen in the ring.

An alternative to heating quaternary ammonium hydroxides to prepare alkenes is the *Cope reaction,* in which an amine oxide is the starting material. This is an example of a *cyclic* elimination mechanism.

Baumgarten *et al.,* (1957) have applied the Cope reaction starting from a carboxylic acid containing an α-methylene group. The first step is the conversion of the acid to an amine by the Schmidt reaction.

$$RCH_2CO_2H \xrightarrow{NH_3} RCH_2NH_2 \xrightarrow{MeI} RCH_2NMe_2$$

$$\xrightarrow{H_2O_2} RCH_2\overset{+}{N}Me_2-O^- \xrightarrow{heat} R^1CH=CH_2$$

Alternatively, the amide is prepared first.

$$RCH_2CO_2H \xrightarrow[(ii)\ NH_3]{(I)\ SOCl_2} RCH_2CONH_2 \xrightarrow{LiAlH_4}$$

$$RCH_2CH_2NH_2 \cdots\rightarrow RCH=CH_2$$

In the former case, oxidation of the alkene gives an acid with two carbon atoms less than the original acid, and in the latter case, one carbon atom less. Thus, either reaction may be used to step down the acid series.

*Methylamine, dimethylamine* and *trimethylamine* may be prepared by any of the general methods, but are conveniently prepared by special methods.

*Methylamine :* By heating ammonium chloride with two equivalents of formaldehyde (in formalinsoln), the yield is 45-51 per cent based on the ammonium chloride.

$$2HCHO + NH_4Cl \longrightarrow CH_3NH_3^+ Cl^- + HCO_2H$$

It is prepared industrially by passing a mixture of methanol and ammonia over alumina at 450°C under pressure and separating the mixture of amines by fractional distillation.

*Dimethylamine :* By heating ammonium chloride with about four equivalents of formaldehyde (in formalin solution).

$$NH_4Cl + 4HCHO \longrightarrow (CH_3)_2NH_2^+ Cl^- + 2HCO_2H$$

It is a gas, b.p. 7°C.

*Trimethylamine :* By heating a solid mixture of ammonium chloride and paraformaldehyde.

$$2NH_4Cl + 9HCHO \longrightarrow 2(CH_3)_3NH^+Cl^- + 3CO_2 \ (89\%)$$

It is a gas, b.p. 3.5°C. It occurs in sugar residues, and is used as a source of methyl chloride.

$$(CH_3)_3N + 4HCl \xrightarrow[\text{pressure}]{\text{heat}} 3CH_3Cl + NH_4Cl$$

**Nitrogen Ylids (Ylides)**

A general method for preparing nitrogen ylids is by the reaction between quaternary ammonium halides and pheynl- or butyl-lithium, *e.g.*, tetramethylammonium chloride gives trimethyl-ammonium methylide.

$$Me_4N^+Cl^- + PhLi \xrightarrow{Et_2O} PhH + [Me_3NCH_2Li]^+Cl^-$$

$$\xrightarrow{-\,LiCl} Me_3\overset{+}{N}CH_2^-$$

Many ylids are of the type

$$\overset{+}{Z}\text{—}\overset{-}{\ddot{C}}\Big\langle \longleftrightarrow Z{=}C\Big\langle$$

where Z is a heteroatom such as N, P, S, etc, attached to various groups. However, on the basis that N cannot expand its octet, N ylids differ from the other ylids in that they are not stabilised by d-orbital conjugation, *i.e.*, the second contributing resonance structure is absent for N ylids.

Ylids undergo reaction with a variety of compounds to form products, many of which may not be readily prepared by other methods, *e.g.*,

$$Me\overset{+}{N}CH_2^- \begin{cases} \xrightarrow{H_2O} Me_4N^+ OH^- \\ \xrightarrow{H_2O_2} [Me_3NCH_2OH]^+OH^- \\ \xrightarrow{I_2} [Me_3NCH_2I]^+I^- \\ \xrightarrow{MeI} [Me_3NCH_2Me]^+I^- \\ \xrightarrow{R_2CO} Me_3\overset{+}{N}(CH_2CR_2)O^- \\ \xrightarrow{CO_2} Me_3\overset{+}{N}CH_2CO_2^- \end{cases}$$

**Enamines**

This name usually refers to α,β-unsaturated, and their most common method of preparation is by reaction between a carbonyl compound which contains at least one a-hydrogen atom and a secondary amine. This may be represented in general terms as.

$$>CH-C=O + HN< \xrightarrow{-H_2O} >\underset{\beta}{C}=\underset{\alpha}{C}-N<$$

Enamines derived from aldehydes are less stable than those from ketones, and both types derived from acyclic secondary amines are far less stable than those from cyclic secondary amines, the most common of which are pyrrolidine, piperidine, and moroph† holine, *e.g.*, cyclohexanone and pyrrolidine form N-cyclohexenopyrrolidine.

$$\text{pyrrolidine (NH)} + O{=}\text{cyclohexanone} \xrightarrow[\text{benzene}]{\text{TsOH}} \left[\text{1-(pyrrolidin-1-yl)cyclohexan-1-ol (OH, H)}\right] \xrightarrow{-H_2O} \text{N-cyclohexenopyrrolidine}$$

Primary amines also form enamines.

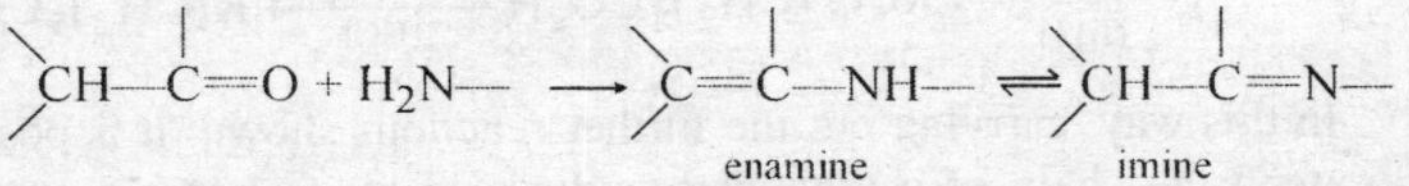

$$\rangle CH-\overset{|}{C}=O + H_2N- \longrightarrow \rangle C=\overset{|}{C}-NH- \rightleftharpoons \rangle CH-\overset{|}{C}=N-$$

enamine imine

However, in this case, enamine-imine tautomerism is possible and the equilibrium lies virtually completely on the imine side, and consequently the enamine cannot be isolated.

Enamines may be represented as resonance hybrids of two canonical structures.

$$\rangle C=\overset{|}{C}-\ddot{N}\langle \longleftrightarrow \rangle \bar{\ddot{N}}-C=\overset{+}{N}\langle$$

Hence, it might be anticipated that the β-carbon atom would be susceptible to electrophilic attack, or alternatively, the β-carbon atoms is a nucleophilic centre. This is so in practice, and consequently enamines are very usueful as intermediates in organic synthesis, *e.g.*,

(i) *Alkylation :* Enamines are alkylated at the β-carbon atom by reactive alkyl halides, and react with α-bromoacetic esters, *e.g.*,

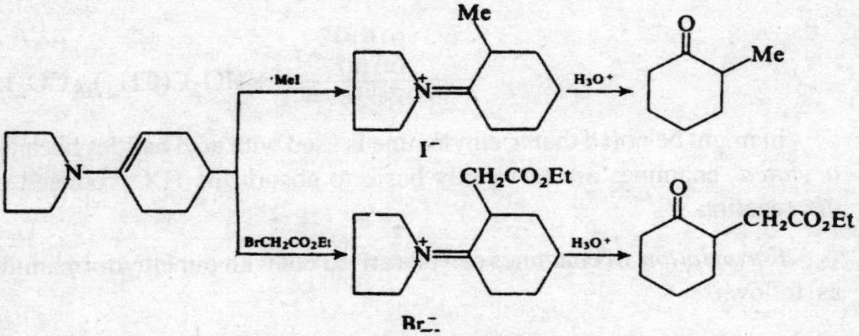

MeI; $I^-$; Me; $H_3O^+$; Me

$BrCH_2CO_2Et$; $CH_2CO_2Et$; $Br^-$; $H_3O^+$; $CH_2CO_2Et$

On advantage of this method over that of the direct action on the ketone (after treatment with sodium ethoxide) is that monosubstituted enamines are relatively difficult to alkylate further, and so can be isolated.

(ii) *Acylation, e.g.*, using N-cyclopentenomorpholine.

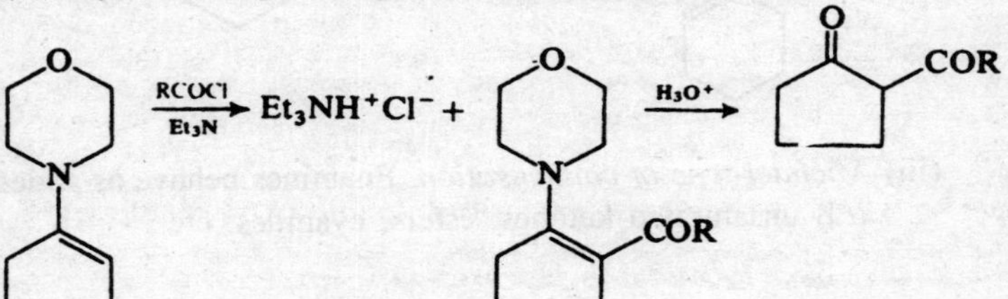

$\xrightarrow[Et_3N]{RCOCl}$ $Et_3NH^+Cl^-$ + (COR) $\xrightarrow{H_3O^+}$ (COR)

$$\xrightarrow[\text{(ii) H}^+]{\text{(i) OH}^-} RCO\,(CH_2)_4CO_2H \xrightarrow[\text{HCl}]{\text{Zn / Hg}} R(CH_2)_5CO_2H$$

In this way, carrying out the further reactions shown, it is possible to extend the chain of a monocarboxylic acid by *five* carbon atoms.

If the corresponding cyclohexeno-enamine is used, the chain can be extended by *six* carbon atoms.

In the same way, it is possible to extend the chain of a dicarboxylic acid by ten or twelve carbon atoms, *e.g.*, [N. B. $(CH_2)n$, $n \geq 5$].

2 [morpholine enamine] + $ClOC(CH_2)_6COCl$ $\xrightarrow{Et_3N}$ [bis-enamine] $CO(CH_2)_6CO$

$$\xrightarrow[\text{(iii) Zn / Hg / HCl}]{\text{(i) H}_3\text{O}^+ \quad \text{(ii) HO}^-} HO_2C(CH_2)_{18}CO_2H$$

In might be noted that triethylamine is used with acyl halides because *acylated* enamines are to weakly basic to absorb the HX produced in the reaction.

*Formylation* of enamines cab be carried out with dimethylformamide as follows.

[morpholine enamine] + $HCONMe_2$ $\xrightarrow[CH_2Cl_2]{POCl_3}$ [2-formylcyclohexanone] CHO

(iii) *Michael-type of condensation.* Enamines behave as addenda to α,β unsaturated ketones, esters, cyanides, etc.,

$$\text{(1-cyclohexenyl)pyrrolidine} \xrightarrow{CH_2=CHCO_2Et} \left[\text{iminium } \overset{+}{N}=\text{C}\ \ CH_2-\bar{C}HCO_2Et \ \ H\right] \longrightarrow$$

$$\text{enamine}-CH_2CH_2CO_2Et \xrightarrow{H_3O^+} \text{2-}(CH_2CH_2CO_2Et)\text{cyclohexanone}$$

(iv) *Introduction of a cyano group.* Enamines are treated with cyanogen chloride in the presence of $Et_3N$.

$$-CH=\overset{|}{C}-N\lt \xrightarrow{ClCN} -\underset{CN}{\underset{|}{CH}}-\overset{|}{C}=\overset{+}{N}\lt \ Cl^- \xrightarrow{Et_3N}$$

$$-\underset{CN}{\underset{|}{C}}H=\overset{|}{C}-N\lt \xrightarrow{H_3O^+} -\underset{CN}{\underset{|}{CH}}-\overset{|}{C}=O$$

(v) *Reduction :* Enamines are reduced catalytically or by sodium borohydride (but not by LAH) to saturated t-amines.

$$\gt C=\overset{|}{C}-N\lt \xrightarrow[\text{or } NaBH_4]{H_2-Pt} \gt CH-\overset{|}{C}H-N\lt$$

## Nitrenes

The parent compound is nitrene, NH (also known as *imidogen, azene,* or *imene*), and is formed when hydrazoic acid is irradiated with ultraviolet light. In the presence of ethylene, nitrene is trapped to form ethylenimine.

$$HN_3 \xrightarrow{h\nu} N_2 + H\ddot{\underset{..}{N}} \xrightarrow{C_2H_4} \text{aziridine (N–H)}$$

Acyl nitrenes, $RCO\ddot{N}$ have been proposed as possible intermediates in the Hofmann. Curtius, and Lossen rearrangements, but recent evidence does not favour this. However, there is evidence that the *photolysis* of acyl azides can lead to the formation of acyl nitrenes. Alkyl nitrence may be prepared by the photolysis of isocyanates.

$$R-N=C=O \xrightarrow{hv} R-\ddot{N} + CO$$

Like methylene, nitrenes can exist in the singlet and the triplet state, the latter being the ground state. Most nitrenes exist in the triplet ground state and behave as electrophiles, undergoing cyclo-addition and insertion reaction.

## DIAMINES

Diamines may be prepared by methods similar to those for the monoamines, using poly-methylene halides instead of alkyl halides.

*Ethylenediamine.* b.p. 118°C. may be prepared by heating, under pressure, ethylene dibromide with a large excess of ammonia.

$$BrCH_2CH_2Br + 2NH_3 \longrightarrow NH_2CH_2CH_2NH_2 + 2HBr$$

It forms *piperazine* when its hydrochloride is heated.

$$Cl^-H_3NCH_2CH_2NH_3Cl^- \longrightarrow Cl^-H_2\overset{+}{N}\langle\rangle\overset{+}{N}H_2Cl^- + 2NH_4Cl$$

Ethylenediamine forms chelate compounds with many metals, *e.g.*, cobalt.

*Putrescine, tetramethylenediamine,* m.p. 27°C, is formed by the putrefaction of proteins (in flesh). It may be prepared as follows.

$$BrCH_2CH_2Br \xrightarrow{KCN} NCCH_2CH_2CN$$

$$\xrightarrow{Na/C_2H_5OH} NH_2(CH_2)_4NH_2$$

A convenient synthesis starts with ethyl adipate and uses the Curtius rearrangement.

$$(CH_2)_4(CO_2Et)_2 \xrightarrow[reflux]{N_2H_4/EtOH} (CH_2)_4(CONHNH_2)_2$$

$$\xrightarrow{HNO_2} (CH_3)_4(CON_3)_2 \xrightarrow[benzene]{heat\ in}$$

$$(CH_2)_4(NCO)_2 \xrightarrow{\text{aq. HCl}} (CH_3)_4(NH_3^+Cl^-)_2 \quad (63\text{-}73\%)$$

When its hydrochloride is heated, *pyrrolidine* is formed.

$$\begin{array}{l} CH_2CH_2{-}NH_3^+Cl^- \\ | \\ CH_2CH_2\ddot{N}H_2 \end{array} \longrightarrow \boxed{\ \ } NH_2^+Cl^- + NH_2$$

*Cadaverine, Pentamethylenediamine,* b.p 178-180°C, is formed by the putrefaction of proteins (in flesh). It may be prepared by an analogous method to that used for putrescine.

$$Br(CH_2)_3Br \xrightarrow{KCN} NC(CH_2)_3CN \xrightarrow{Na/C_2H_5OH} NH_2(CH_2)_5NH_2$$

It is, however, prepared more conveniently by heating pentamethylene dibromide with an excess of ammonia.

$$Br(CH_2)_5Br + 2NH_3 \longrightarrow NH_2(CH_2)_5NH_2 + 2HBr^-$$

The starting material pentamethylene dibromide is readily obtained from piperidine. When its hydrochloride is heated, piperidine is formed.

$$\begin{array}{l} CH_2CH_2NH_3^+Cl^- \\ / \\ CH_2 \\ \backslash \\ CH_2CH_2\ddot{N}H_2 \end{array} \longrightarrow \bigcirc\ NH_3^+Cl^- + NH_3$$

*Hexamethylenediamine* is manufactured by the catalytic hydrogenation of adiponitrile prepared from adipic acid.

$$HO_2C(CH_2)_4CO_2H \xrightarrow[\text{catalyst}]{NH_3} NC(CH_2)_4CN$$

$$\xrightarrow[\text{catalyst}]{NH_3} NH_2(CH_2)_6NH_2$$

Adiponitrile is also manufactured by the electrolytic hydrodimerisation of vinyl cyanide.

$$2CH_2{=}CHCN \xrightarrow[2H_2O]{2e} NC(CH_2)_4CN$$

Hexamethylenddiamine, m.p. 39°C, is used in the manufacture of *nylon.*

*Spemine,* a deliquescent crystalline solid which is isolated from human sperm, is a tetramine.

$$NH_2(CH_2)_3NH(CH_2)_4NH(CH_2)_3NH.$$

Diamines with two amino-groups attached to the same carbon atoms are unknown (c.f., the group $C(OH)_2$. On the other hand, N-substituted derivatives have been prepared, *e.g.*, tetraethyl-methylenediamine

$$(C_2H_5)_2NH + CH_2O + NH(C_2H_5)_2 \longrightarrow (C_2H_5)_2NCH_2N(C_2H_5)_2 + H_2O$$

## UNSATURATED AMINES

The simplest unsaturated amine would be vinylamine (ethenylamine) if it existed. All attempts to prepare it result in the formation of the cyclic compound *ethylenimine (aziridine),* and a good method is to treat 2-aminoethanesulphonic with sulphuric acid and then to heat the product with aqueous sodium hydroxide.

$$NH_2CH_2CH_2OH \xrightarrow{H_2SO_4} \overset{+}{N}H_3CH_2CH_2OSO_3^- \xrightarrow{NaOH} \underset{\text{(ring via NH)}}{CH_2{-}CH_2}$$

Thus tis compound is a three-membered heterocyclic ring. Ethylenimine is a syrupy liquid, b.p. 56°C. It combines with sulphurous acid to form *taurine (2-aminoethanesulphonic* acid), which occurs in human bile.

$$\underset{\text{(ring via NH)}}{CH_2{-}CH_2} + H_2SO_3 \longrightarrow \overset{+}{N}H_3CH_2CH_2SO_3^-$$

*Neurine (trimethylvinylammonium hydroxide)* is found in the brain. It may be prepared.

(i) By boiling choline with barium hydroxide solution.

$$[(CH_3)_3NCH_2CH_2OH]^+OH^- \xrightarrow{Ba(OH)_2} [(CH_3)_3NCH{=}CH_2]^+OH^- + H_2O$$

(ii) By heating trimethylamine and ethylene dibromide (one molecule of each), and subsequently heating the product with silver oxide in water.

$$(CH_3)_3N + BrCH_2CH_2Br \longrightarrow [(CH_3)_3NCH_2CH_2Br]^+Br^- \xrightarrow{AgOH} [(CH_3)_3NC{=}CH_2]^+OH^-$$

Neurine is a very poisonous syrupy liquid, it is a quaternary enamine.

*Allylamine (2-propenylamine),* b.p. 53°C, may be prepared by heating allyl iodide with ammonia.

$$CH_2 = CHCH_2I + NH_3 \longrightarrow CH_2 = CHCH_2NH_2 + HI$$

It is more conveniently prepared by boiling allyl isothiocyanate with dilute hydrochloric acid.

$$CH_2 = CHCH_2NCS + H_2O \xrightarrow{HCl} CH_2 = CHCH_2NH_2 + COS$$

(70-73%)

## AMINOALCOHOLS

The simplest aminoalcohol is *2-aminoethanol* (*cholamine, ethanolamine, 2-hydroxyethylamine).* It occurs in kepahlines, and is best prepared by the action of ethylene oxide on excess of ammonia.

$$\overset{\quad O}{CH_2\!-\!CH} + NH_3 \longrightarrow NH_2CH_2CH_2OH$$

It is a viscous liquid, b.p. 171°C, miscible with water, and is strongly basis. It readily reacts with concentrated hydrobromic acid to form β-bromoethylamine hydrobromide (83 per cent).

*Choline* (*2-hydroxyethyltrimethylammonium hydroxide)* occurs in lecithins, it is best prepared by the action of ethylene oxide on trimethylamine in aqueous solution.

$$\overset{\quad O}{CH_2\!-\!CH_2} + (NH_3)_3N + H_2O \longrightarrow [(CH_3)_3NCH_2CH_2OH]^+OH^-$$

It is a colourless viscous liquid, soluble in water, and is strongly basic. It forms neurine when boiled with barium hydroxide solution. It is present in the vitamin B complex, it is a growth factor in chicks.

## AMINOACIDS

Amino-acids are derivatives of the carboxylic acids in which is hydrogen atom in the carbon chain has been replaced by an amino-group. The amino-group may occupy the α-or β-or γ¯... position, there may also be two or more amino-groups present in the chain.

The three basic classes of foods are, proteins, fats and carbohydrates. The proteins are nitrogenous substances which occur in most cells of the animal body, they also occur in plants. When hydrolysed by strong inorganic acids or by enzymes, proteins yield a mixture of amino-acids, most of which are α-amino-acids. The number of amino-acids obtained form proteins is about twenty-five, of which about ten are essential, *i.e.*, a deficiency in any one prevents growth in young animals, and may even cause death.

The amino-acids are classified in several ways. Table 2.2 shows a convenient classification, the letters g, l and e which follow the name of the acid indicate that the acid indicate that the acid is respectively of general occurrence, lesser occurrence and essential (in man).

In this book we shall consider a detail only the simplest aminoacid, *glycine (aminoacetic acid, glycocoll),* $(CH_2(NH_2)CO_2H$. This acid is found in many proteins, and occurs in certain animal excretions, usually in combination, *e.g.*, *hippuric acid* (in horses' urine), $C_6H_5CONH$ $CH_2CO_2$.

Glycine may be readily prepared by the action of concentrated ammonium hydroxide solution on chloroacetic acid.

$$ClCH_2CO_2H + 2NH_3 \longrightarrow CH_2(NH_2)CO_2H + NH_4Cl \text{ (64-65\%)}$$

$$C_6H_4(CO)_2\bar{N}K^+ + Cl{-}CH_2CO_2H \longrightarrow$$

$$KCl + C_6H_4(CO)_2NCH_2CO_2H \xrightarrow{HCl}$$

$$C_6H_4(CO_2H)_2 + CH_2(NH_2)CO_2H$$

It may also be prepared pure by Gabriel's phthalimide synthesis.

Both methods may also be used to prepare alanine, $CH_3CH$ $(NH_2)CO_2H$, from a-bromo-propionic acid.

Alanine may also be prepared by the *Strecker synthesis* (1850), this is the reaction between an aldehyde-ammonia and hydrogen cyanide, and the resulting amino-nitrile is then hydrolysed.

In practice, the reaction is carried out as follows.

$$MeCHO \xrightarrow[KCN]{NH_4Cl} MeCH(NH_2)CN \xrightarrow[H_2O]{HCl} MeCH(NH_2)CO_2H$$

| *Name* | *Formula* |
|---|---|
| *Neutral amino-acids* (one amino-group and one carboxyl group) | |
| 1. Glycine (g) | $CH_2(NH_2)CO_2H$ |
| 2. Alanine (g) | $CH_3CH(NH_2)CO_2H$ |
| 3. Valine (g. e) | $(CH_3)_2CHCH(NH_2)CO_2H$ |
| 4. Leucine (g.e) | $(CH_3)_2CHCH_2CH(NH_2)CO_2H$ |
| 5. Isoleucine (g. e) | $(C_2H_5)(CH_3)CHCH(NH_2)CO_2H$ |
| 6. Norleucine (*l*)* | $CH_3(CH_2)_3CH(NH_2)CO_2H$ |
| 7. Phenylalanine (g, e) | $CH_2CH(NH_2)CO_2H$ |
| 8. Tyrosine (g) | HO $CH_2CH(NH_2)CO_2H$ |
| 9. Serine (g) | $HOCH_2CH(NH_2)CO_2H$ |
| 10. Cysteine (g) | $HSCH_2CH(NH_2)CO_2H$ |
| 11. Cystine (g) | $(-SCH_2CH(NH_2)CO_2H)_2$ |
| 12. Threonine (g. e) | $CH_3CHOHCH(NH_2)CO_2H$ |
| 13. Methionine (g, e) | $CH_3SCH_2CH_2CH(NH_2)CO_2H$ |
| 14. Di-iodotyrosine or iodogorgic acid (*l*) | I; HO $CH_2CH(NH_2)CO_2H$; I |
| 15. Thyroxine (*l*) | I, I; HO –O– $CH_2CH(NH_2)CO_2H$; I, I |
| 16. Dibromotyrosine (*l*) | Br; HO –$CH_2CH(NH_2)CO_2H$; Br |
| 17. Tryptophan (g, e) | $CH_2CH(NH_2)CO_2H$; N H |

18. Proline (g)

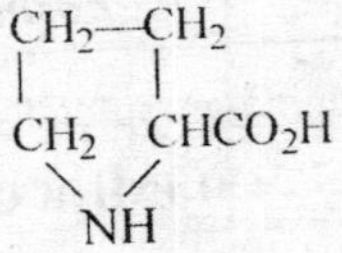

$$\begin{array}{ll} CH_2 & \!\!-\!\!CH_2 \\ | & \quad | \\ CH_2 & CHCO_2H \\ \quad\diagdown & \!\!\diagup \\ & \!\!\!\!\!\!NH \end{array}$$

19. Hydroxyproline (g)

$$\begin{array}{ll} HOCH & \!\!-\!\!CH_2 \\ \quad | & \quad | \\ CH_2 & CHCO_2H \\ \quad\diagdown & \!\!\diagup \\ & \!\!\!\!\!\!NH \end{array}$$

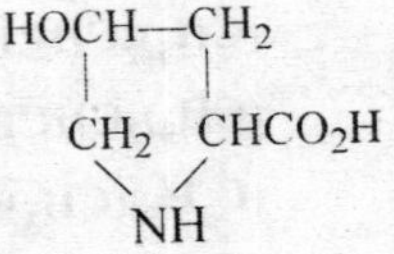

*Acidic amino-acids* (on amono-group and two carboxyl groups)

20. Aspartic acid (g) $HO_2CCH_2CH(NH_2)CO_2H$

21. Glutamic acid (g) $HO_2CCH_2CH_2CH(NH_2)CO_2H$

22. β-Hydroxyglutamic acid (*l*)* $HO_2CCH_2CHOHCH(NH_2)CO_2H$

*Basic amino-acids* (two amino-groups and one carboxyl group)

23. Ornithinet** $NH_2CH_2CH_2CH_2CH(NH_2)CO_2H$

24. Arginine (g, e)

$$\begin{array}{c} NH_2 \\ | \\ HN{=}C{-}NHCH_2CH_2CH_2CH(NC_2)CO_2H \end{array}$$

25. Lysine (g, e) $NH_2CH_2CH_2CH_2CH_2CH(NH_2)CO_2H$

26. Histidine (g, e)

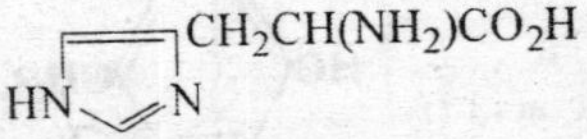

$$\begin{array}{l} \ulcorner\!\!-\!\!\!-\!\!\!\urcorner CH_2CH(NH_2)CO_2H \\ HN \diagdown\!\!\diagup N \end{array}$$

*Occurrence in proteins uncertain.

**Ornithine is probably not present in proteins, but is formed by the hydrolysis, of arginine.

Glycine exists as white prisms which melt, with decomposition, at 289-292°C. It has a sweet taste, and is soluble in water but insoluble in ethanol and ether.

Since it contains an amino-group and a carboxyl group, it combines the properties of a base and an acid, *i.e.*, it is *amphoteric.* The following reactions are typical of all a-amino-acid.

**Reactions Characteristic of the Amino-group**

1. Glycine forms salts with strong inorganic acids, *e.g.*,

$$\bar{C}l\{H_3\overset{+}{N}CH_2CO_2H$$

2. Glycine reacts with acetyl chloride or acetic anhydride to give the acetyl derivative.

$$H_2NHC_2CO_2H + (CH_3CO_2)_2O \longrightarrow CH_3CONHCH_2CO_2H + CH_3CO_2H$$

Similarly, with benzoyl chloride, it forms benzoylglycine (hippuric acid, see above).

These acylated derivatives are acidic, the basic character of the amino-group being effectively eliminated by the presence of the – I group attached to the nitrogen.

3. When glycine is treated with nitrous acid, nitrogen is evolved and glycollic acid is formed.

$$CH_2(NH_2)CO_2H + HNO_2 \longrightarrow HOCH_2CO_2H + N_2 + H_2O$$

4. Nitrosyl chloride (or bromide) reacts with glycine to form chloro- (or bromo-) acetic acid

$$CH_2(NH_2)CO_2H + NOCl \longrightarrow ClCH_2CO_2H + N_2 + H_2O$$

**Reactions Characteristic of the Carboxyl Group**

1. Glycine may be esterified by an alcohol in the presence of inorganic acid⁻, *e.g.*,

$$CH_2(NH_2)CO_2H + C_2H_5OH \xrightarrow{HCl} \bar{C}lH_3\overset{+}{N}CH_2CO_2C_2H_5 + H_2O$$

The ester is liberated from its hydrochloride by alkali.

2. Glycine forms metallic salts when its aqueous solution is warmed with a metallic oxide or hydroxide. These salts are chelate compounds, *e.g.*, the copper salt (deep blue needles) is:

```
O=C—O         H2NCH2
  |    \     /  |
  |     Cu←     |
  |    ↗   \    |
CH2NH2       O—C=O
```

3. Amino-acids may be reduced to aminoalcohols by lithium aluminium hydride (Vogel *et al.*, 1952)

$$RCH(NH_2)CO_2H \xrightarrow{LiAlH_4} RCH(NH_2)CH_2OH$$

### Reactions Due to Both the Amino and Carboxyl Groups

1. When measured in aqueous solution, the dipole moment of glycine is found to have a large value. To account for this large value it has been suggested that glycine exists, in solution, as an *inner salt.*

$$H_2NCH_2CO_2H + H_2O \rightleftharpoons H_3\overset{+}{N}CH_2CO_2^- + H_2O$$

Such a doubly charged ion is known, in addition to an inner salt, as a *zwitterion, ampholyte* or a *dipolar ion.* This dipolar ion structure also accounts for the absence of acidic and basic properties of an amino-acid (the carboxyl and amino-groups of the *same* molecule neutralising each other to form a salt).

The properties of crystalline glycine, *e.g.*, its high melting point and its in solubility in hydrocarbon solvents, also indicate that it exists as the inner salt in the solid state. The presence of the $NH_3^+$ group in glycine has now been confirmed by i.r and NMR spectra studies.

Owing to its amphoteric character, glycine cannot be titrated directly with alkali. When formalin solution is added to glycine, methyleneglycine is formed (reaction probably more complex).

$$N_2NCH_2CO_2H + HCHO \longrightarrow CH_2 = NCH_2CO_2H + H_2O$$

This is a strong acid (the basic character of the amino-group being now suppressed), and can be titrated with sodium hydroxide. This method is known as the *Sorenesen titration.*

2. When heated, glycine forms *diketopiperazine, glycine* esters give a better yield.

$$\begin{matrix} H_2NCH_2CO_2Et \\ + \\ EtO_2CCH_2NH_2 \end{matrix} \longrightarrow \text{(diketopiperazine ring: HN, NH, two C=O)} + 2EtOH$$

*Betaines :* These are the trialkyl derivatives of glycine, which exists as dipolar ions of formula

$$R_3\overset{+}{N}CH_2C\overset{-}{O_2} + 3CH_3I \longrightarrow (CH_3)_3\overset{+}{N}CH_2C\overset{-}{O}_2 + 3HI$$

It is more conveniently prepared by warming an aqueous solution of chloroacetic acid with trimethylamine.

$$(CH_3)_3N + ClCH_2CO_2H \longrightarrow (CH_3)_3\overset{+}{N}CH_2CO_2 + HCl$$

Betaine is a solid, m.p. 300 C (with decomposition), it occurs in nature, especially in plant juices.

It behaves as a base, *e.g.*, with hydrochloric acid it forms the stable crystalline hydrochloride.

$$Cl^-(CH_3)_3\overset{+}{N}CH_2CO_2H\,.$$

## ALIPHATIC DIAZO-COMPOUNDS

The aliphatic diazo-compounds are characterised by the presence of the group $>CN_2$.

*Diazomethane :* May be prepared in various ways, the first is of historical importance, and the others are very convenient methods.

*Method of Von Pechmann* (1894) : Methylamine is treated with ethyl chloroformate to give N-methylurethan which, on treatment with nitrous acid in ethereal solution, forms N-methyl-N-nitroso-urethan.

This, on warming with methanolic potassium hydroxide, decomposes into DIAZOMETHANE, which is collected in cooled ether.

$$CH_3NH_2 + ClCO_2C_2H_5 \longrightarrow CH_3NHCO_2C_2H_5 \xrightarrow{HNO_2}$$
$$CH_3N(NO)CO_2H_5 \xrightarrow{KOH} CH_2N_2 + CO_2C_2H_5$$

An improved method starts with methylurea.

$$MeNHCONH_2 \xrightarrow{HNO_2} MeN(NO)CONH_2 \xrightarrow{KOH}$$
$$CH_2N_2 + KNCO + 2H_2O$$

*Method of McKay* (1948). Methylamine hydrochloride and nitroguanidine are allowed to react in potassium hydroxide solution, the product N-methyl-N′-nitroguanidine treated with nitrous acid, and the N-methyl-N-nitroso-N¢-nitroguanidine so produced is then warmed with potassium hydroxide.

$$CH_3NH_3{}^+Cl + NH_2\overset{\overset{\displaystyle NH}{\|}}{C}NHNO_2 + KOH \longrightarrow$$

$$NH_3 + KCl + CH_3NH\overset{\overset{\displaystyle NH}{\|}}{C}NHNO_2 \xrightarrow{HNO_2}$$

$$CH_3N(NO)\overset{\overset{\displaystyle NH}{\|}}{C}NHNO_2 \xrightarrow{KOH} CH_2N_2$$

*Method of Backer et al.*, (1951). The nitroso-derivative of p-toluene-N-methylsulphonamide is distilled with ethanolic potassium hydroxide,

$$CH_3C_6H_4SO_2N(CH_3)NO + KOH \longrightarrow CH_2N_2 + CH_3C_6H_4SO_3K + H_2O \ (80\text{-}90\%)$$

Diazomethane is a yellow, poisonous gas, liquid diazomethane, b.p. – 24°C, is explosive. The gas is soluble in ether, and since the ethereal solution is fairly to handle, reactions with diazomethane are usually carried out in ethereal solution.

Diazomethane is neutral, it is reduced by sodium amalgam to methylhydrazine.

Diazomethane is best represented as a resonance hybrid derived from linear resonating structures with opposing dipoles.

$$\underset{(I)}{CH_2{=}\overset{+}{N}{=}\overset{-}{\ddot{N}}:} \leftrightarrow \underset{(II)}{\overset{+}{C}H_2{-}\ddot{N}{=}\overset{-}{\ddot{N}}:} \leftrightarrow \underset{(III)}{\overset{-}{\ddot{C}}H_2{-}\overset{+}{N}{\equiv}N:} \leftrightarrow \underset{(IV)}{\overset{-}{\ddot{C}}H_2{-}\ddot{N}{=}\overset{+}{N}:}$$

Under appropriate conditions, diazomethane can behave as an electrophile (II) or a nucleophile (III), as a 1, 3-dipole (IV), or as a source of methylene.

## Reactions

Most of the reactions of diazomethane are readily explained in terms of nucleophilic attack (structure III) on the substrate, followed by loss of nitrogen.

Halogen acids react with diazomethane to form methyl halides.

$$Cl{-}H \ \overset{-}{C}H_2{-}N{\equiv}N \longrightarrow Cl^- \ CH_3{-}\overset{+}{N}{\equiv}N \longrightarrow Cl{-}CH_3 + N_2$$

Diazomethane methylates *acidic* hydroxyl groups very readily, carboxylic acids, sulphuric acids, phenols and enols. A possible mechanism is.

$$RCO{-}O{-}H \; {}^{-}CH_2{-}\overset{+}{N}{\equiv}N \longrightarrow RCO{-}O^{-} \; CH_3{-}\overset{+}{N}{\equiv}N \xrightarrow{-N_2} RCOOCH_3$$

$$RCO{-}O^{-} \; CH_3{-}\overset{+}{N}{\equiv}N \xrightarrow{-N_2} RCO{-}O^{-} + CH_3^{+}$$

Alcohols can also be methylated by diazomethane is the presence of a suitable catalyst, *e.g.*, an aluminium alkoxide. Fluoroboric acid is also a very good catalyst for primary alcohols (Muller, 1958)

$$R^1OH + Al(OR^2)_3 \rightleftharpoons (R^1)(H)\overset{+}{O}{-}\overset{-}{Al}(OR^2)_3 \xrightarrow{CH_2N_2}$$

$$N_2 + (R^1)(H_3C)\overset{+}{O}{-}\overset{-}{Al}(OR^2)_3 \rightleftharpoons R^1OCH_3 + Al(OR^2)_3$$

Catalysts are necessary with alcohols because of the low acidity of the hydroxyl hydrogen.

Diazomethane reacts with aldehydes to form methyl ketones and with ketones to form the higher homologue. In both cases, an epoxide is formed and may be the predominant product.

$$R_2C{=}O \; {}^{-}CH_2{-}\overset{+}{N}{\equiv}N \longrightarrow R{-}C(O^{-})(R){-}CH_2{-}\overset{+}{N}{\equiv}N \xrightarrow{-N_2} R{-}\overset{O}{\overset{\|}{C}}{-}CH_2R$$

$$\longrightarrow R_2C(O^{-}){-}CH_2{-}\overset{+}{N}{\equiv}N \xrightarrow{-N_2} R_2C{-}CH_2 \text{ (epoxide, bridged by O)}$$

Certain cyclic ketones undergo ring expansion when treated with diazomethane.

Diazomethane adds on to ethylenic compounds to form *pyrazoline* derivatives (cf. α,β-unsaturated esters. with ethylene, *pryrazoline* is formed. The addition may be formulated (see also below).

$CH_2{=}CH_2 + \overset{+}{C}H_2{-}N{=}\overset{-}{N} \rightarrow \left[ CH_2{-}\overset{+}{C}H_2,\ CH_2\ N^-,\ N \right] \rightarrow \left[ CH_2{-}\overset{+}{C}H_2,\ CH_2\ N^-,\ N \right] \rightarrow CH_2{-}CH_2,\ CH_2\ N,\ NH$

Diazomethane also adds on to acetylenic compounds, in this case to form *pyrazole* derivatives, with acetylene, *pyrazole* is formed.

$CH_2{\equiv}CH_2 + \overset{+}{C}H_2{-}N{=}N^- \rightarrow \left[ CH{=}\overset{+}{C}H,\ CH_2\ N^-,\ N \right] \rightarrow \left[ CH_2{-}\overset{+}{C}H_2,\ CH_2\ N,\ N \right] \rightarrow CH_2{-}CH,\ CH\ N,\ NH$

Additions such as these, which involve a molecule containing a multiple bond and a molecule containing three atoms in a chain with its terminal atoms carrying a small positive and negative charge, respectively, are referred to as 1.3-*dipolar additions.* It is quite possible that the cyclo- addition occurs in one step rather than two.

Diazomethane is used in the *Arndt-Eistert synthesis* (1935). This is a means of converting an acid (aliphatic, aromatic, alicyclic, or heterocyclic) into the next higher homologue. The acid chloride is treated with diazomethane (2 molecules), the product is a diazoketone, and this loses nitrogen to form a keten by rearrangement when warmed in the presence of silver oxide as catalyst.

$R{-}C({=}O){-} + CH_2{-}N^+{\equiv}N \longrightarrow \left[ R{-}C(O^-)(Cl){-}CH(H){-}N{\equiv}N \right] \xrightarrow{-HCl}$

$R{-}C{-}CH{-}N{\equiv}\overset{+}{N} \longleftrightarrow (O^-)(R)C{=}CH{-}\overset{+}{N}{\equiv}N \xrightarrow{-N_2} O{=}C{=}CHR$

This rearrangement of the diazoketone is usually referred to as the *Wolff rearrangement* (1912).

When the rearrangement of the diazoketone is carried out in the presence of water, a carboxylic acid is formed.

$$RCH = C = O + H_2O \longrightarrow RCH_2CO_2H$$

Diazoketones react with alcohols and ammonia in the presence of silver oxide as catalyst to form esters and amides, respectively (via the keten).

$$R^1CH_2CO_2R^2 \xleftarrow{R^2OH} R^1COCHN_2 \xrightarrow{NH_3} R^1CH_2CONH_2$$

In the absence of a catalyst and in the presence of water and formic acid, diazoketones form hydroxymethyl ketones.

$$RCOCHN_2 + H_2O \xrightarrow{HCO_2H} RCOH_2OH + N_2$$

Diazoketones also react with hydrogen chloride to form chloromethyl ketones.

$$RCOCHN_2 + HCl \longrightarrow RCOCH_2Cl + N_2$$

*Diazoacetic ester (ethyl diazoacetate),* b.p. 141/720 mm, may be readily prepared by treating a cooled solution of the hydrochloride of the hydrochloride of ethyl glycine ester with cold sodium nitrite solution.

$$Cl^-H_3\overset{+}{N}CH_2CO_2C_2H_5 + NaNO_2 \longrightarrow N_2CHCO_2C_2H_5 + NaCl + 2H_2O$$

The stability of diazoacetic ester may be attributed to extended resonance. Thus there will be larger resonance energy then for diazomethane, and so the latter is less stable than diazoacetic ester

$$:\ddot{N}{=}\overset{+}{N}{=}CH{-}\overset{\overset{\displaystyle O}{\|}}{C}{-}OEt \longleftrightarrow :N{\equiv}\overset{+}{N}{-}\overset{-}{C}H{-}\overset{\overset{\displaystyle O}{\|}}{C}{-}OEt \longleftrightarrow :N{=}\overset{+}{N}{=}CH{-}\overset{\overset{\displaystyle O^-}{|}}{C}{=}\overset{+}{O}Et$$

The reactions of diazoacetic ester are similar to those of diazomethane. It is reduced by zinc dust and acetic acid to ammonia and glycine. When boiled with dilute halogen acid, it eliminates nitrogen to form glycollic ester.

$$N_2CHCO_2C_2H_5 + H_2O \longrightarrow HOCH_2CO_2C_2H_5 + N_2$$

When, however, diazoacetic ester is warmed with *concentrated* halogen acid, ethyl halogen-acetate is formed, *e.g.*,

$$N_2CHCO_2C_2H_5 + HCl \longrightarrow ClCH_2CO_2C_2H_5 + N_2$$

Diazoacetic ester reacts with compounds containing an active hydrogen atom, *e.g.*, it forms acetylglycollic ester with acetic acid, and the ethyl ether of glycollic ester with ethanol.

$$CH_3CO_2H + N_2CHCO_2C_2H_5 \longrightarrow CH_3COOCH_2CO_2H_5 + N_2$$

$$C_2H_5OH + N_2CHCO_2C_2H_5 \longrightarrow CH_2(OC_2H_5)CO_2C_2H_5 + N_2$$

It reacts with ethylenic compounds to form *pyrazoline* derivatives, *e.g.*, with ethylene it forms pyrazoline -3-carboxylic ester.

$$\begin{matrix} CH_2 \\ \| \\ CH_2 \end{matrix} + N_2CHCO_2C_2H_5 \longrightarrow \begin{matrix} CH_2-CCO_2C_2H_5 \\ | \qquad \| \\ CH_2 \quad N \\ \diagdown \diagup \\ NH \end{matrix}$$

With acetylenic compounds it forms *pyrazole, e.g.*, with acetylene, it gives pyrazole-3-carboxylic ester.

$$\begin{matrix} CH \\ ||| \\ CH \end{matrix} + N_2CHCO_2C_2H_5 \longrightarrow \begin{matrix} CH-CCO_2C_2H_5 \\ \| \qquad \| \\ CH \quad N \\ \diagdown \diagup \\ NH \end{matrix}$$

In some of its reactions, diazoacetic ester decomposes to carbethoxymethylene, *e.g.*, in the presence of benzene, cycloheptatrienecarboxylic ester is produced (Doering *et al.*, 1956.

$$N_2CHCO_2Et \xrightarrow{hv} N_2 + H-\ddot{C}-CO_2Et \xrightarrow{C_6H_6} C_7H_7-CO_2Et$$

## EXERCISES

1. Write out the structures and names of the isomeric amines of formula $C_4H_{11}N$.
2. Of a group of isomeric amines, the t-amine has the lowest b.p Explain.
3. Deamination of n-$BuNH_2$ with $NANO_2$/HCl gives two butanols, two butenes and two butyl chlorides. Give a possible mechanism for the formation of these products.
4. What products would you expect to get by the application of the Hofmann exhaustive methylation to.

(i) $Me_2CHCH_2CH_2NH_2$, (ii) $Me_2CHCH_2NHMe$,

(iii) $Me_2CHCH_2NHCH_2CH_2Me$. (iv) $EtNHCH_2CH_2Cl$ ?

5. RCONHMe does not undergo the Hofmann amine preparation reaction. Offer an explanation

6. Complete the following equations.

(i) $CH_2{=}CHCN + PhCH_2OH \xrightarrow{H_2SO_4}$

(ii) $PhCH_2COCHN_2 + HCl(g) \xrightarrow{Et_2O}$

(iii) $NH_2CONH_2 + n{-}BuOH \xrightarrow{heat}$

(iv) $MeNH_2$ + ? ⟶ $MeNHCO_2Et$

(v) $HOCH_2CH_2CN \xrightarrow[heat]{aq.HBr}$

(vi) [cyclohexane ring bearing a $CH_2\overset{+}{N}Me_2$–$O^-$ group] $\xrightarrow{heat}$ ?

(viii) $PhCH(CN)CH(CN)CO_2Et \xrightarrow[heat]{aq.HCl}$

(viiii) $+ ClCO_2Et \xrightarrow{Et_3N}$ ? $\xrightarrow{H^+}$

(ix) [pyrrolidine N attached to cyclohexene] $+ CH_2 = CHCN \longrightarrow ? \xrightarrow[warm]{heat}$

(x) $Me_2CHNO_2 + HCHO + EtNH_2 \xrightarrow{HCl}$

7. Discuss the reactions.

(i) $R_2C(OH)CH_2CR_2NH_2 \xrightarrow{HNO_2} R_2CO + R_2C{=}CH_2$

(ii) $PhCH{=}NR \xrightarrow{MeI}$ ? $\xrightarrow{H^+}$ aldehyde + s-amine

(8) Devise an experiment to show whether the Me group in n-$PrNH_2$ undergoes migration during deamination.

(9) The NMR spectrum of $Me_2NNO$ shows two peaks at room temperature and one peak at ~ 200°C. Offer an explanation.

(10) t-$BuNH_2$ and $noePenNH_2$ cannot be prepared by the action of $NH_3$ on the corresponding alkyl bromide. Why not ? Each can be prepared from a carboxylic acid. How ?

(11) Complete the following equations.

(i) $Me_2C(CN)NHNHC(CN)Me_2 \xrightarrow{HOCl} ?$

(ii) $MeNHNHMe \xrightarrow{HgO} ?$

(iii) $Me(CH_2)_2Me \xrightarrow[\sim 500\,C]{HNO_3} ?$

(iv) $RCH{=}CH_2 \xrightarrow{B_2H_6} ? \xrightarrow{ClNH_2}$

(v) Me3N + D-EtMeCHCl ¾® ?

(vi) $PhNCO \xrightarrow{H_2O} ? \xrightarrow{-CO_2} ? \xrightarrow{PhNCO} ?$

(viii) $CH_2(CO_2Et)_2 \xrightarrow[(ii)\ BrCN]{(i)\ EtONa} ?$

12. $MeCOCH_2Cl + CH_2N_2 \longrightarrow$ (A) $C_4H_7ClO$ + (B) $C_4H_7ClO$ (A) is a ketone and (B) is an epoxide. Suggest structures for (A) and (B) and mechanisms for their formation.

13. Complete and comment ($\overset{*}{C} = {}^{13}C$).

$$RMgBr + \overset{*}{C}O_2 \longrightarrow ? \xrightarrow{H^+} ? \xrightarrow[(ii)\ CH_2N_2]{(i)\ SOCl_2} ?$$

$$\xrightarrow{Ag_2O} ? \xrightarrow{H_2O}$$

14. Amine (A), $C_6H_{15}N$, on treatment with MeI and then KOH, gives (B), $C_8H_{20}N^+OH^-$. This on heating produces isobutene and amine (C), $C_4H_{11}N$. What is (A) ?

15. Complete the following equations.

(i) $R_2CO + Ph_3P = NH \longrightarrow ?$

(ii) $Ph_3CCl + RNH_2 \longrightarrow ?$

(iii) $EtNO_2 + 2CH_2O \xrightarrow{KHCO_3} ?$

(iv) $RCH_2NH_2 + 2CH_2O + 2HCO_2H \longrightarrow ?$

(v) $CH_2{=}CO + HCN \longrightarrow ? \xrightarrow{CH_2=CO}$

(vi) $Me_2C{=}CHCOMe + RCH_2NH_2 \longrightarrow ?$ $\xrightarrow{HNO_2} ? \xrightarrow{EtONa}$

(vii) $EtCHO + CH_2N_2 \longrightarrow ?$

(viii) $MeCOEt + CH_2N_2 \longrightarrow$

(ix) $? \xleftarrow{MeBr} Et_3\overset{+}{N}-CH_2^- \xrightarrow{Me_2CO}$

(x) $CH_2 = CHCOMe + CH_2N_2 \longrightarrow ?$

16. Discuss the NMR spectrum of $MeCH_2CH_2NO_2$ and give the relative positions of the signals.

17. Convert : (i) $Me_2CHO_2H$ into $Me_2CH(CH_2)_6CO_2H$, (ii) $(CH_2)_3Br_2$ into $Et_2N(CH_2)_3NH_2$. (iii) $O_2NCH_2CH_2CN$ into $(CH_2)_3(NH_2)_2$, (iv) $O_2NCH_2CH_2CN$ into $O_2NCH_2CH_2CH_2NH_2$.

18. How many N.M.R. signals would you expect for.

(i) $MeCONHCH_2CO_2H$, (ii) $MeCH = CHCN$,

(iii) $MeCH_2CH(NH_2)Me$, (iv) $(MeCO_2)_2NMe$ ?

19. Show the fragmentation paths that would account for the ions m/e.

(i) 30 and 58 from ethyl-n-propylamine,

(ii) 30, 58, and 86 from triethylamine,

(iii) 41 from n-butyl cyanide.

# 3

# ORGANOMETALLIC COMPOUNDS

## INTRODUCTION

Generally speaking, organometallic compounds are those organic compounds in which a metal is directly joined to carbon. However, the ionic character of the carbon-metal bond depends on the nature of the metal, *e.g.*, the ionic character order is. Na > Li > Mg > Al > Zn > Cd > Hg, sodium and potassium alkyls are *salts* (see later). All of these metals are less electronegative than carbon, and the larger the difference, the greater is the ionic character of the carbon-metal bond (the carbon being the *negative* end of the dipole). Also, the greater the ionic character, the more reactive is the metal alkyl.

## THE GRIGNARD REAGENTS

The alkylmagnesium halides, R—Mg—X, or Grignard reagents, introduced by Grignard in 1900, are extremely valuable in organic syntheses. A Grignard reagent is generally prepared by reaction between magnesium (1 atom) and alkyl halide (1 molecule) in dry, alcohol-free ether.

$$RX + Mg \rightarrow RMgX \quad \text{(v.g. –ex.)}$$

The ethereal solution of the Grignard reagent is generally used in all reactions. Other solvents besides ether may be used, *e.g.,* tertiary amines, tetrahydrofuran and the dimethyl ether of ethylene glycol. Tetrahydrofuran is being used increasingly as a solvent, since it has been found that it increases the reactivity of organic halides towards magnesium. Thus, Normant (1953, 1957) has prepared Grignard reagents of vinyl halides and aryl chlorides in THF.

In many cases where the preparation of the Grignard reagent is difficult, the addition of some methyl iodide may result in the formation of the desired Grignard reagent; this is referred to as the *entrainment method.* However, the best entrainment technique is to add ethylene dibromide (3 molecules or more) and to use magnesium in excess (6 molecules or more; *et al.,* 1959).

The ease with which an alkyl halide forms a Grignard reagent depends on a number of factors. It has been found that for a given alkyl group the ease of formation is alkyl iodide > bromide > chloride.

It has also been found that the formation of a Grignard reagent becomes increasingly difficult as the number of carbon atoms in the alkyl group increases, *i.e.,* the ease of formation is $CH_3X > C_2H_5X > C_3H_7X > ...$ . Since tertiary alkyl iodides readily eliminate hydrogen iodide with the formation of an alkene, tertiary alkyl chlorides are used.

Also, in the preparation of the Grignard reagent, a common side-reaction is coupling between two R groups to give R — R.

## Structure of Grignard Reagents

Grignard *et al.,* (1901), in their attempt to isolate the Grignard reagent, showed that the compound contained ether (which they called 'ether of crystallisation'). Although a great deal of work has been done, the structure of the Grignard reagent is still not settled. According to Ashby (1967), the composition of Grignard reagents *in ether* may be represented by the following equilibria.

$$\text{etc.} \rightleftharpoons \text{Dimer} \rightleftharpoons 2\text{RMgX} \rightleftharpoons R_2Mg + MgX_2 \rightleftharpoons \text{Dimer} \rightleftharpoons \text{etc.}$$

The extent of association depends on the concentration, and is extensive above about 0.3 M. Each molecule of Grignard reagent co-ordinates with one molecule of ether, and the halogen atom of one molecule co-ordinates with the magnesium atom of another molecule, *e.g.,* the monomer is (I) and the dimer is (II).

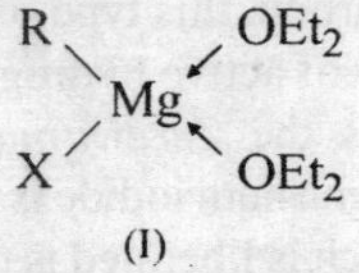

$$\begin{array}{ccccc} R & & OEt_2 \\ & Mg & \\ X & & OEt_2 \end{array} \qquad \begin{array}{ccccc} & R & & & R \\ Et_2O \rightarrow & Mg & —X\rightarrow & Mg & —X \\ & Et_2O & & & OEt_2 \end{array}$$

(I) 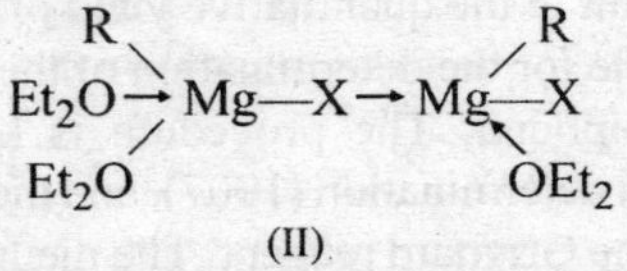 (II)

Crystalline Grignard reagents have been isolated (from ether) as 'di-etherates', *e.g.,* $EtMgBr.(OEt_2)_2$, and X-ray analysis has shown that the four groups are arranged tetrahedrally about the magnesium.

In this book, the Grignard reagent will be designated as RMgX, but as far as ether solutions are concerned, it will occasionally (and more accurately) be designated as $RMgX(OEt_2)_2$.

**Reactions of the Grignard reagents**

When working with Grignard reagents, it is usual to add the other reactant (often in ethereal solution) slowly to the Grignard solution or vice versa, and after a short time, decompose the magnesium complex with water, dilute acid, or aqueous ammonium chloride; the yields are usually *g.-v.g.*

The majority of Grignard reactions fall into two groups,

(i) *Addition of the Grignard reagent to a compound containing a multiple-bond group, e.g.,*

$$>C{=}O; \equiv \quad -C{\equiv}N; \quad >C{=}S; \quad -N{=}O; \quad >S{=}O$$

Addition of the R group of RMgX occurs at the less electronegative and that of the MgX group at the more electronegative atom. This is understandable on the basis that R and MgX are respectively the negative and positive ends of the dipole in RMgX.

(ii) *Double decomposition with compounds containing an active hydrogen atom or a reactive halogen atom.* We shall consider only the former at this stage. As we have seen, an active hydrogen atom is one joined to oxygen, nitrogen or sulphur. When such compounds react with a Grignard reagent, the alkyl group is converted into the alkane, *e.g.,*

$$R^1H + Mg(OH)X \xleftarrow{H_2O} R^1MgX \xrightarrow{R^2OH} R^1H + Mg(OR^2)X$$

Since reactions with compounds containing an active hydrogen atom result in the quantitative yield of hydrocarbon, this type of reaction is valuable for the determination of the number of active hydrogen atoms in a compound. The procedure is known as the *Zerewitinoff active* hydrogen determination (1907), and methylmagnesium iodide is normally used as the Grignard reagent. The methane which is liberated is measured (by volume), one molecule of methane being equivalent to one active hydrogen atom, *e.g.,*

$$RNH_2 + CH_3MgI \longrightarrow CH_4 + RNHMgI$$

$$R_2NH + CH_3MgI \longrightarrow CH_4 + R_2NMgI$$

Only one hydrogen atom in a primary amine reacts at room temperature. At a sufficiently high temperature, the active hydrogen atom in the magnesium derivative of the primary amine will react with a further molecule of methylmagnesium iodide.

$$RNHMgI + CH_3MgI \longrightarrow CH_4 + RN(MgI)_2$$

It is therefore possible to estimate the number of amino- and imino-groups in compounds containing both. It is not possible to get a high enough temperature for the second reaction with ether; a satisfactory solvent for the complete Zerewitinoff determination is pyridine (Lehman and Basch, 1945).

Lithium aluminium hydride also reacts with compounds containing active hydrogen, and so may be used for analytical determinations, *e.g.,*

$$LiAlH_4 + 4ROH \longrightarrow (RO)_4LiAl + 4H_2$$

$$LiAlH_4 + 4RNH_2 \longrightarrow 4H_2 + (RNH)_4LiAl \xrightarrow{LiAlH_4} 2(RN)_2LiAl + 4H_2$$

The enolic form of a compound, since it contains an active hydrogen atom, reacts with a Grignard reagent, *e.g.,* acetoacetic ester and nitroethane,

$$CH_3COCH_2CO_2C_2H_5 \rightleftharpoons CH_3\overset{\overset{\displaystyle OH}{|}}{C}{=}CHCO_2C_2H_5 \xrightarrow{CH_3MgI} CH_3\overset{\overset{\displaystyle OH}{|}}{C}{=}CHCO_2C_2H_5 + CH_4$$

$$CH_3CH_2NO_2 \rightleftharpoons CH_3CH{=}NO_2H \xrightarrow{CH_3MgI} CH_3CH{=}NO_2MgI + CH_4$$

In both cases the methane will not be liberated immediately, but at a rate depending on the speed of the conversion of keto into enol.

The hydrogen atom in the $\equiv CH$ group is also active with respect to a Grignard reagent (to be expected, since it is acidic). Thus, when acetylene is passed through an ether solution of ethylmagnesium bromide, ethynylbis (magnesium bromide) is formed,

$$C_2H_5MgBr + C_2H_2 \longrightarrow HC \equiv CMgBr + C_2H_6 \xrightarrow{C_2H_5MgBr} BrMgC \equiv CMgBr$$

The second step can be avoided by adding the Grignard solution to tetrahydrofuran saturated with acetylene (Jones *et al.,* 1956).

*Order of reactivity of functional groups.* In most cases of syntheses in which a Grignard reagent is used, the other compound has only one functional group, and consequently the reaction takes place in one direction only. Occasionally it is necessary to carry out a synthesis with a compound containing two (or possibly more) functional groups.

If an excess of Grignard reagent is used, then both groups react as would be expected. It has been found experimentally that the reactivity of different groups is not equal, and hence when one equivalent of Grignard reagent is added, two competitive reactions take place simultaneously, but at different rates, resulting in two products in unequal amounts.

Experiments have shown that an active hydrogen reacts very much faster than any other group; so much so, in fact, that a compound containing an active hydrogen and another group, reacts with one equivalent of a Grignard reagent as if it had only one reactive group, the active hydrogen. In general, the order of reactivity of the oxo group is as shown, and in all cases it is more reactive than halogen of the alkyl halide type.

$$—CHO > \rangle CO > —COCl > —CO_2R > —CH_2X$$

## SYNTHETIC USES OF THE GRIGNARD REAGENTS

*Hydrocarbons.* When a Grignard reagent is treated with any compound containing active hydrogen, a hydrocarbon is produced ; in practice, water or dilute acid is used.

$$RMgBr + H_2O \xrightarrow{H^+} RH + Mg(OH)Br$$

Since alkyl halides are readily prepared from alcohols, it becomes a relatively simple matter to convert an alcohol (saturated or unsaturated) into the corresponding hydrocarbon. It has already been mentioned that coupling to give $R_2$ usually occurs during the preparation of the Grignard reagent. On the other hand, Kharasch *et al.,* (1941) showed that the addition of a small amount of cobaltous chloride to the Grignard reagent increases the yield of coupling product.

According to Wilds *et al.,* (1949) and Hey *et al.,* (1965, 1969), the reaction proceeds via dialkylcobalt intermediates.

$$2RMgX + CoCl_2 \longrightarrow 2MgXCl + R_2Co \longrightarrow R_2 + Co$$

$$2RMgX + Co \longrightarrow Mg + MgBr_2 + R_2Co \longrightarrow R_2 + Co, \text{etc.}$$

Many transition-metal halides may be used besides cobaltous chloride, *e.g.,* AgBr, $CuCl_2$, $CrCl_3$, $MnCl_2$, $FeCl_3$, etc.

Coupling between a Grignard reagent and alkyl halide containing a reactive halogen atom can be effected directly; this reaction is probably $S_N2$.

$$XMg—R^1 \quad R^2—X \longrightarrow R^1—R^2 + MgX_2$$

The yield is very good if $R^2$ is allyl, t-butyl, or benzyl. This reaction also works well for α-halogeno-ethers (see later). A good yield $R^1 - R^2$ can be obtained for any $R^2$ by using its tosyl ester.

$$R^1MgI + TsOR^2 \longrightarrow R^1 - R^2 + TsOMgI$$

Unreactive halides react with Grignard reagents in the presence of cobaltous chloride, etc., preferably in benzene solution. The main products are $R_2^1$ and $R_2^2$; the crossed product is usually obtained in poor yield. The formation of $R_2^1$ (from $R^1MgBr$) has been described above, the other products are possibly formed as follows.

$$R1Co + 2R2Br \longrightarrow R_2^1 + R_2^2 + CoBr_2$$

$$R^2Br + Mg \longrightarrow R^2MgBr$$

$$R^1MgBr + R^2MgBr + CoCl_2 \longrightarrow R^1CoR^2 \longrightarrow R^1 - R^2 + Co$$

*Primary alcohols.* A Grignard reagent may be used to synthesis an alcohol by treating it with dry oxygen and decomposing the product with acid.

$$RMgX \xrightarrow{O_2} RO_2MgX \xrightarrow{RMgX} 2ROMgX \xrightarrow{H_3O^+} 2ROH \quad (g.\text{–}v.g.)$$

This method (which can be used for all three classes of alcohols) is little used in practice since an alkyl halide may be converted into the corresponding alcohol by simpler means. The method, however, is useful for converting aryl halides into phenols.

When a Grignard reagent (in ethereal solution) is treated with formaldehyde gas, or when a Grignard reagent *ir, e.g.,* di-n-butyl ether (b.p. 141°C), is refluxed with paraformaldehyde, a primary alcohol is obtained by decomposing the magnesium complex with dilute acid. In view of the fact that mechanisms of reactions involving Grignard reagents are not known with certainty, and what is known leads to a complicated sequence (tertiary alcohols, below), it is customary to use *simplified* equations for Grignard reactions. Thus,

$$\begin{matrix} R\text{—}MgI \\ CH_2{=}O \end{matrix} \longrightarrow \begin{matrix} H \quad R \\ C \\ H \quad OMgI \end{matrix} \xrightarrow{H_3O^+} RCH_2OH \text{ (g.)}$$

A primary alcohol containing two carbon atoms more than the Grignard alkyl group can be prepared by adding one molecule of ethylene chlorohydrin to two molecules of Grignard reagent.

$$RMgBr + ClCH_2CH_2OH \longrightarrow RH + ClCH_2CH_2OMgBr$$

$$\xrightarrow{RMgBr} CH_2CH_2OMgBr$$

$$\xrightarrow{H_2O^+} RCH_2CH_2OH \quad \textit{(g.-v.g.)}$$

Two molecules of Grignard reagent are not necessary if ethylene oxide is used instead of ethylene chlorohydrin.

$$\begin{matrix} CH_2 \\ | \quad \rangle O \\ CH_2 \end{matrix} + RMgX \longrightarrow \begin{matrix} CH_2 \\ | \quad \rangle O \\ CH_2 \end{matrix} \rightarrow Mg\begin{matrix} R \\ X \end{matrix} \longrightarrow$$

$$RCH_2CH_2OMgX \xrightarrow{H_3O^+} RCH_2CH_2OH \text{ (g.)}$$

Similarly, a primary alcohol containing *three* more carbon atoms can be prepared from the Grignard reagent by using trimethylene oxide.

*Secondary alcohols.* When a Grignard reagent is treated with any aldehyde other than formaldehyde, a secondary alcohol is formed.

$$R^1CHO + R^2MgX \longrightarrow R^1CH\begin{matrix} OMgX \\ R^2 \end{matrix} \xrightarrow{H_3O^+} R^1CHOHR^2 \textit{ (f.g.–g.)}$$

It can be seen that the secondary alcohol, $R^1CHOHR^2$, is obtained whether we start with $R^1CHO$ and $R^2MgX$, or $R^2CHO$ and $R^1MgX$. Which pair we use is generally a matter of their relative accessibility.

Secondary alcohols may also be prepared by interaction of a Grignard reagent (2 molecules) and ethyl formate (1 molecule) (see also t-alcohols, below).

$$\mathrm{HC(=O)OEt} \xrightarrow{\mathrm{RMgX}} \mathrm{H(R)C(OMgX)(OEt)} \xrightarrow{\mathrm{RMgX}}$$

$$\mathrm{R(H)C(OMgX)R} \xrightarrow{\mathrm{H_3O^+}} \mathrm{RCHOHR}$$

*Tertiary alcohols.* A tertiary alcohol may be prepared by the action of a Grignard reagent on a ketone.

$$\mathrm{R^1R^2C{=}O + R_3MgX \longrightarrow R^1R^2C(OMgX)R^3 \xrightarrow{H_3O^+} R^1R^2C(OH)R^3} \quad (g.)$$

By this means a tertiary alcohol with three different alkyl groups may be prepared, and the starting materials may be any of the following pairs of compounds. $R^1COR^2$ and $R^3MgX$, $R^1COR^3$ and $R^1MgX$, or $R^2COR^3$ and $R^1MgX$ (*cf.* secondary alcohols). It might be noted here that because t-alcohols are readily dehydrated to alkenes by acids, the complex is often broken up with aqueous ammonium chloride or buffered acid solution.

The mechanism of the reaction between a Grignard reagent and a carbonyl compound has been studied in great detail. The following mechanism is that proposed by Ashby (1967), who examined the reaction between phenylmagnesium bromide and benzophenone (R = Ph).

$$\mathrm{R_2C = O + RMgX(OEt_2)_2 \rightleftharpoons R_2C = O \longrightarrow MgRX(OEt_2)}$$

This is followed by combination with another molecule of Grignard reagent to form an intermediate alkoxymagnesium alkyl, which then reacts as shown.

$$\mathrm{R_2C{=}O \rightarrow Mg(R)(X) \leftarrow OEt_2 \xrightarrow{-Et_2O} [R_2C{-}O{\rightarrow}Mg(R)(X)(\leftarrow OEt_2)\cdots R{-}Mg(X)(\leftarrow OEt_2)] \xrightarrow{-2Et_2O}}$$

$$\mathrm{R_3C{-}OMgR + MgX_2 \rightleftharpoons [R_3C{-}O{-}Mg{-}R \cdots X{-}Mg{-}X] \longrightarrow}$$

$$\mathrm{R_3C{-}OMgX + RMgX \xrightarrow{H_3O^+} R_3COH}$$

This mechanism is in keeping with many of the observed facts and, in particular, explains the observed decrease in rate of reaction when half of the Grignard reagent has been used up. Further support comes from the fact that a mixture of alkoxymagnesium alkyl and magnesium bromide produces the alkoxymagnesium halide (Ashby *et al.,* 1967).

Tertiary alcohols containing at least two identical alkyl groups' may be prepared by the reaction between a Grignard reagent (2 molecules) and any ester (1 molecule) other than formic ester.

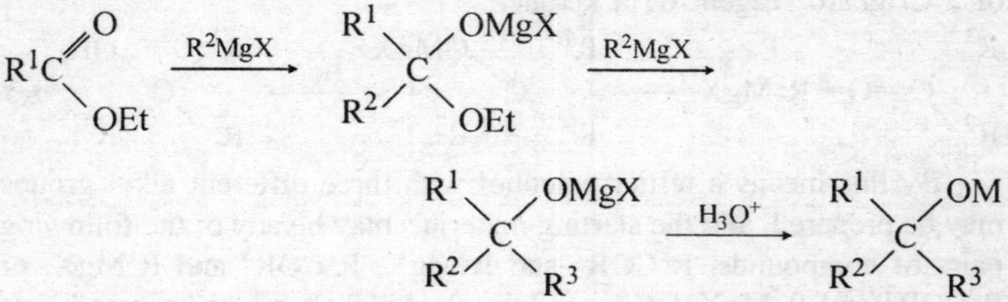

$$R^1C(=O)OEt \xrightarrow{R^2MgX} R^1R^2C(OMgX)(OEt) \xrightarrow{R^2MgX}$$

$$R^1R^2C(OMgX)R^3 \xrightarrow{H_3O^+} R^1R^2C(OM)R^3$$

There is very little evidence for the mechanism of this reaction, but by analogy with that for Grignard/ketone reactions (see above), the steps may be (ether molecules are omitted).

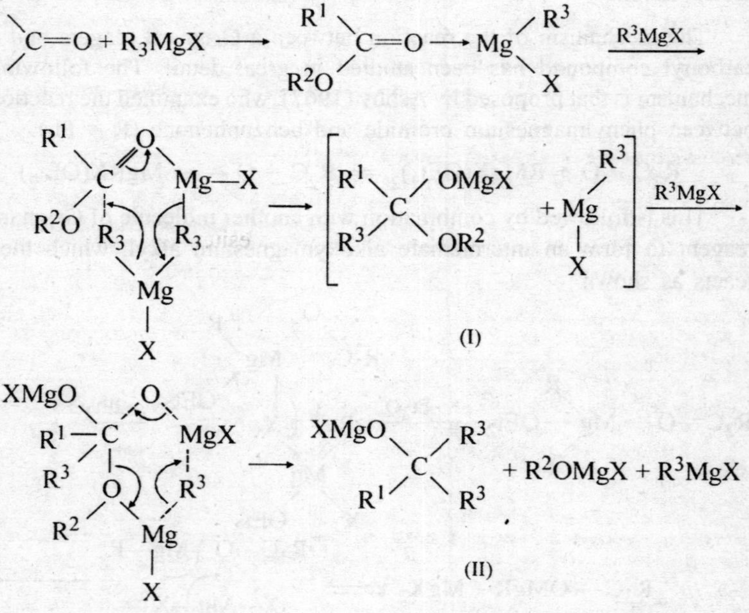

$$>C{=}O + R_3MgX \longrightarrow R^1(R^2O)C{=}O \rightarrow Mg(R^3)X \xrightarrow{R^3MgX}$$

$$R^1(R^2O)C{=}O \cdots Mg{-}X,\ R_3 \cdots R_3,\ Mg{-}X \longrightarrow \left[ R^1R^3C(OMgX)(OR^2) + Mg(R^3)X \right] \xrightarrow{R^3MgX}$$

(I)

$$XMgO(R^1)(R^3)C{-}O(R^2) \cdots MgX,\ R^3 \cdots Mg{-}X \longrightarrow XMgO(R^1)C(R^3)_2 + R^2OMgX + R^3MgX$$

(II)

For most esters, it appears that the step leading to (II) is much faster than that leading to (I). Thus, treatment of an ester with one equivalent of a Grignard reagent generally gives t-alcohol and recovery of half of the ester.

If however, the Grignard is added to ester *(reverse addition)* and the reaction is carried out at low temperatures, then reasonable yields of ketone may be obtained. If the Grignard reagent (or the ester) is sterically hindered, ketone and not alcohol is the major product.

Tertiary alcohols may also be prepared by using an acid and a slight excess of Grignard reagent, *i.e.,* more than 3 molecules (Huston *et at.,* 1946).

$$R^1C(=O)OH \xrightarrow{R^2MgI} R^2H + R^1C(=O)OMgI \xrightarrow{R^2MgI}$$

$$R^1C(OMgI)_2R^2 \xrightarrow{R^2MgI} R^1C(OMgI)R^2_2 \xrightarrow{H_3O^+} R^1R^2_2COH$$

Tertiary alcohols containing three identical alkyl groups may be prepared by the reaction between a Grignard reagent (3 molecules) and ethyl carbonate (1 molecule).

$$3RMgX + (C_2H_5O)_2CO \longrightarrow R_3COMgX + 2Mg(OC_2H_5)X$$

$$\xrightarrow{H_3O^+} R_3COH \quad \text{(f.g. - g.)}$$

*Ethers.* Ethers may be prepared by reaction between a Grignard reagent and an α-monochloroether (this is an example of double decomposition).

$$R^1OCH_2Cl + R^2MgX \longrightarrow R^1OCH_2R^2 + MgXCl \quad \text{(g.)}$$

Chloroethers of the above type are readily prepared, and using monochlorodimethyl ether with Grignard reagents, it is possible to step up both the ether and alcohol series, *e.g.,*

$$MeOH + CH_2O + HCl \longrightarrow MeOCH_2Cl + H_2O$$

$$ROM + HBr \longrightarrow RBr \xrightarrow{Mg} RMgBr \xrightarrow{MgOCH_2Cl}$$

$$RCH_2OMe \xrightarrow{HBr} RCH_2Br \xrightarrow{OH^-} RCH_2OH$$

*Aldehydes.* An aldehyde may be prepared by the reaction between a Grignard reagent (1 molecule) and ethyl formate (1 molecule).

If the Grignard reagent is in excess, a secondary alcohol is formed (see above). Hence, to avoid this as much as possible, *the Grignard reagent is added to the ester.*

$$\mathrm{(H)(EtO)C{=}O + RMgX \longrightarrow (H)(EtO)C(OMgX)(R) \xrightarrow{H_3O^+} RCHO}$$

If ethyl orthoformate is used instead of ethyl formate, a better yield of aldehyde is obtained, since the formation of secondary alcohol *is* prevented by the formation of an acetal.

$$HC(OC_2H_5)_3 + RMgX \longrightarrow RCH(OC_2H_5)_2 + Mg(OC_2H_5)X$$

$$\xrightarrow{H_3O^+} RCHO + 2C_2H_5OH$$

Aldehydes may also be prepared by the reaction between a Grignard reagent and hydrogen cyanide or formamide (see ketones, below).

*Ketones.* These are not usually prepared by reaction between a Grignard reagent (1 molecule) and any ester (1 molecule) other than formic ester (see tertiary alcohols, above).

They may be prepared, however, as ketals, by using any orthoester other than orthoformic ester *(cf.* aldehydes).

$$R^1C(OC_2H_5)_3 + R^2MgX \longrightarrow R^1R^2C(OC_2H_5)_2$$

$$+ Mg(OC_2H_5)X \xrightarrow{H_3o^+} R^1COR^2$$

Ketones may also be prepared by adding an alkyl cyanide to a Grignard reagent.

$$R^1CN + R^2MgX \longrightarrow R^1R^2C = NMgX \xrightarrow{H_3o^+}$$

$$[R^1R^2C = NH] \xrightarrow{H_3o^+} R^1COR^2 + NH_3$$

The starting materials may be either $R^1CN$ and $R_2MgX$ or $R_2CN$ and $R^1MgX$. If hydrogen cyanide is used instead of an alkyl cyanide, and aldehyde is formed.

Acyl chlorides (1 molecule) react rapidly with Grignard reagents (1 molecule) to form ketones.

$$R^1(Cl)C{=}O + R^2MgX \longrightarrow R^1(Cl)C(OMgX)R^2 \xrightarrow{H_3O^+} R^1COR^2$$

$$R^1(Cl)C(OMgX)R^2 \xrightarrow{R^2MgX} R^1C(OMgX)(R^2)_2 \xrightarrow{H_3O^+} R^1C(OH)(R^2)_2$$

Tertiary alcohol is also formed.

By analogy with that for Grignard/ketone reactions, the mechanism may be.

$$R^1COCl + R^2MgX \rightleftharpoons R^1(Cl)C{=}O \rightarrow Mg(R^2)X \xrightarrow{R^2MgX}$$

$$\text{cyclic transition state: } R^1(Cl)C{=}O\rightarrow Mg{-}X,\ R^2,\ R^2\text{–}Mg\text{–}X \longrightarrow R^1(Cl)C(OMgX)R^2 + R^2MgX \xrightarrow{H_3O^+} R^1COR^3$$

Acid anhydrides also form ketones, the reaction being best carried out at about – 70°C.

$$(R^1CO)_2O + R^2MgX \xrightarrow{-70^\circ C} R^1COR^2 + R^1CO_2MgX \text{ (g. – v.g).}$$

Amides and N-substituted amides may also be used.

$$R^1C(=O)NH_2 + R^2MgX \longrightarrow R^2H + R^1C(=O)NHMgX \xrightarrow{R^2MgX}$$

$$R^1C(OMgX)(R^2)NHMgX \xrightarrow{H_3O^+} R^1COR^2$$

Formamide gives rise to the formation of an aldehyde.

*Acids.* When a Grignard reagent is treated with solid carbon dioxide and the complex decomposed with dilute acid, a monocarboxylic acid is obtained.

$$RMgX + O{=}C{=}O \longrightarrow R{-}C({=}O)OMgX \xrightarrow{H_3O^+} RCO_2H \quad (f.g.\text{–}g.)$$

On the other hand, a good yield of acid may also be obtained by passing carbon dioxide into the Grignard solution cooled to 0°C.

The above method is particularly useful for preparing acids of the type $R_3CCO_2H$, which usually cannot be prepared by the cyanide synthesis using a tertiary alkyl halide.

It may also be noted that this method offers a means of ascending the acid series.

γ-Keto-acids may be prepared by the action of a Grignard reagent on a cyclic anhydride.

*Esters.* When a Grignard reagent (1 molecule) reacts with ethyl chloroformate (1 molecule) an ester is formed, and to avoid reaction with the carbethoxy group, the Grignard reagent is added to the ester (reverse addition).

$$(Cl)(EtO)C{=}O + RMgX \longrightarrow (Cl)(EtO)C(OMgX)(R) \xrightarrow{H_3O^+} RCO_2Et$$

Esters of t-alcohols are conveniently prepared as follows.

$$R^1OH + CH_3MgI \longrightarrow CH_4 + R^1OMgI \xrightarrow[\text{(ii) } H_3O^+]{\text{(i) } R^2COCl} R^2CO_2R^1$$

*Alkyl cyanides.* Cyanogen (1 molecule) reacts with a Grignard reagent (1 molecule) to form an alkyl cyanide.

$$RMgX + (CN)_2 \longrightarrow RCN + Mg(CN)X$$

Alkyl cyanides are also formed, together with alkyl chloride, when a Grignard reagent (1 molecule) is added to an ethereal solution of cyanogen chloride (1 molecule), the latter should always be in excess, since the alkyl cyanide produced tends to react with the Grignard reagent (*cf.* aldehydes and esters).

$$RMgCl + ClCN \longrightarrow RCN + MgCl_2$$

$$RMgCl + ClCN \longrightarrow RCl + Mg(CN)Cl$$

This reaction is probably the best method for preparing tertiary alkyl cyanides.

*Primary amines.* A satisfactory method for preparing pure primary amines is the reaction between O-methylhydroxylamine and a Grignard reagent which may be either the alkyl-magnesium chloride or bromide, but *not* iodide (Brown and Jones, 1946).

$$2RMgCl + CH_3ONH_2 \longrightarrow RNHMgCl + RH + Mg(OCH_3)Cl \xrightarrow{H_3O^+} RNH_2 \quad (40\text{–}90\%)$$

Chloramine has also been used to prepare amines, but is unpleasant to handle, and yields are reasonable only for the t-alkyl group.

$$RMgX + ClNH_2 \longrightarrow RNH_2 + MgXCl$$

*Organometallic and organo-non-metallic compounds.* These may be prepared by interaction of a Grignard reagent and an inorganic halide.

*Alkyl halides.* Bromides and iodides are formed when a Grignard reagent is treated with bromine or iodine.

$$RMgX + I_2 \longrightarrow RI + MgXI$$

*Thioalcohols.* These may be prepared by the action of sulphur on a Grignard reagent *(cf.* alcohols).

$$RMgX + S \longrightarrow RSMgX \xrightarrow{H_3O^+} RSH$$

*Sulphinic acids.* When sulphur dioxide is passed into a well-cooled Grignard solution, a sulphinic acid (as its magnesium complex) is formed.

$$RMgX + S(=O)_2 \longrightarrow R{-}S(=O){-}O \xrightarrow{H_3O^+} RSO_2H$$

*Dithioic acids.* These may be prepared by the action of carbon disulphide on a Grignard reagent *(cf.* carboxylic and sulphinic acids).

$$RMgX + C(=S)_2 \longrightarrow R{-}S(=S){-}SMgX \xrightarrow{H_3O^+} RCS_2H$$

## ABNORMAL BEHAVIOUR OF GRIGNARD REAGENTS

In certain cases a Grignard reagent does not react with compounds containing a functional group which is normally capable of reaction. Generally, branching of the carbon chain near the functional group prevents reaction; the cause is probably the steric effect, *e.g.,* methylmagnesium bromide or iodide does not react with hexamethylacetone, $(CH_3)_3CCOC(CH_3)_3$.

It has also been found that if the Grignard reagent contains large alkyl groups, reaction may *be prevented; e.g.,* isopropyl methyl ketone reacts with methylmagnesium iodide but not with t-butyl-magnesium chloride.

In other cases, *abnormal reaction* may take place, *e.g.,* when isopropyl-magnesium bromide is added to di-isopropyl ketone, the expected tertiary alcohol is *not* formed; instead, the secondary alcohol, di-isopropylcarbinol, is obtained, resulting from the *reduction* of the ketone.

$$(CH_3)_2CHCOCH(CH_3)_2 \xrightarrow{(CH_3)_2CHMgBr} (CH_3)_2CHCHOHCH(CH_3)_2$$

This abnormal reaction may be explained by the transfer of a hydride ion from the Grignard reagent via a cyclic transition state *(cf.* the M..P.V. reduction).

$$\begin{array}{c} \text{Br} \\ | \\ \text{Mg} \\ \diagup \quad \diagdown \\ \text{MeCH} \quad \text{O} \\ \| \qquad \| \\ \text{CH}_2 \quad \text{CPr}^i_2 \\ \diagdown \quad \diagup \\ \text{H} \end{array} \longrightarrow \begin{array}{c} \text{MeCH} \\ \| \\ \text{CH}_2 \end{array} + \begin{array}{c} \text{MgBr} \\ | \\ \text{O} \\ | \\ \text{CHPr}^i_2 \end{array}$$

In some cases, aldehyde and the Grignard reagent interact to form ketone and a primary alcohol, *e.g.,*

$$R^1CHO + R^2MgBr \longrightarrow R^1R^2CHOMgBr$$

$$\xrightarrow{R^1CHO} R^1R^2O + R^1CH_2OMgBr$$

This also occurs by hydride ion transfer via a cyclic T.S.

Grignard reagents often reduce nitro groups (to nitroso or hydroxylamine),. and hence mixtures of products may be obtained. However, in certain cases, normal reaction will proceed if carried out at low temperatures.

Compounds which are capable of enolisation liberate hydrogen with Grignard reagents. α,β,-Unsaturated carbonyl compounds react with a Grignard reagent in the 1,2- or 1,4-positions.

**1, 2-addition**

$$\overset{4}{R^1—CH}=\overset{3}{CH}—\overset{2}{CR^2}=\overset{1}{O} \xrightarrow{R^2MgX}$$

$$R^1—CH=CH—CR^2R^3—OMgX \xrightarrow{H_3O^+}$$

$$R^1—CH=CH—R^2R^3—OH$$

**1, 4-addition**

$$\overset{4}{R^1—CH}=\overset{3}{CH}—\overset{2}{CR^2}=\overset{1}{O} \xrightarrow{R^2MgX}$$

$$R^1R^3CH—CH=CR^2—OMgX \xrightarrow{H_3O^+}$$

$$\underset{\text{enol}}{[R^1R^3CH—CH=CR^2OH]} \longrightarrow \underset{\text{keto}}{R^1R^3CH—CH_2—CR^2=O}$$

The predominant factor in deciding the course of addition is steric hindrance, and when both $R^1$ and $R^2$ are large, $-R^2$ has greater influence than $R^1$ (the first step in 1,2-addition-is addition to the carbonyl group).

On the other hand, 1,4-addition can be made to predominate by carrying out the reaction in the presence of, *e.g.*, cuprous chloride, cupric acetate, etc. (the mechanism is now free-radical).

The reaction of dihalides of the type $Br(CH_2)_nBr$ with magnesium depends on the value of *n*. For $n$ = 1-3, no Grignard reagent is formed, *e.g.*, ethylene dibromide ($n$ = 2) gives ethylene; 1, 3-dibromopropane ($n$ = 3) gives propene, cyclopropane, and other products. When $n \geq 4$, the dimagnesium compound is obtained, but the yields are usually of the order of about 30 per cent., *e.g.*, n = 5.

$$Br(CH_2)_5Br + 2Mg \longrightarrow BrMg(CH_2)_5MgBr$$

The reactions of these di-Grignards depend on the nature of the added substrate.

## THE HALOGEN DERIVATIVES OF ALKANES

**Introduction :** The *halogen derivatives* of the alkanes are divided into mono, di-, tri, etc, substitution products according to the number of halogen atoms in the molecule.

## GENERAL REACTIONS OF THE ALKYL HALIDES

(1) Alkyl halides are hydrolysed to alcohols very slowly by water, but rapidly by silver oxyl suspended in boiling water, or by boiling aqueous alkalis (see also below).

$$RX + KOH \longrightarrow ROH + KX.$$

This type of reaction is an example of nucleophilic substitution ($S_N$) since the attacking reagent is nucleophilic regent. It has been assumed that this type of heterolytic reaction in solution can take place by two different mechanisms, unimolecular or bimolecular.

*Unimolecular mechanism* : This is a *two-stage process.* The first stage consisting to slow heterolysis of the compound to form a carbonium ion, followed by rapid combination between the carbonium ion and the substituting nucleophilic reagent. Since the rate-determining step is the first one, and since in this step only one molecule is undergoing a covalency change, this type of mechanism is called *unimolecular* and is labelled $S_N1$ (Ingold *et al.* 1928. 1933). Thus, if the hydrolysis of an alkyl halide is an $S_N1$ reaction it may be written.

$$R\text{—}X \xrightleftharpoons{\text{slow}} X^- \text{—} R^+ \xrightarrow[\text{fast}]{OH^-} ROH$$

Heavy metal ions, particularly silver ions, catalyse $S_N1$ reactions and according to Kevill *et al.* (1967), the mechanism involves a fast pre-equilibrium step (the silver ion has an empty orbital).

$$R\text{—}X + Ag^+ \xrightleftharpoons{\text{heat}} [R\text{—}X\text{—}Ag]^+ \xrightarrow{\text{heat}} R^+ + AgX$$

*Bimolecular mechanism* : This is a *one-stage process,* two molecules simultaneously undergoing covalency change in the rate-determining step. This type of mechanism is called *bimolecular* and is labelled $S_N2$. Since the rate-determining step in this reaction is the formation of the transition state, the hydrolysis of an alkyl halide by an $S_N2$ reaction may be written.

$$HO\ R\text{—}X \xrightarrow{\text{slow}} \overset{\delta-}{HO}\cdots R \cdots \overset{\delta-}{X} \xrightarrow{\text{fast}} HO\text{—}R + X^-$$

A number of differences exist between $S_N2$ and $S_N1$ reactions, *e.g.*, (i). When both reactants are present in small and controllable concentrations. $S_N2$ reactions are second-order and $S_N1$ reactions are first-order. If one reactant is present in large excess *e.g.*, one of the reactants is the solvent, than an $S_N2$ mechanism will lead to first-order kinetics. (ii) The $S_N2$ mechanism always leads to inversion of the product, whereas with the $S_N1$ mechanism the configuration of the product depends on various factors, (iii) No rearrangement is possible for the $S_N2$ mechanism, but is possible (and usually occurs) for the $S_N1$, (iv) The rate constant of an $S_N2$ reaction with a given substrate depends on the nature of the nucleophilic and for a given nucleophile. on the nature of the leaving group in the substrate (see above).

**Stereochemistry of Substitution Reactions**

(i) *The $S_N2$ mechanism.*

$$HO^- \; R—X \longrightarrow \overset{\delta-}{HO} \cdots R \cdots \overset{\delta-}{X} \longrightarrow HO—R + X^-$$

In the transition state, the groups OH and X are collinear and on opposite sides of the attacked carbon atom. Furthermore, the line joining OH and X is prependicular to the plane containing the other three groups a, b, and d. In the original molecule *Cabd* X, the four groups are arranged tetrahedrally. Hence, to achieve a planar configuration of *Cabd* in the transition state, the carbon atom changes from tetrahedral to trigonal hybridisation, the remaining $p_x$ orbital being used (by means of its two loves) to hold the groups OH and X by 'half-bonds'. When X is ejected, the carbon atom returns to its state of tetrahedral hybridisation.

The above reaction is a three-centre-reaction, proceeding by an end-one approach.

$$HO^- + C(a)(b)(d)X \longrightarrow \overset{\delta-}{HO} \cdots C(a)(b)(d) \cdots \overset{\delta-}{X} \longrightarrow (HO)(a)(b)(d)C + X^-$$

Hughes *et al.*, (1935) studied. (a) the interchange reaction of (+)-2iodo-octane with radio active iodine (as NaI*) in acetone solution, and (b) the racemisation of (+)-2-iodo-octane by ordinary NaI under the same conditions. These reactions were shown to occur by the $S_N2$ mechanism, and the result showed that every halide-halide* displacement was always accompanied by inversion. Thus this experiment leads to the *assumption that an* $S_N2$ *reaction always gives inversion,* this is fully supported by other experimental work. Hence, if it can be shown that the $S_N2$ mechanism is operating, then inversion must have occurred.

(ii) *The $S_N1$ mechanism*

$$R—X \underset{}{\overset{\text{slow}}{\rightleftharpoons}} X^- + R^+ \xrightarrow[\text{fast}]{OH} ROH$$

When the reaction proceeds by this mechanism, then inversion and retention (recemisation) will occur, the amount of each depending on various factors. The carbonium ion is flat (trigonal hybridisation), and hence attack by nucleophilic regents can take place equally well on either side, *i.e.*, equal amounts of (+)- and (–)-forms are produced, this is recemisation. Complete recemisation, however, can be expected only if the carbonium ion is sufficiently long-lived, this is favoured by low reactivity of the carbonium ion and low concentration of the nucleophilic reagent.

On the other hand, during the actual ionisation, the retiring negative group will shield attack on that side, this encourages end on attack *on the other side*, there by leading to inversion. In general, the $S_N1$ mechanism is accompanied by inversion and racemisation, but in some cases there may be complete inversion.

It will be seen from the above discussion that if heterolysis of RX is the rate-determining step, than the mechanism, is $S_N1$. However, circumstances might arise where the *second* step is the rate-determining one. In this case, Hughes and Ingold (1954) have labelled the reaction $S_N2$ (C*) [substitution, nucleophilic, bimolecular, with a rapidly formed carbonium ion.

Another $S_N$ reaction is the SN1 type (substitution, nucleophilic, internal), *e.g.*, Hughes. Ingold *et al* (1937) have shown that optically active α-phenylethanol (C6H5CHMe – = R) reacts with thionyl chloride to give a-phenylethyl chloride with almost complete retention of configuration. The mechanism proposed in $S_N1$ via a chlorosulphite.

$$\text{R—O(H)} + \text{Cl—S(=O)—Cl} \xrightarrow{\text{fast}} \text{HCl} + \text{R—O—S(=O)—Cl} \xrightarrow{\text{fast}} \text{RCl} + SO_4$$

Now inversion occurs in either stage (both stages are four-centre type). The situation arises with *arylmethanols,* but in the presence of pyridine, the chloride now has the *inverted* configuration. This has been explained on the basis that the $S_N2$ mechanism is operating, but now involves a pyridine complex.

$$\text{ROSOCl} + C_5H_5N \xrightarrow{\text{fast}} \text{ROSONC}_5H_5 + \xrightarrow[S_N2]{\text{slow}}$$

$$\overset{\delta-}{\text{Cl}}\ldots\text{R}\ldots\overset{\delta-}{\text{O}}\ \text{SONC}_5H_5 \xrightarrow{\text{fast}} \text{Cl} - \text{R} + SO_2 + C_5H_5N\,.$$

Any factor that affects the energy of activation of a given type of reaction will affect the rate and/or the mechanism. The following discussion is largely qualitative, and because of this, one cannot be sure which are the predominant factors in deciding the value of the energy of activation (see also later).

Let us first consider the nature of R. There are two effects to be considered, the polar factor and the steric factor.

*polar factor.* Consider the series of ethyl, isopropyl, and t-butyl halides. Since the methyl group has a + 1 effect, the larger the number of methyl group on the carbon atom of the C—X the greater will be the electron density on this carbon. If we use one arrow-head to represent (qualitatively) the electron-repelling effect of a methyl group, then we have the following state of affairs.

$$Me \rightarrow CH_2 \rightarrow X \qquad \begin{matrix} Me \searrow \\ \quad CH \twoheadrightarrow X \\ Me \nearrow \end{matrix} \qquad \begin{matrix} Me \searrow \\ Me \rightarrow CH \Rrightarrow X \\ Me \nearrow \end{matrix}$$

This increasing electron density on the central carbon atom increasingly opposes attack at this carbon atom by a negatively charged nucleophilic reagent.. Thus, the formation of the transition state for the $S_N2$ mechanism becomes increasingly difficult. It can therefore be anticipated that the $S_N2$ mechanism is made more difficult in passing from EtX to t-BuX. On the other hand, space the $S_N1$ mechanism involves ionisation as the first step, then it can be expected that as the electron density increases on the central carbon atom, the bonding pair in C—X becomes more and more displaced towards the X atom, and consequently ionisation of X as a negative ion will become easier. It therefore follows that the tendency for the SN1 mechanism should increase from EtX to t-BuX. These anticipated results have been realised in practice. Hughes, Ingold *et al*, (1935-1940) examined the hydrolysis of alkyl bromides in alkaline aqueous ethanol and showed that MeBr and EtBr undergo hydrolysis by the $S_N2$ mechanism, isoPrBr by *both* $S_N2$ and SN1 mechanism, and t-BuBr by the $S_N1$ mechanism only. It has also been shown, however, that the actual position where the mechanism changes over from $S_N2$ to SN1 in a given graded polar uses (such as the one above) is one fixed, but also depends on other factors such as the conformation of the nucleophilic reagent and the nature of the solvent.

An alternative explanation for the effect of the nature of R is as follows. When the $S_N1$ mechanism is favoured, it means that the activation energy for this path is less than that for the $S_N2$ and vice versa. Since the $S_N1$ reaction proceeds through a carbonium ion, the more stable this ion, the lower is the activation energy and the more favoured will be this mechanism. Now the order of stabilities of carbonium ions is p < s < t, and hence the tendency for the SN1 mechanism to operate will be t-BuX > iso rX > EtX, and the reverse tendency for the $S_N2$ mechanism.

*Steric effects.* The steric effect was originally thought to be a spatial effect brought into play be mechanical interference between groups, and was described as *steric hindrance*. The term steric hindrance considered the geometry of the *reactant* molecule. When however a molecule undergoes chemical reaction, it does so via a transition state. Consequently the geometry of both the initial and transition state must be taken into consideration. Thus steric forms may affect the speed and the mechanism of a reaction. It is however, often very difficult is distinguish between steric and polar factors.

Nevertheless, the effects of the steric factor can often be assessed in some sort of qualitative manner. When steric effects slow down a reaction, the reaction is said to be subject to steric hindrance (*or* retardation), and when they speed up a reaction, that reaction is said to be subject to steric Halogen Derivatives of the Alkanes.

In an $S_N2$ reaction there will be five groups 'attached' to the carbon atom at which reaction groups. Thus there will be 'crowding' in the transition state, and the bulkier the groups, the greater will be the *compression energy,* and consequently the reaction will be hindered specially.

Let us consider the following example. Hughes *et al.* (1946) examined the rate of exchange of made ions in acetone solution with MeBr, EtBr, isoPrBr and t-BuBr under conditions where only the $S_N2$ mechanism operated. The relative reactivites were found to be 10,000; 65; 0-50;

$$\bar{I} \; R—Br \longrightarrow \overset{\delta-}{I}\cdots Br \cdots \overset{\delta-}{Br} \longrightarrow I—R + Br^-$$

Thus, increasing the number of methyl groups on the central carbon atom increases the step retardation.

The problem is some what different for the SN1 mechanism Here the transition state do not contain more than four groups attached to the central carbon atom , and hence one word expect steric hindrance to be less important in the SN1 mechanism If, however, the molecular contains bulky groups, then by ionising, the molecule can relieve the steric strain, since carbonium ion-produced is flat (trigonal hybridisation) and so there is more room to accomodate the three alkyl groups. Thus, in such $S_N1$ reactions, there will be steric Halogen Derivatives of the Alkanes. Brown *et al* (1949) showed that the solvolysis of tertiary halides is subject to steric accelerate (*solvolysis is a* nucleophilic substitution reaction in which the *solvent* is the nucleophilic reagent

$$R_3C{-}X \xrightarrow{H_2O} R_3C^{+}{-}X^{-}$$

(tetrahedral; large strain) (trigonoal; planar; small strain)

These authors showed that as R increases in size, the rate of solvolysis increases. However, a larger the R groups are, the more slowly will the carbonium ion be expected to react with slowly molecules, since steric strain will be reintroduced into the molecule. In cases like this, the bonium ion tends to undergo an elimination reaction to form an alkene, and Brown *et al*. (1958) have shown that this elimination process increases as the alkyl groups become larger.

It should be noted that a fundamental part of the $S_N1$ mechanism is the postulate of the trances existence of carbonium ions. Triarylmethyl carbonium ions have been isolated as their salts, *e.g.*, phenylmethyl carbonium ion, $Ph_3C^*$, as its perchlorate and borofluoride (Dauben *et al*., 1960). The stability of ions such as these is attributed to resonance. The order of stability of a carbonium ions is t > s > p, and so one is more likely to expect success in the isolation of the Olah *et al*. (1964) have prepared, e.g, $Me_3C^*$ $SbF_6^-$, and their infra-red spectra studies have to stantiated be planar $sp^2$ hybridised structure of simple alkyl carbonium ions.

Now let us consider the effect of the nature of the halogen atom. Experimental work be shown that the nature of X has very little effect, if any, on *mechanism,* but it does affect the rat of reaction for a given mechanism, *e.g.*, it has been found that in both $S_N2$ and $S_N1$ reactions the rate follows the order RI > RBr > RCI. Experimental work has shown that many $S_N2$ reactions of alkyl chlorides and bromides are catalysed by the presence of iodide ions. This is be explained on the basis that the activation energies of *both* steps in reaction (i) are lower that the activation energy of the one-step reaction (ii).

$$\text{(i)}\ I^- + R{-}Cl \longrightarrow I{-}R + Cl^- \xrightarrow{OH^-} ROH + I^- + Cl^-$$

(ii) $HO^- + R—Cl \longrightarrow HO—R + Cl^-$

The more pronounced the nucleophilic ·· tivity reagent, then more the $S_N2$ mechanism will be favoured, since in the $S_N1$ mechanism the reagent does not enter into the rate-determining step of ionisation. However, it can also be expected that as the nucleophilic that the mechanisms will change from $S_N2$ to $S_N1$. Hughes. Inglod , *et al.*, (1935) examined the mechanism and rate of decomposition of various trimethylsulphonium salts ($Me_3S^+X$) and showed that for X = OH, the mechanism is $S_N2$, and X = I, Cl or Br, the mechanism is $S_N1$.

Mechanism and rate of reaction are very affected by the nature of the solvent. It has been found that the ionising power of a solvent depends on its dielectric constant and its power of solvation, and it appears that the later is more important than the former. Solvation is due to the 'attachment' between solvent and solute molecules, and one important contributing factor attraction of a charge for a dipole.

Since electrostatic work is done in the process of solvation, energy is lost by the system, and consequently the system is more stable. Furthermore, the greater the polarity (dipole moment) of the solvent, the greater is the attraction between solute and absent molecules and consequently the more stable is the system.

Thus, increasing the dielectric constant of the solvent increases the ionising potentiality of the solute molecules, and the higher the polarity of the solvent the greater is the amount of solvation. Solvation, however, may also be partly due to hydrogen bonding of the solvent (if this is protic) with the solute (if this can from hydrogen bonds). Some common solvents and their dielectric constants and dipole comments (D) are: water (81 : 1 ; 2– 3); formic acid (48.0; 1. 5), nitrobenzene (35.7; 4.0), ethanol (258, 1.7); acetone (21.3; 3.0), acetic acid (7.1, 1 – 1 : 5); chloroform (4.6; 1.1); ether (4.3;1.25); become (2.3; 0); carbon tetrachloride (2.2; 0).

When solute molecules have ionised, the oppositely charged pair of ions may become enclosed in a 'cage' of the surrounding solvent molecules, and may therefore recombine before they can escape from the cage. Such a complex is known as an *ion-pair,* and their recombination is known *is internal return.* It has now been shown that many organic reactions proceed via ion-pairs rather than dissociated ions. Thus we have the possibility of the following steps.

$$RX \rightleftharpoons \underset{\text{ion-pair}}{R^+X^-} \rightleftharpoons \underset{\text{dissociated ions}}{R^+ \quad + X^-}$$

Many attempts have been made to correlate reaction rates and nature of the solvent. Hughes and Ingold (1935, 1948) proposed the following qualitative theory of solvent effects.

(i) Ions and polar molecules, when dissolved in polar solvents, tend to become solvated.

(ii) For a given solvent, solvation tends to increase with increasing magnitude of charge on the solute molecules or ions.

(iii) For a given solute, solvation tends to increase with the increasing dipole moment of the solvent.

(iv) For a given magnitude of charge solvation decreases as the charge is spread over a larger volume.

(v) The decrease of solvation due to the dispersal of charge will be less than that due to its deutuction.

Let us consider the following $S_N1$ reaction.

$$R—X \xrightleftharpoons{\text{slow}} \overset{\delta+}{R}...\overset{\delta-}{X} \xrightarrow{\text{fast}} R^+ + X^- \xrightarrow[\text{fast}]{OH^-} R—OH \cdot$$

Since the transition state has a larger charge than the reactant molecule, the former will be more solvated than the latter (rule ii). Thus the transition state is more stabilised that the reactant molecule, *i.e.*, the energy of activation is lowered, and so the reaction proceeds faster than had there been no (or less) solvation.

The rates of $S_N2$ reactions are also affected by the polarity of the solvent.

$$H\bar{O} \; R—X \rightleftharpoons \overset{\delta-}{HO}\cdots R \cdots \overset{\delta-}{X} \longrightarrow HO—R + X^-$$

A solvent with a high dipole moment will solvate both the reactant ion (nucleophilic reagent) and the transition state, but more so the former than the latter, since in the latter the charge, although unchanged in magnitude (δ – = – 1/2), is more dispersed than in the former. Thus fore solvation tends to stabilise the reactants more than the transition state (rule iv). Thus a activation energy is increased, and so the reaction is retarded.

The rate of the following $S_N2$ reaction (Menschutkin reaction 1890) is increased as the polarity of the solvent increases.

$$R_3\ddot{N}\ \ R{-}X \rightleftharpoons \overset{\delta-}{R_3}{-}N\cdots\overset{\delta-}{R}\cdots X \longrightarrow R_4N^+X^-$$

Here the charge on the transition state is greater than that on the reactants. Hence the transist state more solvated than the reactants and consequently stabilised, and so the activation is lowered.

The polarity of the solvent may *change* the mechanism of a reaction, *e.g.*, Olivier (1995) showed that the alkaline hydrolysis of benzyl chloride in 50 per cent, aqueous acetone process by both $S_N2$ and $S_N1$ mechanisms. When water was used as solvent, the mechanism was mainly $S_N1$. The dielectric constant of water is greater than that of aqueous acetone, an ionisation of the benzyl chloride is facilitated.

The above discussion has been purely qualitative, but it is also of interest to consider the problem a quantitative way. This involves the use of the following thermochemical cycle.

RX (solvated) —1 (leassion)→ R⁺ (solvated) + X⁻ (solvated)

2 (desolvation) ↓ ; 6 ↑ (desolvation) ; 7 ↑ (desolvation)

R⁺ (vapour) X⁻ (vapour)

6 ↑ (desolvation) ; 7 ↑ (desolvation)

RX (vapour) —3 (dessestion)→ R· (vapour) + X· (vapour)

The, if the enthalpy change for each step is represented by the number (of the step) as a subscript follows from Hess's law that

$$\Delta H_1 = \Delta H_2 + \Delta H_3 + \Delta H_4 + \Delta H_5 + \Delta H_6 + \Delta H_7$$

Where $\Delta H_4$ is the ionisation potential of R. and $\Delta H_7$ is the electron affinity of X.

All of these values except $\Delta H_2$ have been calculated from suitable experimental data, and the rat of $\Delta H_2$ has been estimated. The values given in the following discussion are those used by Frater is Singer (see References).

First let us consider the problem of reactions in the gas phase and in solution. In the gas phase, and reaction is homolytic, then ΔH for step 3 in the above cycle is $\Delta H_3$. If we use methyl chloride and example, than $\Delta H_3$ is + 334.7 kJ. If however, the reaction is heterolytic, then

$$\Delta H = \Delta H_3 + \Delta H_4 + \Delta H_5 = +\ 334.7 + 970.7 - 364.0 = 941.4\ \text{kJ}$$

It can be seen that heterolytic reaction in the gas phase is very unfavourable energetically. If the reaction is carried out in water, then

$$\Delta H_1 = +\ 33.47 + 334.7 + 970.7 - 364.0 - 343.1 - 288.7$$
$$= 343.1 \text{ kJ}$$

This much more favourable value is due to the large contribution of the heats of solvation ($\Delta H_6$ , $\Delta H_7$).

As we have already seen, in the $S_N1$ mechanism (with all ions and molecules solvate).

$$RX_{(aq)} \underset{}{\overset{\text{slow}}{\rightleftharpoons}} R^+_{(aq)} + X^-_{(aq)} \xrightarrow[\text{fast}]{H_2O} ROH_{(aq)} + H^+_{(aq)}$$

The rate order is t-Bu > isoPr > Et > Me. The following Table gives the values of $\Delta H_1$ the ionisation in kJ $mol^{-1}$) for various alkyl halides.

| *Group* | Cl | Br | 1 |
|---|---|---|---|
| Me | 343.1 | 343.1 | 364.0 |
| Et | 238.5 | 213.4 | 238.5 |
| isoPr | 104.6 | 96.23 | 115.1 |
| t-Bu | 71.13 | 62.76 | 75.31 |

It can be seen in going in a downward direction for a given halogen atom, $\Delta H_1$ decreases rapidly. Moreover, calculations have shown that $\Delta H_1$ and E (activation energy) are almost identical. Hence the rate of the $S_N1$ reaction will increase in the downward direction *i.e.*, t-Bu >> isoPr > Me.

It has been shown experimentally that the rate of the $S_N1$ reaction (and that of the $S_N2$) for a given alkyl group is RI > RBr > RCI. The above Table indicates the order RBr > RCl > Rl. This again illustrates the point that even though a quantitative approach to a problem is always desirable, unless all the necessary values are known (reasonably) accurately, estimates may lead to some, wrong conclusions. This is more the case with reactions in solution than with gas-phase reactions.

2. Alkyl halides are reduced by dissolving metals to alkanes:

$$RX \xrightarrow{eH^+} RH + HX$$

Primary and secondary alkyl chloride and bromides are nor affected in natural of acidic media, whereas prim., s-and t-iodides, t-chlorides and t-bromides are converted into alkanes by *catalytic hydrpgenolysis, i.e.,* hydrogenation in the presence of a catalyst. The best catalyst is Pd—C, but Raney nickel is also effective provided it is used in large amounts.

$$RX + H_2 \xrightarrow{Pd-C} RH + HX$$

Reduction to alkane can also be carried out with lithium aluminium hydride or sodium borohydride.

Johnson *et al.* (1964) have shown that primary and secondary alkyl iodides are oxidised by dimethyl sulphoxide to aldehydes or ketones, respectively. The mechanism is uncertain, but a possibility is:

$$RCH_2I + Me_2{=}O \longrightarrow I^- + R{-}\underset{H}{CH}{-}O{-}SMe_2 \longrightarrow RCHO + Me_2S + H^+$$

3. Alkyl halides undergo the Wurtz reaction to form alkanes. On the other hand alkyl halides, when treated with certain metallic alloys, form *organometallic* compounds, *e.g.*, ethyl chloride heated with a sodium lead alloy under pressure gives tetraethyl-lead.

$$4C_2H_5Cl + 4Na/Pb \longrightarrow (C_2H_5)_4Pb + 4NaCl + 3Pb.$$

4. When an alkyl halide is heated at about 300°C, and at a lower temperature in the presence of aluminium chloride as catalyst, the alkyl halide undergoes rearrangement, *e.g.*,

$$CH_3CH_2CH_2Cl \xrightarrow{-Cl^-} CH_3CH_2CH_2^+ \longrightarrow CH_3\overset{+}{C}HCH_3 \xrightarrow{Cl^-} CH_3CHClCH_3.$$

If there is no hydrogen atom on the carbon adjacent to the C—X group, an alkyl group migrates, *e.g.*, neopentyl chloride rearranges to 2-chloro-2-methylbutane.

$$(CH_3)_3CCH_2Cl \longrightarrow (CH_3)_2CClCH_2CH_3$$

5. When alkyl halides are boiled with ethanolic potassium hydroxide, alkenes are obtained, *e.g.*, propyl bromide gives propene.

$$CH_3CH_2CH_2Br + KOH \xrightarrow{ethanol} CH_3CH = CH_2 + KBr + H_2O$$

This is an example of the 1,2- or β-elimination reactions, in which two groups are lost from *neighbouring* atoms to form a double bond. Other elimination types (x, g, etc) are also known but these are much less common.

There are two important mechanism which operate with alkyl halides, E1 and E2, both of which are *polar* (ef, cyclic eliminations.

E1 mechanism (unimolecular elimination). This mechanism is parallel to the $S_N1$ mechanism in that the rate-determining step is ionisation to a carbonium ion, but the second step is now elimination of a β-proton to form an alkene.

$$HCR_2{-}CR_2{-}Z \xrightleftharpoons{slow} HCR_2{-}CR_2^+ + Z^-$$

$$B\ H{-}CR_2{-}CR_2 \xrightleftharpoons{fast} BH + CR_2{=}CR_2$$

The reaction is carried out in the presence of a base and the proton in transferred to this, *e.g.*, t-butyl bromide in aqueous ethanolic potassium hydroxide gives isobutene.

$$(CH_3)_3CBr \xrightleftharpoons{slow} (CH_3)_3C^+ + Br^-$$

$$HO^-\ H{-}CH_2{-}C(CH_3)_2 \xrightarrow{fast} H_2O + CH_2{=}C(CH_3)_2$$

Evidence for the E1 mechanism has been obtained in several ways, *e.g.*, the rate low for many eliminations is first order (alkyl halide only), *i.e.*, rate = k [R.X]. This is consistent with the mechanism given.

E2 mechanism (bimolecular elimination). The general reaction may be written.

$$B\ H{-}CR_2{-}CR_2{-}Z \longrightarrow BH{-}CR_2{=}CR_2 + Z^-$$

An example of this type is the action of ethanolic potassium hydroxide on isopropyl bromide.

$$EtO\ H{-}CR_2{-}CH(CR_3){-}Br \longrightarrow EtOH + CH_2{=}CHCH_3 + Br^-$$

Evidence for the E2 mechanism comes from the fact that the rate law for may eliminations is seconds order, *i.e.*, rate = k [RX] [B]. There is, however, a difficulty here. The second-order rate law is consistent with the one-step mechanism given above, but it is also consistent with the following two-step mechanism (Ph = $C_6H_5$).

$$EtO + PhCH_2CH_2Br \xrightleftharpoons{fast} EtOH + PhCHCH_2Br$$

$$Ph\bar{C}H—CH_2—Br \xrightarrow{slow} PhCH{=}CH_2 + Br^-$$

A simplified derivation of the rate law for this reaction is as follows. Since the reaction is carried out in ethanol, the concentration of the *ethanol produced* in the first step may be neglected. Hence.

$$K = [Ph\bar{C}HCH_2Br]/([PhCH_2CH_2Br][OEt^-])$$

$$rate = k = [Ph\bar{C}HCH_2Br] = kK\ [PhCH_2CH_2Br][OEt^-].$$

Since k and K are both constants, their product may be replaced by a constant, and the rate law

$$rate = k\ [PhCH_2CH_2Br]\ [OEt^-]$$

*i.e.*, the rate law is second-order. Thus, kinetics cannot distinguish between these two possible elimination mechanisms.

In the second mechanism, a slow unimolecular elimination occurs in the conjugate base (cB) of the reactant, and hence this mechanism is called the E1 cB or carbanion mechanism. Since the first step must be reversible (acid-conjugate base equilibrium), if ethanol containing EtOD is used as solvent,

would be expected that the original bromide would incorporate deuterium. Skell at al. (1945) examined this reaction and found that after the reaction and found that after the reaction was half-completed, the recovered bromide did not contain deuterium. Hence *this* mechanism is not E1cB, but is E2.

It appears that definite examples of the E1cB mechanism are rate.

Which mechanism, E1 or E2 operates depends on the nature of the alkyl halides and the conditions used. Before discussing this, let us consider the situation where two alkanes may be formed by dehydrohalogenation of an alkyl halide. Originally, the direction of elimination was described empirically by Saytzeff's rule (1875). This rule may be sated in two ways.

The predominant product is the most substituted alkene, *i.e.*, the one carrying the largest number of alkyl substituents; or Hydrogen is eliminated preferentially from the carbon atom joined to the heat number of hydrogen atoms. There is also another rule which describes empirically the direction elimination in quaternary ammonium hydroxide. This rule, the Hofmann rule (1881), is discussed later, but may be sated here as follows. The predominant alkene (from an onium hydroxide) is the least substituted alkene. An illustration of these rules is the dehydro bromination of 2-bromobutane.

$$\underset{\text{(I)}}{CH_3CH=CHCH_3} \longleftarrow CH_3CH_2CHBrCH_3 \longrightarrow CH_3CH_2CH=CH_2$$

(I) is the Saytzeff product and (II) is the Hofmann product, and since we are dealing with elimination from an alkyl halide, (I) will be expected to be the predominant product,. Ingold *et al.* (1994) have offered the following explanation for the Saytzeff rule. If (I) and (II) are produced by an E2 mechanism, their formation may be written as follows.

(i) $HO^- \; H—CH(CH_3)—CH(CH_3)—Br \longrightarrow H_2O + CH(CH_3)=CH(CH_3) + Br^-$ (I)

(ii) $HO^- \; H—CH_2—CH(CH_2CH_3)—Br \longrightarrow H_2O + CH_2(CH_3)—CH=CH_2 + Br^-$ (II)

In the transition state of (i), the $CH_3$ group can enter into hyperconjugation with the partly formed double bond, there by lowering the energy of the transition state. In (ii), however, the $CH_3$ group cannot enter into hyperconjugation with the partly formed double bond, and so the energy of this transition state is higher than the of (i), and consequently the latter path is favoured.

On the basis that the hyperconjugative effect determines the stability of the alkene produced, it can be deduced that the ease of formation of alkene form an alkyl halide by the E2 mechanism should be t—R > s—R > p—R, *e.g.*, t—Bu, (2 Me groups) > isoPr (1 Me group) > Et (No Me groups.

$$H—CH_2(—C\cdots H\cdots \overset{\delta-}{OEt})—C(Me)(Me)\cdots Br^{\delta-} \qquad H—CH_2(—C\cdots H\cdots \overset{\delta-}{OEt})—C(Me)(H)\cdots Br^{\delta-} \qquad H—CH(—C\cdots H\cdots \overset{\delta-}{OEt})—C(H)(H)\cdots Br^{\delta-}$$

These predictions are borne out in practice. Also, since the most substituted alkene is general the most stable isomer, it therefore follows that Saytzeff's rule leads to the predominant of the most stable product. A further point in this connection is that when the alkene is capable of existing as geometrical isomers, the *trans*-isomer generally predominates over the *cis*.

Saytzeff's rule also applies when the mechanism is E1. In this mechanism the formation of a carbonium ion is the rate-determining step (as for the $S_N1$ mechanism), and then the carbonium eliminates a proton to form an alkene, the stability of which will be largely determined by hyperconjugative effect.

Stereochemistry of elimination reactions. Many examples in the literature show that *trans* elimination occurs more readily than *cis*., chlorifumaric acid undergoes dehydrochlorinating about fifty times as fast times as fast as chloromaleic acid.

$$\underset{\text{(HO}_2\text{C)(Cl)C=C(H)(CO}_2\text{H)}}{} \xrightarrow[(-HCl)]{NaOH} HO_2C\text{-}C\equiv C\text{-}CO_2H \xleftarrow[(-HCl)]{NaOH} \underset{\text{(Cl)(CO}_2\text{H)C=C(H)(CO}_2\text{H)}}{}$$

Experiments of this type (involving 'rigid' systems) let Hughes and Ingold to propose the bimolecular elimination reactions take place when the two groups to be eliminated are *trans* and lie in one plane with the two carbon atoms to which they are attached, *i.e.*, E2 reactions an stereoselectively *trans*.

$$B^{\frown}\ H\text{-}C\text{-}C\text{-}Z \longrightarrow \rangle C{=}C\langle + BH + Z$$

Evidence for E2 *trans*-elimination has also been obtained from acyclic systems, *e.g.*, Crist *et. al.* (1952) showed that the alkenes produced by the base-catalysed dehydrpbromination of the diastereoisomeric 1-bromo-1,2-diphenylpropanes (I and (II) can only arise by *trans*-eliminates.

Later work has shown that bimolecular eliminations may, in certain circumstances, proceed by *cis*-elimination.

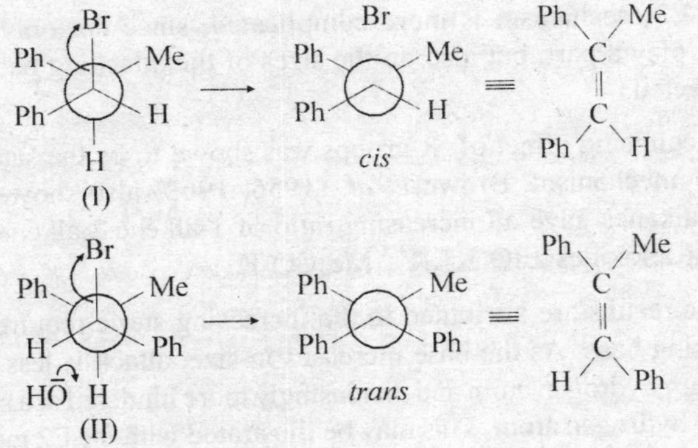

(I) ... *cis*

(II) ... *trans*

## STERIC EFFECTS IN ELIMINATION

Let us consider the E1 reaction with $RCH_2CMe_2Br$. Accepting that elimination of HBr is selectively *trans*, then we may represent the two directions of elimination as follows.

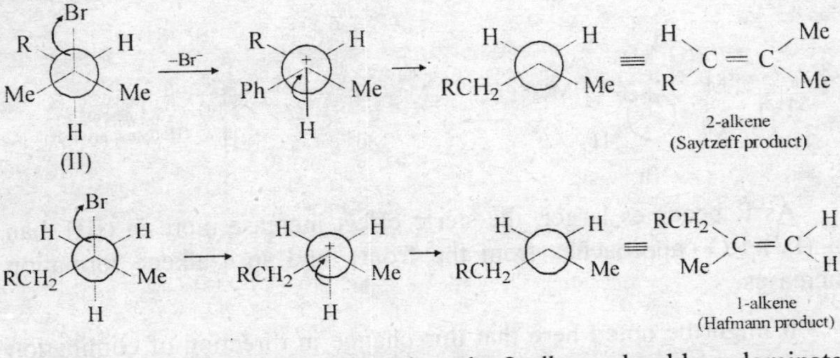

(II)

2-alkene
(Saytzeff product)

1-alkene
(Hafmann product)

On the basis of hyperconjugation, the 2-alkene should predominate (Saytzeff's rule). However, Brown *et al.* (1955) showed that as R increased in size, the ration 1-alkene/2-alkene also increased.

As R gets bigger, the steric repulsion in the transition state leading to the 2-alkene increases (due to eclipsing of R and M), whereas steric repulsion is minimised in the T.S. Leading to the 1-alkene (Hofmann rule).

Thus, in this case, the 1-alkene is more stable than the 2-isomer. Hence, in the E1 mechanism, the structure of the alkene is determined by hyperconjugation and steric effects.

The E2 mechanism is more complicated, since not only does the size of R play a part, but also do the sizes of the attacking base and the group ejected.

The eclipsing effect of R groups was shown to be the same as that in the E1 mechanism. Brown *et al.* (1956, 1968) also showed that 2-halogenoalkanes gave an increasing ratio of 1-alkene.2-alkene with the potassium alloxides $EtO^-K^+$,$K^+$, $Me_3CO^-K^+$.

These results are attributed to the increasing steric requirements of the attacking base. As the base increases in size, attack is less hindered co a primary hydrogen atom and increasingly more hindered at a secondary or tertiary hydrogen atom. This may be illustrated with the E2 mechanism using $Me_2CHCBrMe_2$ as substrate.

(III) RO⁻ H; Me, Me, Me, Me; Br ⟶ $Me_2C=CMe_2$ 2-alkene Saytaff product)

(IV) RO⁻ H; Me, Me, $CHMe_2$, H; Br ⟶ $Me_2CH(Me)C=CH_2$ 1-alkene (Hofmann product)

As R becomes larger, the steric effect increase more in (III) than in (IV) $RO^-$ approaches from the front), and so 1-alkene formation increases.

It might be noted here that this change in direction of elimination may be used to shift a double bond in an alkene, *e.g.*,

$$Me_2C=CMe_2 \xrightarrow{HCl} Me_2CHCMe_2Cl \xrightarrow{Et_5CO^-K^-}$$

$$Me_2CMCMe-CH_2 \qquad (97\%)$$

Brown *et al.* (1956) showed that as the size of the leaving group increases, the yield of 1-alkene also increases (for a given base). Here again, there is less steric hindrance in (IV) than in (III) as Z gets larger (replace Br by Z).

Now let us consider the conditions which favour the formation of alkenes from alkyl halides. In practice, the situation is complicated by the fact that substitution and elimination often occur together,. However, it has been shown that under E1/$S_N$1 conditions, substitution is more favoured than under E2/SN2 conditions. Thus, for a given alkyl halide, the alkene yield is higher for the E2 mechanism than for the E1. Since a high concentration of the base with high nucleophilic power favours the E2 mechanism, alkene preparation is best carried out in solvents with ionising power. Thus, ethanolic sodium, hydroxide solution (solvent of low ionising power and a very strong base, $OEt^-$) gives better alkene yield than does aqueous sodium hydroxide (solvent of high ionising power and a base. $OH^-$, which is weaker than $OEt^-$.). Also, a base of and nucleophilic power will have little affinity for hydrogen, and consequently alkene formation will be decreased and substitution encouraged. The acetate ion a very weak base, and hence its use will given an ester as the main product rather than alkene. Thus, the conversion of an alkyl halide into the alcohol is often best carried out via the ester, which is then hydrolysed.

A point of interest here is that even under the most favourable conditions, ethyl halides give only 1 per cent, of ethylene, the main reaction leads to substitution products.

The eliminations discussed above have dealt with only alkyl halides as substrates.

However, many other types of alkyl compounds also undergo E1 and E2 reactions, *e.g.*, in RZ, Z may be $R_3N^+$ $R_2S^+$, $R_3P^+$, OTs, OCOR.etc., and the base may be $OH^-$, $OR^-$, $NH_2^-$, etc.

6. When alkyl halides are heated with ethanolic ammonia under pressure, a mixture of *ammonias i.e.*, substituted ammonias, is obtained, *e.g.*, .

$$RX + NH_3 \longrightarrow RNH_2 + HX \longrightarrow RNH_2.HX \text{ or } [RNH_3]^+X^-$$

7. When alkyl halides are heated with aqueous ethanolic potassium cyanide, alkyl *cyanides* are obtained.

$$RX + KCN \longrightarrow RCN + KX$$

When aqueous ethanolic silver cyanide is used, cyanides and *isocyanides* are formed.

$$RX + AgCN \longrightarrow RCN + AgX$$

$$RX + AgCN \longrightarrow RCN + AgX$$

8. Alkyl halides form esters when heated with an ethanolic solution of a silver salt.

$$R^1CO_2Ag + R^2X \longrightarrow R^1CO_2R^2 + AgX$$

The acid may be organic or inorganic. If the salt is silver nitrite two isomeric compounds are obtained, the *nitrite* (ester) and the *nitro-* compound.

$$R{-}O{-}N{=}O \longleftarrow RX + Ag{-}O{-}N{=}O \longrightarrow R{-}\overset{+}{N}\begin{matrix}{/\!\!/}O \\ \diagdown O\end{matrix}$$

9. When alkyl halides are heated with sodium alkoxides, *ethers* are obtained.

$$R^1ONa + R^2X \longrightarrow R^1OR^2 + NaX$$

10. Alkyl halides heated with aqueous ethanolic sodium hydrogen sulphide form *thioalcohols* and *thioethers,* when mercaptides are used.

$$RSH \xleftarrow{NaSH} RX \xrightarrow{RSNa} RSR$$

With sodium sulphite, a *sulphonate* is formed.

$$RX + Na_2SO_3 \longrightarrow RSO_3Na + NaX$$

11. Alkyl halides may be used in the *Freidel-Crafts* reaction and to prepare Grignard reagents.

Methyl chloride is prepared industrially by chlorinating methane with chlorine diluted with nitrogen, the ration by volume of $CH_4:Cl_2:N_2$ being 8:1:80. The reactions is carried out in the presence of partly reduced cupric chloride as catalyst. All four chloromethanes are obtained, the methyl chloride comprising 90 per cent of the chlorine used.

$$CH_4 + Cl_2 + (N_2) \longrightarrow CH_3Cl + HCl + (N_2) + (CH_2Cl_2 + CHCl_3 + CCl_4)$$

By adjusting the ration of chlorine to methane, each chloromethane can be obtained as the main product.

Methyl chloride, b.p – 24°C, is used in the manufacture of aniline dyes, silicone rubbers, tetramethyl, methylation of cellulose, as a refrigerating agent, local anesthetic and as a fire extinguisher.

Methyl bromide, b.p. 45°C, is manufactured by passing the vapours of methanol and hydrobromic acid iodide is prepared industrially.

Methyl iodide is prepared industrially.

(i) By warming a mixture of methanol and red phosphorus with iodine.

$$6CH_3OH + 2P + 3I_3 \longrightarrow 6CH_3I + 2H_3PO_3$$

(ii) By the action of methyl sulphate on potassium iodide solution in the presence of calcium carbonate.

$$KI + (CH_3)_2SO_4 \longrightarrow CH_3I + K\,(CH_3)SO_4 \quad (90\text{–}94\%)$$

Methyl iodide is a sweet-smelling liquid. b.p. 42.5°C. Since it is liquid, it is easier to handle than methyl chloride, and so is used a great deal as a methylating agent in laboratory organic syntheses, but the chloride is used industrially because it is cheaper.

Ethyl chloride is prepared industrially.

(i) By the chlorination of ethane at about 400°C.

$$C_2H_6 + Cl_2 \longrightarrow C_2H_5Cl + HCl$$

(ii) By the addition of hydrogen chloride to ethylene (from cracked petroleum) in the presence of aluminium chloride as catalyst.

$$C_2H_4 + HCl \xrightarrow{ACl_3} C_2H_5Cl$$

Ethyl chloride, b.p. 12.5°C, is used in the preparation of tetraethyl lead, etc, and as a refrigerating agent.

Ethyl bromide, b.p. 38.5°C, is manufactured by irradiating a mixture of ethylene and hydrogen bromide with γ-radiation from a cobalt—60 source.

$$HBr \xrightarrow{\gamma\text{-rays}} H\cdot + Br\cdot$$

$$C_2H_4 + Br\cdot \longrightarrow \cdot CH_2 - CH_2Br \xrightarrow{HBr} CH_3CH_2Br + Br\cdot$$

It is used for, preparing barbiturate drugs.

## DERIVATIVES OF DIHALOGEN

### General Methods of Preparation

*1,1-Dihalides may be prepared*

(i) By the action of phosphours pentahalides on aldehydes or ketones, *e.g.*, acetone gives isopropylidene dichloride when treated with phosphorus pentachloride.

$$CH_3COCH_3 + PCl_5 \longrightarrow CH_3CCl_2CH_3 + POCl_3$$

(ii) By the addition of halogen acids to alkynes, *e.g.*, ethylidene dibromide from acetylene and hydrogen bromide.

$$CH \equiv CH + HBr \longrightarrow CH_2 = CHBr \xrightarrow{HBr} CH_3CHBr_2$$

1,2,-Dihalides may be prepared by the addition of halogen to alkenes, *e.g.*, propene dibromide from propene and bromine.

$$CH_3CH{=}CH_2 + Br_2 \longrightarrow CH_3CHBrCH_2Br$$

The method of preparing an αω-dihalide is special to the particular halide, *e.g.*, trimethylene dibromide may be prepared by the addition of hydrogen bromide to alkyl bromide at low temperatures.

$$CH_2{=}CHCH_2Br + HBr \longrightarrow CH_2BrCH_2CH_2Br$$

Oldham (1950), however, has found that αω-dibromides may be prepared by the action of bromine in carbon tetrachloride on the silver salt of a dicarboxylic acid .

$$(CH_2)_n(CO_2\ Ag)_2 + 2Br \longrightarrow Br(CH_2)_nBr + 2AgBr + 2CO_2$$

The yields are very good provided n is 5 or more.

## ALKYL HALIDES

The alkyl halides have the general formula $C_nH_{2n+1}X$ or R.X, where X denotes chlorine bromine or iodine. Fluorine is not included, since fluorides do not behave like the other halides.

### General Methods of Preparation

1. The method most widely used is the replacement of the hydroxyl group of an alcohol by an X atom. Halogen acid is the most common reagent used with alcohols, but the conditions depend on both the structure of the alcohol and the particular acid.

$$ROH + HX \xrightarrow{heat} RX + H_2O$$

Hydrogen chloride is used with zinc chloride (*Groves' process)* for primary and secondary alcohols, but tertiary alcohols readily react with concentrated hydrochloric acid in the absence of zinc chloride.

Zinc chloride as a Lewis acid and consequently can co-ordinate with the alcohol.

$$ROH + ZnCl_2 \longrightarrow R-\overset{+}{\underset{H}{\underset{|}{O}}}-\overline{Z}nCl_2 \xrightarrow{S_N1} R+ + [HO-ZnCl_2]$$

The R—O bond is weakened and so the complex readily forms a carbonium ion ($S_N1$) mechanism. If the nature of $R^+$ is such that it can undergo rearrangement, the product will be a mixture of isomeric alkyl chlorides. When the concentration of zinc chloride is low, the reaction is still catalysed, but no rearrangement occurs ($S_N2$ mechanism).

$$ZnCl_2 + HCl + ROH \longrightarrow$$

$$ZnCl_3^- + ROH_2^+ \xrightarrow{S_N2} ZnCl + RCl + H_2O$$

Fridine and dimethylamine also catalyse the reaction between alcohols and hydrochloric acid without rearrangement.

The formation of the various halides by rearrangement is one example of 1, 2-shifts. In this type of rearrangement a group migrates from a *carbon* atom (the *migration origin)* to an *adjacent* atom (the *migration terminus(* which is usually carbon or nitrogen. The rearrangement may be expressed in the following general form.

$$\overset{Z}{\overset{|}{A}}-B-Y \longrightarrow \overline{Y} + \overset{Z}{\overset{|}{A}}-B^+ \longrightarrow \overset{+}{A}-\overset{Z}{\overset{|}{B}} \xrightarrow[\text{rageet}]{\text{nucleopbilic}} \text{products}$$

When Y ionises, B is left with only six valency electrons (open sextet), and consequently carries a positive charge Z migrates *with its bonding pair of electrons,* and A now has an open sexted Attack now takes place at A by $Y^-$ or some other nucleophilic reagent to form the products. Then mechanism was first suggested by Whit more (1932) and is often referred to as the *Whit more mechanism.*

The nature of Z is quite varied, the 'key' atom (*i.e.*, the atom joined to the bonding pair which migrates) may be carbon, halogen, oxygen, etc., If Z has a lone pair of electrons, *e.g.*, halogen, then this pair may be used in the migration to give a cyclic intermediate, *i.e.*, a bridged or non chemical carbonium ion.

$$\underset{\displaystyle A\text{—}B^+}{\overset{Z}{|}} \longrightarrow A\overset{\overset{+}{Z}}{\diagup\diagdown}B \longrightarrow \text{product}$$

If Z is an aryl group, then this group can supply an electron pair to form a bridged ion. If Z is an alkyl group, then there are o lone pairs or multiple bonds, but again it is believed that a bridged ion is possible, *i.e.*,

$$-\overset{R}{\underset{|}{\overset{|}{C}}}-\underset{|}{\overset{|}{C^+}} \longrightarrow \;\diagdown\!\overset{R}{C\text{=====}C}\!\diagup$$

This type of bridged ion contains three *partial bonds formed form one pair of electrons. The* formation of these bridged ions in 1,2-shifts is an example of *nighbouring group participation,* and when the rate is increased because of this effect, the rearrangement is said to be *an chemically assisted* (Winstein *et al.*, 1953).

A very important point about 1,2-shifts is that they are *intramolecular*', *i.e.*, Z is never actually free. If the mechanism were such that Z was free during is migration, then the rearrangement would be called *intermolecular.* The intramolecular nature of the 1,2-shifts has been established by using a group Z which contains an asymmetric carbon atom attached to A, i.e, have type of C—A. If Z actually separated during the migration, then it means that the free migrating that is carbanion, *i.e.* abC:$^-$. Examination of such carbanions has shown that none ever mains its configuration, it recemises, *i.e.*, half the molecules have their configuration inverted, and so the product is no longer optically active. Thus if Z moves with its bonding pair and the configuration is *retained* in the product, than Z can never been free, *i.e.*, the rearrangement is intramolecular.

We can now illustrate the foregoing discussion with neopentyl alcohol as our example. Whit more *et al*. (1932) showed that the reactions undergone by neopentyl alcohol and neopentyl halides are of two types. (i) replacement reactions, which occur very slowly (if at all) and produce neopentyl compounds, (ii) replacement (and elimination) reactions, which occur very fast and produce t-amyl compounds.

Hughes and Ingold (1996) have shown that the slow reactions take place by the $S_N2$ mechanism and consequently without rearrangement, *e.g.*, neopentyl bromide reacts with ethanolic sodium ethoxide to give ethyl neopentyl ether.

$$\underset{BrCH_2}{\overset{Me_3C}{|}} + OEt^- \rightleftharpoons \underset{Br\cdots CH_2\ldots OEt^-}{\overset{Me_3C}{\delta-|\quad\delta-}} \overset{Br^-}{\rightleftharpoons} \underset{CH_2OEt}{\overset{Me_3C}{|}}$$

On the other hand when neopentyl alcohol reacts with anhydrous hydrogen bromide, the main product is t-amyl bromide (72 per cent), together with primary and secondary bromides. The arranged bromides are the result of 1,2-shifts (methyl and hydride)

$$Me_3CCH_2OH \overset{HBr}{\rightleftharpoons} Me_3CCH_2-\overset{+}{O}H_2 \rightleftharpoons Me_3C-\overset{+}{C}H_2 + H_2O \xrightarrow{Br^-} Me_3CCH_2Br$$

$$Me_3C-\overset{+}{C}H_2 \xrightarrow{Me\ shift} Me_2\overset{+}{C}CH_2Me \xrightarrow{Br^-} Me_2CBrCH_2Me$$

$$Me_2\overset{+}{C}CH_2Me \xrightarrow{H^-\ shift} Me_2CH\overset{+}{C}HMe \xrightarrow{Br^-} Me_2CHCHBrMe$$

The 'deriving force' of the rearrangement is probably due to the stabilities of carbonium being tertiary > secondary > primary.

Any alkyl halide may be prepared by the action of a phosphorus halide on the alcohol Phosphorus pentachloride gives variable yields depending on the alcohol.

$$ROH + PCl_5 \longrightarrow RCl + HCl + POCl_3$$

Straight- chain primary alcohols react with phosphorus trihalides to give unrearranged halides but many secondary alcohols undergo rearrangement.

$$PX_3 + 3ROH \longrightarrow (RO)_3P + 3HX \longrightarrow (RO)_3$$

$$PH \xrightarrow[S_N2]{XH} RX + (RO)_2P(H)O \xrightarrow[S_N2]{HX}$$

$$RX + (RO)P(H)(OH)O \xrightarrow[S_N2]{HX} RX + H_3PO_3$$

Thionyl chloride reacts with straight-chain primary alcohols without rearrangement in the presence or absence of pyridine (Darzens procedure. 1911), the reaction proceeds via a chlorosulphite.

$$ROH + SOCl_2 \longrightarrow HCl + ROSOCl$$

$$\longrightarrow Cl^- + ROSO^+ \xrightarrow{S_N2} RCl + SO_2$$

2. By direct halogenation *i.e.*, substitution reactions of the alkanes with halogen, and this ml be brought about by light, catalysts or heat.

Photohalogenation is carried out by treating the alkane with chlorine (or bromine) at ordinate temperature in the presence of light. The reaction takes place by a free- radical chain mechanism the initiation being brought about by the formation of chlorine atoms by absorption of lights

$$Cl_2 \xrightarrow{hv} 2Cl\cdot$$

$$Cl. + H—CH_3 \longrightarrow Cl...H...CH_3 \longrightarrow HCl + CH_3$$

$$CH_3. + Cl—Cl \longrightarrow CH_3...Cl...Cl \longrightarrow CH_3Cl + Cl.$$

$$Cl. + H—CH_2Cl \longrightarrow Cl...H...CH_2Cl \longrightarrow HCl + CH_2Cl.$$

$$ClCH_2\cdot + Cl—Cl \longrightarrow ClCH_2...Cl...Cl \longrightarrow CH_2Cl_2 + Cl. \text{ etc.}$$

The termination of the chain reaction may take place by adsorption of the chlorine atom on the walls of the containing vessel, of by two chlorine atoms combining with each other form a chlorine molecule.

There appears to be no evidence to show that $CH_3$, $CH_2Cl$ radicals combine to terminate the chain reaction, but another termination step is the combination of R, with Cl. (to give RCl).

Thermal halogenation. Thermal chlorination has been studied in great detail by Hass. Mass and their co-workers (1935 onwards), and as a result of their work they suggested a number rules for chlorination.

## GENERAL METHODS OF PREPARATION AND PHYSICAL PROPERTIES OF HALOGEN DERIVATIVES OF HYDROCARBONS

### General Methods of Preparation of Haloalkanes and Haloarenes

General methods of preparation of haloalkanes and haloarenes are described below.

### From Hydrocarbons

Halo derivatives of hydrocarbons can be prepared by various methods. Some of these methods are described below.

*By free radical halogenation* : Direct chlorination or bromination of alkanes in the presence of heat or ultraviolet radiation proceed via a free radical mechanism.

Usually a mixture of isomeric mono- and poly haloalkanes is obtained. Therefore, halogenation of alkanes by free radical mechanism is not suitable for obtaining a pure haloalkane.

$$\underset{\text{bu tan e}}{CH_3CH_2\,CH_2CH_2} + Cl_2 \xrightarrow{UV/heat} \underset{\text{isomeric monochlorobu tan es}}{CH_3CH_2CH_2CH_2Cl,\ CH_3CH_2CH(Cl)CH_3}$$

$$\left.\begin{array}{ll} CH_3CH(Cl)CH(Cl)CH_3, & CH_3CH_2CH(Cl)CH_2Cl \\ CH_3CH(Cl)CH_2CH_2Cl\ , & CH_3CH_2CH_2CHCl_2 \\ CH_3CH_2CCl_2CH_3\ , & ClCH_2CH_2CH_2CH_2Cl \end{array}\right] + \underset{\text{derivatives}}{\text{Higher}}$$

isomeric dichlorobu tan es

The direct halogenation method, however is useful in certain cases. For example,

(i) When no isomeric product is formed and the mixture can be separated into different products by fractional distillation due to difference in the boiling points of the products formed

$$\underset{\text{methane}}{CH_4} \xrightarrow[-\,HCl]{+\,Cl_2\,,\,hv} \underset{\text{chloromethane}}{CH_3Cl} \xrightarrow[-\,HCl]{+\,Cl_2} \underset{\text{dichloromethane}}{CH_2Cl_2}$$

$$\xrightarrow[-\,HCl]{+\,Cl_2} \underset{\text{trichloromethane}}{CHCl_3} \xrightarrow[-\,HCl]{+\,Cl_2} \underset{\substack{\text{tetrachloromethane} \\ \text{(carbon tetrachloride)}}}{CCl_4}$$

$$\underset{\substack{\text{2,2-dimethylpropane} \\ (\textit{neo}\text{-pentane})}}{CH_3-\overset{\overset{CH_3}{|}}{\underset{\underset{CH_3}{|}}{C}}-CH_3} + Cl_2 \xrightarrow{heat/UV} \underset{\substack{\text{1-chloro-2,2-dimethylpropane} \\ (\textit{neo}\text{-pentyl chloride})}}{CH_3-\overset{\overset{CH_3}{|}}{\underset{\underset{CH_3}{|}}{C}}-CH_2Cl} + HCl$$

In halogenation reactions,

(a) Chlorine reacts with alkanes readily, but the reaction with bromine is solver. The reaction with iodine is reversible. Thus, the reactivity of halogens follow the order.

$$Cl_2 > Br_2 > I_2$$

Reaction with fluorine is highly exothermic and violent. It leads to the formation of polyfluoro compound and cleavage of the carbon-carbon bonds.

(b) During halogenation reactions, the tendency of the replacement of hydrogen atom of an alkane by a halogen is tertiary alkane > secondary alkane > primary alkane

(c) By this method, generally a mixture of haloalkane and polyhaloalkanes is obtained. These can be separated by fractional distillation.

(ii) When pure compounds are not required, such as those used as industrial solvents. The mixture of different products can be enriched in the desired product either by fractional distillation or by adjusting the ration of the reactants. For example, when methane is chlorinated in the presence of excess of methane, the major product formed is chloromethane

$$\underset{\text{excess}}{CH_4\,(g)} + Cl_2\,(g) \xrightarrow{\text{heat / UV}} \underset{\text{chloromethane}}{CH_3Cl} + HCl$$

The unreacted methane is separated and recycled.

Chloromethane and bromomethane are obtained industrially by this method.

(iii) When difference in the reactivity of different types of hydrogens in the hydrocarbons molecule is so large that only one product is formed under suitable conditions.

The order of reactivity for different types of hydrogens is

$$\underbrace{\text{Benzylic} \approx \text{Allyic}}_{\text{high reactivity}} > \text{Tertiary} > \text{Secondary} > \text{Primary} > \underbrace{\text{Vinylic} \approx \text{Aryl}}_{\text{these hydrogens do not usually participate in free radical halogenation reactions}}$$

Thus *benzylic and allylic hydrogens can be replaced by halogens without any side reaction.* For example,

$$\underset{\text{propane}}{CH_2 = CH - CH_3} + Cl_2 \xrightarrow[\text{gas phase}]{500\text{–}600^\circ C} \underset{\text{3-chloropropene}}{CH_2 = CH - CH_2Cl_2} + HCl$$

$$\underset{\text{toluene}}{C_6H_5CH_3} + Br_2 \xrightarrow[\text{reflux}]{\text{light}} \underset{\text{benzyl bromide}}{C_6H_5CH_2Br} + HBr$$

The high reactivity of allylic and benzylic hydrogens is due to the stabilization of the allylic and benzylic free radicals formed during the reactions by resonance.

$$CH_2=CH—\dot{C}H_2 \xleftrightarrow{} CH_2—\dot{C}H=CH_2$$

resonance in allyl radical

$CH_3$ $CH_2$ $CH_2$ $CH_2$ $\dot{C}H_2$

resonance in benzyl radical

1. *By electrophilic substitution :* The chloro and bromo derivatives of arenes can be obtained by electrophilic substitutions of arenes with chlorine or bromine in the presence of iron or iron (III) halide or any Lewis acid as catalyst.

Low temperature, and presence of a halogen carrier favour nuclear substitution. Chlorination and bromination may be very conveniently carried out at room temperature in the presence of a Lewis acid such as, $FeCl_3$, FeBr, $AlCl_3$, etc.

The Lewis acid acts as a *halogen carrier*.

In actual practice, chlorination and bromination are done in the presence of iron. Iron gets converted into the corresponding halide by the halogen being used.

$$\underset{\text{arene}}{ArH} + \underset{(X=Cl, Br)}{X_2} \xrightarrow{\text{Lewis acid}} \underset{\text{haloarene}}{ArX} + HX$$

$$\underset{\text{benzene}}{C_6H_6} + Cl_2 \xrightarrow[\text{room temp.}]{\text{Fe or FeCl}_3} \underset{\text{chlorobenzene}}{C_6H_5Cl} + HCl$$

$$\underset{\text{toluene}}{C_6H_5CH_3} + Cl_2 \xrightarrow[\text{room temp.}]{\text{Fe or FeCl}_3} \underset{o\text{-chlorotoluene}}{o\text{-}ClC_6H_4CH_3} + \underset{p\text{-chlorotoluene}}{p\text{-}ClC_6H_4CH_3} + HCl$$

The *ortho* and *para* isomers can be separated due to large difference in their melting points. This method is applicable to the preparation of chloroarenes and bromoarenes only. Fluoro and iodo compounds cannot be prepared by this method.

With iodine, the reaction is reversible. So, iodination is carried out in the presence of an oxidising agent like $HIO_3$ or $HNO_3$, $HIO_3$ oxidises HI produced in the reaction to $I_2$. This shifts the equilibrium to the right.

$$C_6H_6 + I_2 \xrightarrow{HIO_3} C_6H_5I + HI$$

$$5HI + HIO_3 \longrightarrow 3I_2 + 3H_2O$$

2. *By electrophilic addition :* Alkenes and alkynes undergo electrophilic addition of halogens ($Cl_2$, $Br_2$) and hydrogen halides (HCl, HBr, HI) to form halogen derivatives.

$$\underset{\substack{\text{ethene}\\\text{(ethylene)}}}{CH_2=CH_2} + HX \longrightarrow \underset{\substack{\text{haloethane}\\\text{(ethyl halide)}}}{CH_3CH_2X}$$

$$\underset{\text{but-2-ene}}{CH_3-CH=CH-CH_3} + HX \longrightarrow \underset{\text{2-halobu tan e}}{CH_3-\underset{\substack{|\\H}}{CH}-\underset{\substack{|\\X}}{CH}-CH_3}$$

$$R-C\equiv CH + HX \longrightarrow \underset{\text{monohalo}}{R-\underset{\substack{|\\X}}{C}=\underset{\substack{|\\H}}{CH}} \xrightarrow{+HX} \underset{\text{dihalo}}{R-\overset{\substack{X\\|}}{\underset{\substack{|\\X}}{C}}-\overset{\substack{H\\|}}{\underset{\substack{|\\H}}{C}}-H}$$

In symmetrical alkenes, the carbon atoms across the double bond are equivalent. So, the addition of an acid molecule to symmetrical alkenes gives only one product.

The addition of halogen acid to unsymmetrical alkenes proceeds in accordance with the *Markownikoff's rule,* and secondary and tertiary haloalkanes are obtained.

*Markownikoff's rule :* According to this rule, the *negative part of the molecule to be added across a double or a triple bond gets attached to the carbon atom containing lesser number of hydrogen atoms.*

$$\underset{\text{unsymmetrical alkene}}{R—CH=CH_2} + HX \longrightarrow \underset{\text{2-haloalkane}}{\underset{X}{\underset{|}{R—CH}}—\underset{H}{\underset{|}{CH_2}}} \text{ or } R—\underset{X}{\underset{|}{CH}}—CH_3$$

$$\underset{\text{propene}}{CH_3CH=CH_2} + HBr \longrightarrow \underset{\text{2-bromopropane}}{CH_3—\underset{Br}{\underset{|}{CH}}—\underset{H}{\underset{|}{CH_2}}} \text{ or } CH_3—\underset{Br}{\underset{|}{CH}}—CH_3$$

*Peroxide effect :* The addition to a double bond in unsymmetrical haloalkanes takes place in accordance with the Markownikoff's rule. However, in the presence of peroxides *e.g.*, benzoyl peroxide ($C_6H_5$ — CO — O — O — CO — $C_6H_5$), the addition of hydrogen halide to an unsymmetrical alkene takes place contrary to the Markownikoff's rule. This is called *peroxide effect, or Kharasch effect.*

$$\underset{\text{propene}}{CH_3—CH=CH_2} + HBr \xrightarrow{\text{Peroxide}} \underset{\text{1-bromopropane}}{CH_3—CH_2—CH_2Br}$$

The presence of peroxide, the negative part of the molecule to be added goes to the carbon atom containing larger number of hydrogen atoms. This is also called *Anti-Markownikoff's Rule.*

This is because in the presence of a peroxide, the addition takes place by a free-radical mechanism.

## FROM ALCOHOLS

This is the most widely used method for the preparation of haloalkanes. The hydroxyl group of alcohols may be replaced by halogen atom by using hydrogen halides (HX), phosphorus halides ($PCl_5$, $PCl_3$) or thionyl halides, ($SOCl_2$).

1. *By the action of hydrogen halide (halogen acids) :* Action of a hydrogen halide on an alcohol in the presence of a dehydrating agent such as, anhydrous zinc chloride, *(Groove's process),* or conc., sulphuric acid, gives haloalkanes. Hydrobromic acid may be generated in the reaction mixture ( *in situ)* by the reaction of NaBr or KBr with conc. $H_2SO_4$.

$$\underset{\text{alcohol}}{ROH} + HX(g) \xrightarrow{ZnCl_2 \text{ anhyd}} \underset{\text{haloalkane}}{R.X} + H_2O$$

$$ROH + NaBr + H_2SO_4 \xrightarrow{\Delta} RBr + NaHSO_4 + H_2O$$

Reactivites of acids follow the order.

HI > HBr > HCl

Reactivites of alcohols follows the order.

Tertiary > Secondary > Primary

$$\underset{\text{ethyl alcohol}}{C_2H_5OH} + HCl(g) \xrightarrow{ZnCl_2\ \text{anhyd.}} \underset{\text{chloroethane}}{C_2H_5Cl} + H_2O$$

$$C_2H_5OH + HBr(g) \xrightarrow{\Delta, H_2SO_4} \underset{\text{bromoethane}}{C_2H_5Br} + H_2O$$

$$C_2H_5OH + HI(g) \xrightarrow{\Delta} \underset{\text{iodoethane}}{C_2H_5I} + H_2O$$

$$\underset{\text{propan-1-ol}}{CH_3-CH_2-CH_2OH} + HCl(g) \xrightarrow{ZnCl_2\ \text{anhyd}}$$

$$\underset{\text{1-chloropropane}}{CH_3-CH_2-CH_2Cl} + H_2O$$

$$\underset{\text{propane-2-ol}}{CH_3-\underset{\underset{OH}{|}}{C}-CH_3} + HCl(g) \xrightarrow{ZnCl_2\ \text{anhyd.}} \underset{\text{2-chloropropane}}{CH_3-\underset{\underset{Cl}{|}}{CH}-CH_3} + H_2O$$

Tertiary alcohols are very reactive. So, tertiary alcohols react with conc. HCl even in the absence of zinc chloride.

$$\underset{\text{propane-2-ol}}{CH_3-\overset{\overset{CH_3}{|}}{\underset{\underset{CH_3}{|}}{C}}-OH} + \text{con. HCl} \longrightarrow \underset{\substack{\text{2-chloro-2-methylpropane}\\ \text{(tert-butyl chloride)}}}{CH_3-\overset{\overset{CH_3}{|}}{\underset{\underset{CH_3}{|}}{C}}-Cl} + H_2O$$

2. *By the action of thionyl chloride* : When thionyl chloride (SOCl2) and alcohol are heater in the presence of pyridine, haloalkane is obtained.

   *This method is preferred beacause the other products of the reaction (HCl and SO2) being escape out leaving behind pure haloalkane.*

$$ROH + SOCl_2 \xrightarrow{\text{pyridine}} RCl + HCl(g) + SO_2(g)$$

***Example:***

$$CH_3CH_2OH + SOCl_2 \xrightarrow{\text{pyridine}} C_2H_5Cl + HCl(g) + SO_2(g)$$

Haloarenes cannot be prepared by similar reaction using phenols. This is because due to resonance, the carbon—oxygen bond in phenols develops a double bond character and is therefore, stronger than the carbon—oxygen single bond in alcohols. As a result, the carbon-oxygen bond in phenols is not cleaved easily.

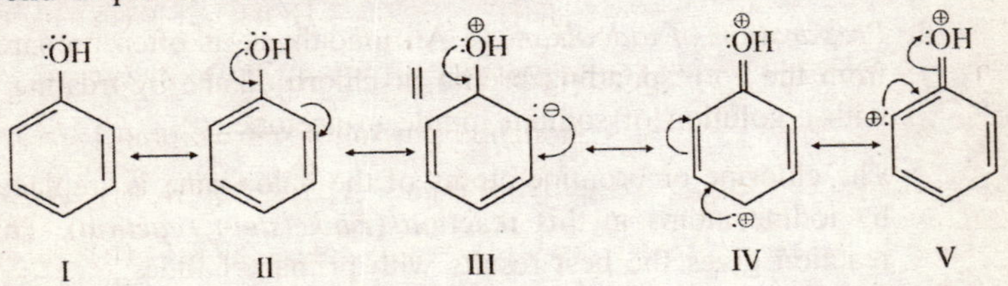

:ÖH ⟷ :ÖH ⟷ ⊕ÖH ⟷ ⊕ÖH ⟷ ⊕ÖH

I II III IV V

Resonance structures of phenol

3. *By the action of phosphorus halides :* Alcohols on refluxing with phosphorus penta or trihalides give haloalkanes. For preparing chloroalkanes, $PCl_5$ or $PCl_3$ is used.

   To prepare a bromoalkane, or an iodoalkane, the alcohol is heated with red phosphorus and bromine ($P + Br_2$) or red phosphorus and iodine ($P + I_2$) respectively.

$$\underset{\text{alcohol}}{ROH} + PCl_5 \xrightarrow{\Delta} \underset{\text{chloroalkane}}{RCl} + POCl_3 + HCl$$

$$\underset{\text{alcohol}}{3ROH} + PCl_3 \xrightarrow{\Delta} \underset{\text{chloroalkane}}{3RCl} + H_3PO_3$$

***Example:***

$$\underset{\text{propanol}}{CH_3CH_2CH_2OH} + PCl_3 \xrightarrow{\Delta} \underset{\text{1-chloropropane}}{CH_3CH_2CH_2Cl} + H_3PO_3$$

$$\underset{\text{propan-1-ol}}{CH_3CH_2CH_2OH} + PCl_5 \xrightarrow{\Delta} \underset{\text{1-chloropropane}}{CH_3CH_2CH_2Cl} + POCl_3 + HCl$$

$$\underset{\substack{\text{propane-2-ol} \\ \text{(secondary alcohol)}}}{CH_3\text{—}\underset{OH}{\underset{|}{CH}}\text{—}CH_3} + PCl_3 \longrightarrow \underset{\text{2-chloropropane}}{CH_3\text{—}\underset{Cl}{\underset{|}{CH}}\text{—}CH_3} + POCl_3 + HCl$$

$$\underset{\text{ethanol}}{6CH_3-CH_2OH} \xrightarrow{2P+2Br_2} \underset{\text{bromoethane}}{6CH_3-CH_2Br} + 2H_3PO_3$$

$$\underset{\text{ethanol}}{6CH_3-CH_2OH} \xrightarrow{2P+3I_2} \underset{\text{iodoethane}}{6CH_3-CH_2I} + 2H_3PO_3$$

$$\underset{\text{propan-1-ol}}{6CH_3-CH_2-CH_2OH} \xrightarrow{2p+3I_2} \underset{\text{1-iodopropane}}{6CH_3-CH_2-CH_2I} + 2H_3PO_3$$

1. *Preparation of iodoalkanes* : An iodoalkane is often prepared from the corresponding bromo or chloro alkane by treating it with a solution of sodium iodide in acetone.

   The chlorine or bromine atoms of the haloalkane is replaced by iodine atoms in this reaction (*Finkelstein reaction*). This reaction gives the best results with primary halides.

$$\underset{(X=Cl,\ Br}{R-X} + NaI \xrightarrow{\text{acetone, reflux}} R-I + NaX(s)$$

$$\underset{\text{chloroethane}}{CH_3CH_2Cl} + NaI\ (\text{in acetone}) \xrightarrow{\text{reflux}} \underset{\text{iodoethane}}{CH_3CH_2I} + NaCl(s)$$

   NaBr, and NaCl are less soluble in acetone than NaI. So they precipitate from the solution and thus preventing the backward reaction to occur.

2. *Preparation of fluoro derivatives of alkanes* : Fluoro derivatives of alkanes can be obtained by heating a chloroalkane or bromoalkane with a metallic fluoride such as AgF, $Hg_2F_2$, $CoF_3$ or $SbF_3$, (*Swarts reaction*). For example,

$$\underset{\text{bromomethane}}{CH_3Br} + AgF \xrightarrow{\text{heat}} \underset{\text{fluoromethane}}{CH_3F} + AgBr$$

   For di or tri halo derivatives with halogen atoms on the same carbon atom (in the alkane) commonly $SbF_3$ or $CoF_3$ is used

$$\underset{\text{2,2-dichloropropane}}{3CH_3CCl_2CH_3} + \underset{\text{antimony trifluoride}}{2SbF_3} \xrightarrow{\text{heat}} \underset{\text{2,2-difluoropropane}}{3CH_3CF_2CH_3}$$

3. *Preparation of Haloarenes From Diazonium Salts*

   Diazonium salts are wonderful strating materials for many aromatic compounds. These will be taught to you in class XII. Here we provide a brief account of their use in preparative chemistry of haloarenes. Haloarenes can be very conveniently prepared from diazonium salts. Diazonium salts are very unstable compounds. Therefore, freshly, prepared solution of a diazonium salt is used for these preparations.

A diazonium salt is obtained by treating an aromatic primary amine with nitrous acid at temperature below 5°C. This reaction is known as *diazotisation reaction.*

In actual practice, an ice-cold aqueous solution of $NaNO_2$ (sodium nitrite) is slowly added into an ice-cold solution of aniline in HCl (maintained at < 5°C). The reactions taking place are,

$$NaNO_2 + HCl \xrightarrow{<5°C} HNO_2\,(aq) + NaCl$$

$$\underset{\text{aniline}}{C_2H_5 - NH_2} + HNO_2\,(aq) + HCl(aq) \xrightarrow{<5°C}$$

$$\underset{\text{benzenediazonium chloride}}{C_6H_5 - N^+ \equiv NCl^-} + 2H_2O$$

The ice-cold solution of benzendiazonium salt is used as such.

4. *From Chloroalkanes, Bromoalkanes By Halogen Exchange Method*

Fluro and iodoalkanes cannot be prepared by direct fluorination and iodination. For preparing fluoro and iodo derivatives of alkanes, generally halogen exchange method is used. Such methods are described below.

## PHYSICAL PROPERTIES OF HALOALKANES AND HALOARENES

Some general physical properties of haloalkanes are described below.

(i) *Physical state and odour :* The lower members *e.g.*, bromomethane and chloroethane are gases. Iodomethane, and the majority of the higher members are sweet-smelling liquids. But still higher homologues are odourless solids.

Haloarenes are colourless liquids or crystalline solids.

(ii) *Colour :* Haloalkanes (alkyl halides) are colourless when pure, but bromides and particularly iodides develop colour when exposed to light.

(iii) *Boiling points :* The boiling points of halogen derivatives of hydrocarbons are influenced by the polarity of the C—X bond. Due to the greater polarity as well as higher molecular mass as compared to the parent hydrocarbon, the intermolecular forces of attraction are stronger in the halogen derivatives relative to that in hydrocarbons.

The boiling points of halogen derivatives are considerably higher than those of the corresponding hydrocarbons.

(i) If high temperature is avoided, no carbon selection rearrangement occur in either thermal or photochemical chlorination.

(ii) Every possible monochloride is formed, and over-chlorination, *i.e.*, chlorination beyond mono-substitution, may be suppressed by controlling the ratio of chlorine to alkane (this rule also holds good for photochemical chlorination).

(iii) The order of ease of substitution is tertiary hydrogen > secondary > primary. At 3000°C, with reaction in the vapour phase, the relative rates of substitution of primary secondary and tertiary hydrogen atoms are 1.00 : 3.25 : 4.43.

(iv) As the temperature rises above 300°C the relative rates of substitution tend to become equal, *i.e.*, 1 : 1 : 1. Increased pressure causes in the relative rate of primary substitution.

(v) According to Tender et at. (1960), the effect of the halogen atom already present in the molecule a to retard substitution at a β-carbon atom, and also affects substitution at an α-carbon atom.

It is of interest here to consider the method of calculation of rates of substitution. Let $r_r$ $r_s$ and $r_t$ be the *relative* rates of substitution of a primary, secondary and tertiary hydrogen atom, respectively. Hass found that the thermal chlorination of propane at 300°C gave 48 per cent of the 1-chloro- and 52 percent of the 2-chloro-isomer. Since there are six $H_p$ atoms, and statistically each one has an equal chance of being replaced, their total rate of replacement is 6r. Similarly for two $H_s$ atoms their total rate of replacement is 2rs. Since the amount of each isomer formed depends on its rate of formation it therefore follows that

$$\frac{2r_s}{6r_p} = \frac{52}{48} \therefore \frac{r_q}{r_p} = \frac{3.25}{1}$$

Hass also found that chlorination of isobutane gave 67 per cent of the 1-chloro and 33 per cent of the 2-chloro-isomer. Since isobutane contains $9H_p$, atoms and $H_t$ atom,

$$\frac{r_t}{9r_p} = \frac{33}{67} \quad \therefore \quad \frac{r_t}{r_p} = \frac{4.43}{1}$$

Thus the relative rates of substitution of $H_p$, $H_s$ and $H_t$ are 1 : 3. 25 : 4.43 (these are the figures given in (iii) above).

With these values it is possible to calculate the proportions of the various monochloro isomers for any alkane, *e.g.*, isopentane.

$$\overset{1}{C}H_3\diagdown\overset{2}{C}H\overset{3}{C}H_2\overset{4}{C}H_3 \quad (\text{with a second } CH_3 \text{ on C-2})$$

| | % | 1 | 2 | 3 | 4 |
|---|---|---|---|---|---|
| | calc. | 30 | 22 | 33 | 14 |
| | obs. | 33.5 | 22 | 38 | 16.5 |

There are $9H_p$, $2H_s$ and $1H_t$. Thus the total rate is.

$$9 \times 1 + 2 \times 3.25 + 1 \times 4.43 = 19.93$$

$$\therefore \quad r_p \times 100 = \frac{9 \times 100}{19.93} = 45\%, \quad r_s \times 100 = \frac{6.5 \times 100}{19.93} = 33\%,$$

$$r_t \times 100 = \frac{4.43 \times 100}{19.93} = 15\% \cdot$$

Since there are two $CH_3$ groups labelled 1 and one $CH_3$ group labelled 4, the rate of formation of the 1-chloro- derivative will be twice that of the 4-chloro. Thus 30 per cent will be the 1- and 15 per cent the 4- chloro isomer. These calculated values compare favourably with the values observed by Hass.

Bromination takes place with greater difficulty than chlorination, and there is less tendency for poly substitution. Furthermore, Roberts *et al.* (1952) have shown that bromination is more relective, bromine has much greater tendency than chlorine to replace tertiary hydrogen in preference to secondary or primary. It is the study of reactions such as these where the more reactive reagent shows less selectivity that has given rise to the empirical *reactivity selectivity principle.*

Pease *et al.* (1931) investigated the thermal chlorination of methane and found that the reaction was retarded by the presence of oxygen. This is strong evidence for a free-radical chain reaction and the authors proposed the following mechanism (cf. photohalogenation).

$$Cl_2 \xrightarrow{\text{heat}} Cl^{\cdot} + Cl^{\cdot}$$

$$Cl. + CH_4 \longrightarrow HCl + CH_3\text{,etc..}$$

Estimations have shown that in thermal chlorination of some alkanes, one chlorine atom may propagate a chain reaction involving about $10^7$ hydrocarbon molecules.

Pease *et al*. also had proposed an alternative mechanism.

$$Cl. + CH_4 \longrightarrow CH_3Cl + H.$$
$$H^{\cdot} + Cl \longrightarrow HCl + Cl\text{, etc.}$$

When we consider the enthalpy changes in these two chain reactions, we find (using bond energies in .

(i) (a) $Cl^{\cdot} + Me\text{—}H\ (+\ 414.2) \longrightarrow H\text{—}Cl\ (-\ 426.8) + Me^{\cdot}$

$\Delta H^{\theta} = -\ 12.6$ kJ

(b) $Me. + Cl\text{—}Cl\ (242.7) \longrightarrow Me\text{—}Cl\ (-\ 326.4) + Cl.$

$\Delta H^{\theta} = -\ 83.7$k J

(ii) (a) $Cl^{\cdot} + Me\text{—}H\ (+\ 414.2) \longrightarrow Me\text{—}Cl\ (-\ 326.4) + H^{\cdot}$

$\Delta H^{\theta} = +\ 87.8$ kJ

(b) $H^{\cdot} + Cl\text{—}Cl\ (+\ 242.7) \longrightarrow H\text{—}Cl\ (-\ 426.8) + Cl^{\cdot}$

$\Delta H^{\theta} = -\ 184.1$ kJ

The overall enthalpy change for both mechanism is – 96.3 kJ. This of course, is to be expected since the overall equation for both mechanism (obtained by adding the steps of the reaction) is the same (Hess's law), viz.,

$Cl\text{—}Cl\ (1 + 242.7) + Me\text{—}H\ (+\ 414.2) \longrightarrow Me\text{—}\ Cl\ (-\ 326.4)$

$+\ H\text{—}Cl\ (-\ 426.8)$

$$\Delta H^{\oplus} = -\ 96.3 \text{ kJ.}$$

Now let us consider the entropy change for the overall reaction values

($Jk^{-1}$ $mol^{-1}$) are. $Cl_2$ 223.0, $CH_4$, 186.2, $CH_3Cl$. 234.3. HCl, 186.6.

$\therefore \quad \Delta S^{\oplus} = (234.3 + 186.6) - (223.0 + 186.2) = 11.7$ kJ

Hence there is a very small increase in entropy.

$$\Delta G^{\oplus} = \Delta H^{\oplus} - T\Delta S^{\oplus} = -96.3 - (298 \times 11.7/1000)$$

$$= -96.3 - 3.49 = -99.79 \text{ kJ}$$

$$\therefore \quad \text{RT in } 99.79. \log K = 99.79/2.3 \text{ RT}$$

$$= 99.79/5.7 = 17.4 \text{ K}$$

$$= -10^{17} \text{ at } 25°\text{C}.$$

Thus, the overall reaction (which is independent of mechanism) is thermodynamically highly favourable.

In the foregoing account, $\Delta H^{\oplus}$ has been *estimated* from bond energies. This usually gives a reason ably good estimate. Entropy changes have been evaluated from a knowledge of absolute entropy values of the reactants and products. Unfortunately, may of these values are not available. Since their estimation is very difficult, one can assume that the entropy changes are small at 25°C and so may be neglected in comparison with $\Delta H^{\oplus}$ values if these are reasonably high (say, 40 kJ, or more). In this ease, then, $\Delta H^{\oplus}$ may be taken as a measure of $\Delta G^{\oplus}$

Although the above reaction is thermodynamically highly favourable, it is immeasurably slow in the dark at room temperature. As we have seen, rates depend on activation energy values. Since the light) must be supplied from outside the system because the value required, 242.kJ, is far too high in be reached through thermal collision at room temperature. In mechanism (i) both (a) and (b) are exothermic, and so is thermodyanamically highly favourable. Now in an endothermic reaction, *E* must be at equal to $\Delta H^{\theta}$. Thus a minium value of *E* for (ii) is 87.86.k J.

This step is a therefore energetically highly unfavorable, and consequently, for a chain mechanism, (i) is involved rather than (ii). There is, however, one further point that should also be considered. It can be shown, from thermodynamic considerations, that the value of the equilibrium constant, K, decreases will rise in temperature for exothermic reactions and increases for endothermic reactions. This, of course follows from Lc Chatelier's principle. Thus, exothermic reactions become thermodynamically less favourable and endothermic reactions more favourable as the temperature rise, but in both cases the equilibrium position is reached much more quickly. Hence, endothermic reactions can give reasonable yields at a fast rate when the temperature is raised sufficiently. From this it follows that at some particular temperature, mechanism (ii) could compete with mechanism (i).

It is also of interest to consider the other halogens, fluorine, bromine and iodine.

| | *F* | *Cl* | *Br* | *I* |
|---|---|---|---|---|
| $X_2 \longrightarrow 2X.$ | + 150.6 | +212.7 | +188.3 | +150.6 |
| (ia) $X^{\cdot} + MeH \longrightarrow HX + Me$ | – 146.4 | –12.55 | + 50.21 | +117.2 |
| (ib) $Me^{\cdot} + X_2 \longrightarrow MeX + X.$ | – 334.7 | –83.68 | –96.23 | –62.76 |
| $\Delta H^{\theta}$ (overall) | – 481.1 | –96.23 | –46.02 | +54.44 |

The rate of each step will depend on the relevant activation energy. E for fluorination (ia) is apparently less than that for chlorination, and because the reaction is so strongly exothermic, fluorination is often explosive. E for bromination (ia) is about 74.48. kJ. The weaker the bond (that is broken), the lower is the necessary activation energy, and so the rate of substitution will be $\tau > s > p$. For bromination, the activation energy is reached for t-hydrogen, and so this is replaced almost exclusively, A might also be noted that order of replacement, $\tau > s$ p,. is what we might expect from the consideration of the stabilities of the free radicals produced in step (ia). Finally, as the temperature rise above 300°C (in thermal chlorination), the chlorine and alkane molecules acquire so much energy that reaction occurs at every collision respective of the type of C—H bond. Under these conditions, the replacement rates will be the same.

2. *Hunsdiecker or Brodine-Hunsdiecker reaction* : Hunsdiecker *et al.* (1935) found that silver and of the carboxylic acids in carbon tetrachloride solution are decomposed by chlorine of bromine to form the alkyl halide, *e.g.*,
3. Direct chlorination of alkanes may be effected by means of sulphuryl chloride. Sulphuryl chloride in the absence of light and catalysts does not react with alkanes even at their boiling points, but in the presence of light and a trace of an organic peroxide the reaction is fast. The mechanism is probably.

$$(R^1CO_2)_2 \longrightarrow 2R^1CO_2. \longrightarrow 2R^1 + 2CO_2$$

$$R^{1.} + SOlCl_2 \longrightarrow R^1Cl + SO_2Cl \longrightarrow SO_2 + Cl.$$

$$Cl: + R^2H \longrightarrow HCl + R^2. \xrightarrow{SO_2Cl_2} R^2Cl + SO_2Cl$$

In practice, chlorination with sulphuryl chloride is best carried out in the presence of αα–azobisisobutyronitrile (this readily produces free radicals).

$$\underset{\displaystyle CN}{\underset{|}{Me_2C}}—N{=}N—\underset{\displaystyle CN}{\underset{|}{CMe_2}} \longrightarrow N_2—2Me_2C(CN)\cdot$$

$$Me_2C(CN)\cdot + SO_2Cl_2 \longrightarrow Me_2C(CN).$$

Sulphuryl chloride is less selective than chlorine, and Russell (1958) has shown this is due to the nature of the solvent (when one is used).

## CHEMICAL REACTIONS OF HALOGEN DERIVATIVES OF HYDROCARBONS

### Reactivity of Haloalkanes

Haloalkanes are very reactive compounds. That is why these are used as the starting materials the preparation of a large variety of organic compounds. The high reactivity of haloalkanes is due to the polarity in the C—C bond.

The polarity of the C—X bond is due to the difference in the electronegativities of C and X atoms. The halogen atom (X) being more electronegative than X atom, pulls the shared pair of electrons towards itself. This causes a partial positive charge on the C atom, and a partial negative charge on X atom. This charge-separation is called *polarisation.* The polarity in the C—X bond is shown in Fig. 3.6. The extend of polarisation depends upon the nature of the halogen atom X.

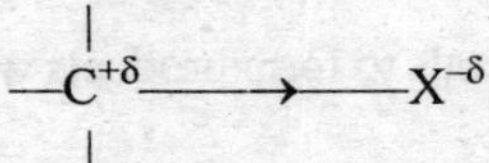

The polarity in the C—X bond is haloalkanes. This polarisation of bond is due to the difference between the electronegativities of C and X (halogen) atoms.

The reactivity of haloalkanes depends upon the following factors.

1. *Nature of the halogen atom :* Reactivity of a haloalkane depends upon the bond energy of the C—X bond. Lower is the bond energy, greater is the reactivity of the haloalkane. Since, the bond energy of the C—X bond follows the order.

| | C – Cl | > | C – Br | > | C – I |
|---|---|---|---|---|---|
| Bond energy, kJ, mol$^{-1}$ | 326.4 | | 284.5 | | 213.5 |

Hence, the reactivity of haloalkanes follows the order,

Iodoalkane > Bromoalkane > Cloroalkane.

This appears logical because greater is the size of the halogen atom, greater is the ease with which it can be lost by the haloalkane.

2. *Nature of the alkyl group viz, primary, secondary and tertiary:* Haloalkanes are polar in nature — the alkyl group at the positive and the halogen at the negative end of the dipole. The alkyl groups are electron-releasing. So, the presence of an alkyl group increases the polarisation of the C—X bond.

$$\underset{\text{primary}}{R\rightarrow \overset{\overset{H}{|}}{\underset{\underset{H}{|}}{C}}\rightarrow X} \qquad \underset{\text{secondary}}{R\rightarrow \overset{\overset{H}{|}}{\underset{\underset{H}{\uparrow}}{C}}\rightarrow\rightarrow X} \qquad \underset{\text{tertiary}}{R\rightarrow \overset{\overset{H}{\downarrow}}{\underset{\underset{H}{\uparrow}}{C}}\rightarrow\rightarrow\rightarrow X}$$

Large the number of alkyl groups attached to a carbon, greater is the repulsion of the electron-pair shared between C and X atoms towards the halogen atom. As a result, a tertiary haloalkane is more reactive than secondary haloalkane, and secondary haloalkane is more reactive than primary haloalkane. Thus the order of reactivity of haloalkanes is,

Tertiary haloalkane > Secondary haloalkane > Primary haloalkane

## REACTIONS WITH METALS

Haloalkanes react with active metals to form products which depend upon the nature of the metal used.

(a) *With sodium metal :* Haloalkanes when heated with metallic sodium in ether solution given alkanes. This reactions is known as *Wurtz reaction*. This reactions is used for obtaining symmetrical alkanes. This reaction, however, fails with tertiary haloalkanes and usually a mixture of alkanes is obtained.

$$\underset{\text{haloalkane}}{2RX} + 2Na \xrightarrow{\text{ether}} \underset{\text{alkane}}{R-R} + 2NaX$$

$$RX + 2Na + R'X \xrightarrow{\text{ether}} R-R + R-R' + R'-R' + 2NaX$$

***Example:***

$$\underset{\text{chloroethane}}{2C_2H_5Cl} + 2Na \xrightarrow{\text{ether}} \underset{\text{n-butane}}{C_2H_5-C_2H_5} + 2NaCl$$

Haloarene when heated with sodium metal in presence of dry ether, the halogen atom of haloarene is replaced by an aryl group. For example, chlorobenzene with sodium in ether gives biphenyl. This reaction is called *Fitting reaction.*

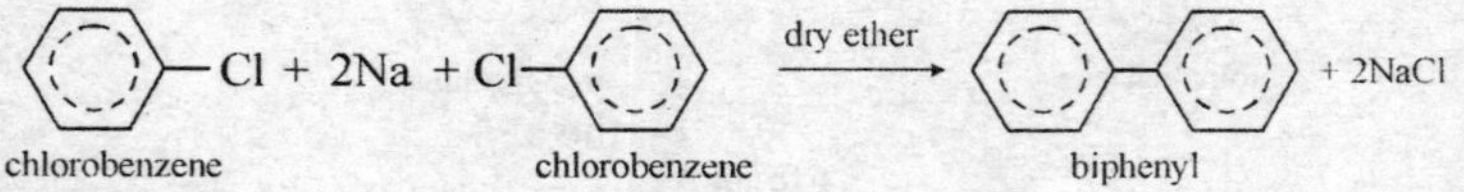

$$C_6H_5-Cl + 2Na + Cl-C_6H_5 \xrightarrow{\text{dry ether}} C_6H_5-C_6H_5 + 2NaCl$$

chlorobenzene chlorobenzene biphenyl

**Fig. 3.1**

When a haloarene is heated with a haloalkane and sodium in ether, the halogen atom of haloarene is replaced by the alkyl group. This reaction is called *Wurtz-Fitting reaction.* For example,

**Wurtz-Fitting reaction**

$$C_6H_5-Cl + 2Na + Cl-CH_3 \xrightarrow{\text{dry ether}} C_6H_5-CH_3 + 2NaCl$$

chlorobenzene (haloarene) chloromethane (haloalkane) toluene

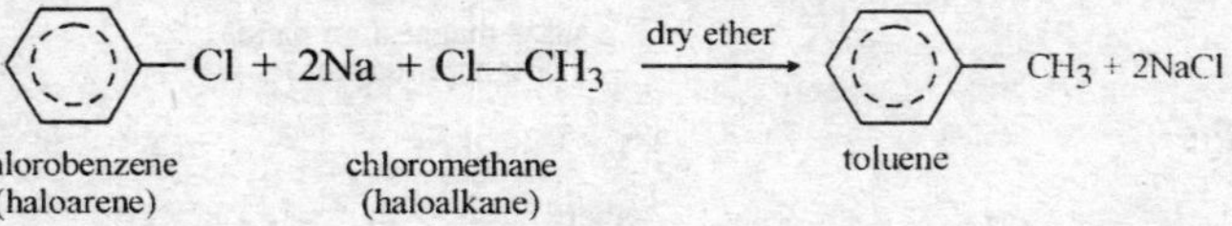

**Fig. 3.2**

(b) *When other active metals :* Haloalkanes also react with some other active metals such as, lithium, aluminium, zinc, and lead etc., in dry ether to form the corresponding organometallic compounds. For example,

$$\underset{\text{bromoethane}}{CH_3-CH_2Br} + 2Li \xrightarrow{\text{dry ether}} \underset{\text{ethyl lithium}}{CH_3-CH_2Li} + LiBr$$

$$\underset{\text{2-bromopropane}}{CH_3-\underset{\underset{Br}{|}}{CH}-CH_3} + 2Li \xrightarrow{\text{dry ether}} \underset{\text{2-propyl lithium}}{CH_3-\underset{\underset{Li}{|}}{CH}-CH_3} + 2LiBr$$

$$2\underset{\text{bromoethane}}{CH_3-CH_2Br} + 2Zn \xrightarrow[\Delta,\text{ inert atm}]{\text{dry ether}} \underset{\text{diethyl zinc}}{(CH_3-CH_2)_2Zn} + ZnBr_2$$

$$4CH_3-CH_2Br + 4Na-Pb \longrightarrow (CH_3-CH_2)_4Pb + 4NaBr + 3Pb$$

teraethyl lead (TEL)
(used as an antiknocking agent in gasoline)

Like haloalkanes, haloarenes also react with certain active metals.

Bromoarenes and iodoarenes react with lithium in the presence of dry ether to give phenyl lithium.

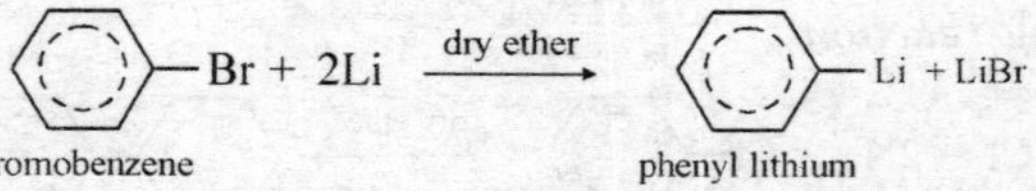

$$C_6H_5-Br + 2Li \xrightarrow{\text{dry ether}} C_6H_5-Li + LiBr$$

bromobenzene — phenyl lithium

**Fig. 3.3**

(c) *With magnesium :* Haloalkanes on reaction with magnesium in dry ether form alkyl magnesium halides commonly known as the *Grignard reagent,* (R – Mg – X). *Grignard reagent is an organometallic compound.*

$$RX + Mg \xrightarrow[\Delta]{\text{ether}} R-Mg-X$$

alkyl magnesium halide (Grignard reagent)

***Example:***

$$\underset{\text{iodomethane}}{CH_3I} + Mg \xrightarrow[\Delta]{\text{ether}} \underset{\text{methyl magnesium iodide}}{CH_3-Mg-I}$$

$$C_3H_7Br + Mg \xrightarrow[\Delta]{\text{ether}} C_3H_7MgBr$$

Bromo and iodoarenes react with dry magnesium power in the presence of dry ether to yield alymagnesium halides *(Grignard's reagent).*

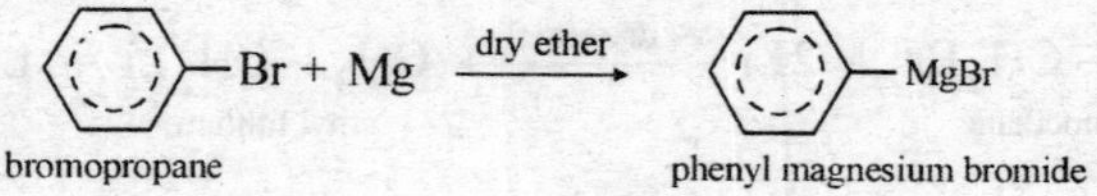

$$C_6H_5-Br + Mg \xrightarrow{\text{dry ether}} C_6H_5-MgBr$$

bromopropane — phenyl magnesium bromide

**Fig. 3.4**

Chloroarenes react with magnesium in the presence of tetrahydrofuran (THF) solvent to give phenylmagnesium chloride.

$$C_6H_5-Cl + Mg \xrightarrow{\text{THF}} C_6H_5-MgCl$$

bromopropane — phenyl magnesium chloride

**Fig. 3.5**

Grignard reagents, (named after their discover Victor Grignard, a French chemist) are very useful preparative reagents. They react with any protic compound to form hydrocarbons.

$$R - Mg + X + ZH\ (Z = OH, RO, RNH, RC \equiv C) \longrightarrow RH + MgZX$$

## REDUCTION REACTIONS

Haloalkanes on reduction give the corresponding alkanes. Reduction of haloalkanes can be carried out by using,

(a) Zn/HCl,

or Sn/HCl

(b) Zinc-copper couple + Ethanol, or Sodium + Ethanol

(c) Lithium aluminium hydride ($LiAIH_4$)

$$\underset{\text{haloalkane}}{RX} + 2H \xrightarrow{Zn/HCl} \underset{\text{alkane}}{RH} + HX$$

***Example:***

$$\underset{\text{bromoethane}}{C_2H_5Br}\ 2H \xrightarrow{Zn-Cu/HCl} \underset{\text{ethane}}{C_2H_6} + HBr$$

Iodoalkanes may be reduced by heating it with HI and red phosphorus at 150°C.

$$\underset{\text{iodoethane}}{C_2H_5I} + HI + \underset{\text{red}}{P} \xrightarrow{150^\circ C} \underset{\text{ethane}}{C_2H_6} + I_2$$

Haloarenes are reduced by nickel-aluminium alloy in alkali, or by sodium amalgam and aqueous alcohol in the presence of alkali, to produce the corresponding hydrocarbons. For example,

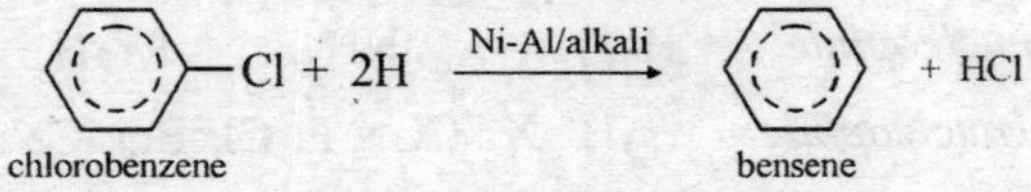

$$\underset{\text{chlorobenzene}}{C_6H_5-Cl} + 2H \xrightarrow{\text{Ni-Al/alkali}} \underset{\text{bensene}}{C_6H_6} + HCl$$

**Fig. 3.6**

## REACTIVITY OF HALOARENES RELATIVE TO HALOALKANES

Haloalkanes and haloarenes are polar compounds. The polarisation of C – X bond takes place towards the halogen atom due to its higher electronegativity, as shown below.

$$-\overset{|}{\underset{|}{C}}{}^{\delta+} \longrightarrow X^{\delta-}$$

As a result, the halogen atom becomes reactive. However, *haloarenes are less reactive than haloalkanes because of their stabilsation due to resonance.*

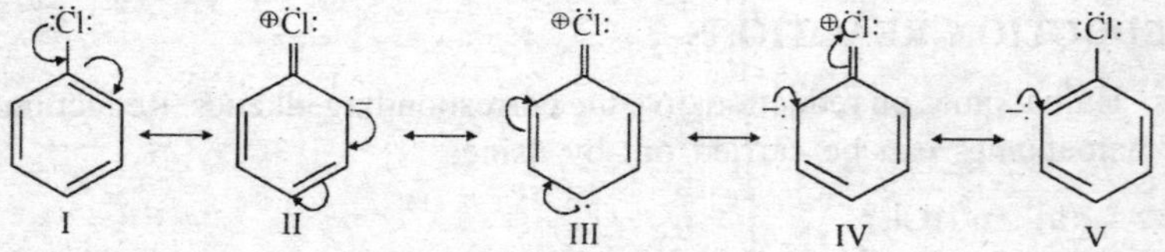

Resonating structures of chlorobenzene.

The contributions of the structures II, III and IV impart some double bond character to the carbon-chlorine bond. As a result, *the C – Xl bond length in haloarenes is expected to be shorter than the C – Cl bond length in haloalkanes.* This is found to be so. The C – Cl bond length in chlorobenzene is found to be 169 pm, as compared to the C – Cl bond length of 180 pm in chloroethane. This shortening of C – Cl bond length in haloarenes imparts stability to haloarenes chloroethane. This shortening of C – Cl bond length in haloarenes imparts stability to haloarenes relative to haloalkane. As a result, therefore, haloarenes are less reactive than haloalkanes.

## CHEMICAL REACTIONS OF HALOGEN DERIVATIVES OF NUCLEOPHILIC SUBSTITUTION REACTIONS

Nucleus-loving, or electron-donating species are called *nucleophiles.* Nucleophiles may be natural molecules ($Nu^-$) with a lone-pair of negatively charged species ($\ddot{N}u$). For example,

*Neutral nucleophile:* $H_2\ddot{O}:$ $:NH_3$, $R\ddot{O}H$, $R\ddot{N}H_2$

*Negative nucleophile:* $:\ddot{O}H^-$ $X^-$ (X = F, Cl, Br), $CN:^-$,

Carbanions $-\overset{|}{\underset{|}{C}}:^-$

*Nucleophiles* attack the substrate at a location where the electron density is low. The reactions in which a strong nucleophile displaces a weaker nucleophile from a compound are called *nucleophilic substitution (SN) reactions. A* nucleophilic substitution reaction is represented as,

$$R - X \quad \underset{\text{stronger nucleophile moves in}}{Y^{\ominus}} \longrightarrow R - Y + \underset{\text{weaker nucleophile goes out}}{X^{\ominus}}$$

The nucleophiles that have two nucleophilic centre of attack are called *ambident nucleophiles.* Such nucleophiles can react through either of these centres. However, the reaction may take place predominataly at one of these centres depending upon the substrate (reagent) and the reaction conditions.

Nucleophilic substitution reactions are used to prepare a variety of organic compounds.

*Nucleophilic substitution reactions of haloalkanes :* Nucleophilic substitution reactions given by haloalkanes are described below.

*Nucleophilic substitution reactions of haloarenes :* In haloarenes, the halogen atom is attached to the benzene ring, and it cannot be easily displaced by a nucleophilic agent (*e.g.*, $OH^-$, $NH_3$, $CN^-$ etc.–.).

Thus, haloarenes are less reactive reactions of chlorobenzene are summarized below.

$C_6H_5$–Cl (chlorobenzene) $\xrightarrow[-NaCl]{NaOH,\ 300°C/200\ atm}$ $C_6H_5$–ONa (sod. phenoxide) $\xrightarrow{HCl}$ $C_6H_5$–OH (phenol) + NaCl

$C_6H_5$–Cl (chlorobenzene) $\xrightarrow[200°C/under\ pressure]{NH_3/Cu_2O}$ $C_6H_5$–$NH_2$ (aniline) + 1/2 $Cu_2Cl_2$ + 1/2$H_2O$

$C_6H_5$–Cl (chlorobenzene) $\xrightarrow[200°C]{Cu_2(CN)_2/pyridine}$ $C_6H_5$–CN (phenyl cyanide) + 1/2 $Cu_2Cl_2$

**Fig. 3.7**

*Effect of substitutents on the nucleophilic substitution :* The presence of other substituents on the ring in haloarenes affects the nucleophilic substitution reactions appreciably.

For example, the presence of electron withdrawing group such as, – $NO_2$, – CN, – COOH etc., at *ortho–* and *para–* positions to the halogenation greatly activates the halogen atom for nucleophilic substitution reactions.

The effect of the presence of – $NO_2$ group on the reaction conditions for nucleophilic substitution reactions is illustrated below.

The presence of group, such as – $NO_2$, – CN, – COOH etc. on the ring makes the nucleophilic substitution easier.

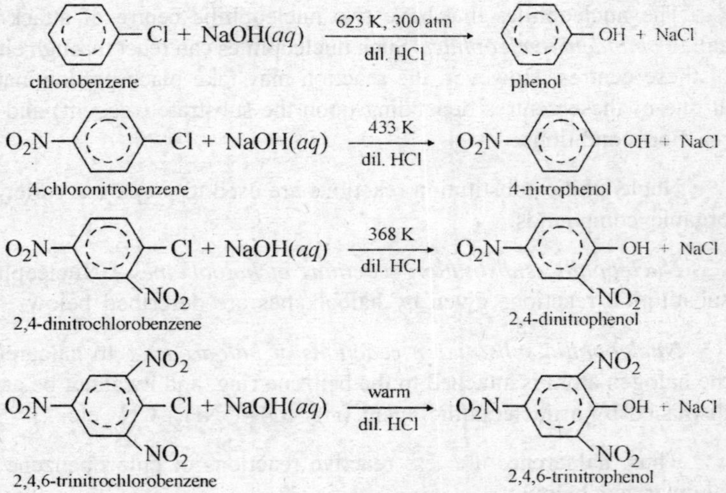

$$C_6H_5-Cl + NaOH(aq) \xrightarrow[\text{dil. HCl}]{623\ K,\ 300\ atm} C_6H_5-OH + NaCl$$

chlorobenzene → phenol

$$O_2N-C_6H_4-Cl + NaOH(aq) \xrightarrow[\text{dil. HCl}]{433\ K} O_2N-C_6H_4-OH + NaCl$$

4-chloronitrobenzene → 4-nitrophenol

$$O_2N-C_6H_3(NO_2)-Cl + NaOH(aq) \xrightarrow[\text{dil. HCl}]{368\ K} O_2N-C_6H_3(NO_2)-OH + NaCl$$

2,4-dinitrochlorobenzene → 2,4-dinitrophenol

$$O_2N-C_6H_2(NO_2)_2-Cl + NaOH(aq) \xrightarrow[\text{dil. HCl}]{\text{warm}} O_2N-C_6H_2(NO_2)_2-OH + NaCl$$

2,4,6-trinitrochlorobenzene → 2,4,6-trinitrophenol

**Fig. 3.8**

## MECHANISM OF NUCLEOPHILIC SUBSTITUTION

The nucleophilic substitution in halogen derivatives containing $C_{sp^3} - X$ bond can occur by any of the two mechanisms.

- Substitution nucleophilic bimolecular, ($S_N2$)
- Substitution nucleophilic unimolecular ($S_N1$).

Molecularity of the reaction is the number of chemical species taking part in the slowest stop.

1. *$S_N2$ Mechanism :* In this mechanism, the nucleophile, $Nu^-$ attacks the partially positive carbon attached to the halogen atom from a direction opposite to the C – X bond.

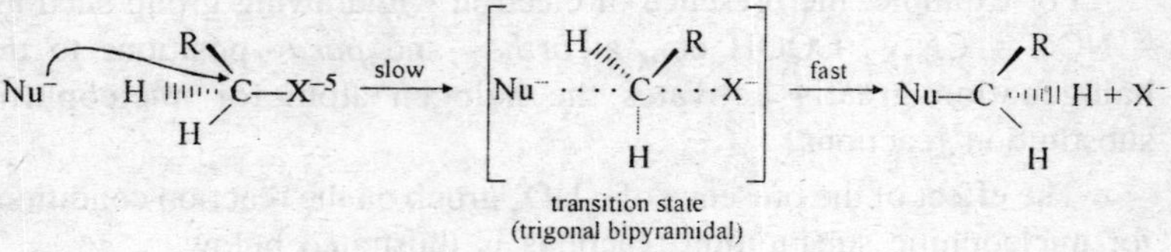

$$Nu^- \quad H\cdots C(R)(H)-X^{-5} \xrightarrow{\text{slow}} \left[ Nu^- \cdots C(H)(R)(H) \cdots X^- \right] \xrightarrow{\text{fast}} Nu-C(R)(H)\cdots H + X^-$$

transition state (trigonal bipyramidal)

**Fig. 3.9**

2. *$S_N1$ Mechanism* : In this mechanism, the reaction proceeds through the heterolytic fission of the C – X bond leading to the formation of a carbocation, and a halide ion. The nucleophile then attacks the carbocation to form the substituted product. In this mechanisms, the first step is slow and the second step is fast.

   The $S_N1$ mechanism is facilitated by polar solvents such as water, alcohol or aqueous–organic solvents.

$$R_3C—X \underset{}{\overset{\text{slow}}{\rightleftharpoons}} R_3C^+ + X^-$$

carbocation (planar)

$$R_3C^+ + Nu^- \xrightarrow{\text{fast}} R_3C—Nu$$

carbocation

**The order $S_N1$ reactivity is**

$$CH_2—X < R—CH_2—X < R—CH(R)—X < R—C(R)(R)—X$$

or primary < secondary < tertiary

**The order $S_N2$ reactivity is**

$$R—C(R)(R)—X < R—CH(R)—X < RCH_2—X < CH_3—X$$

or tertiary < secondary < secondary

3. *$S_N1$ Mechanism in Primary Allylic and Primary Benzylic Halides:* In the $S_N1$ mechanism, the carbocation formed in the first step immediately reacts with the nucleophile to complete the reaction. More easily the carbocation is formed, easier is the overall reaction to occur. Thus, the tendency of the primary haloalkanes towards SN1 mechanism is quite low. However, the primary allylic and benzylic halides show greater reactivity than other

primary haloalkanes in $S_N1$ reactions due to greater stabilization of the allylic and benzylic carbocations by resonance as shown below.

$$CH_2=CH-{}^{+}CH_2 \longleftrightarrow {}^{+}CH_2-CH=CH_2$$

**resonance in allyl carbocation**

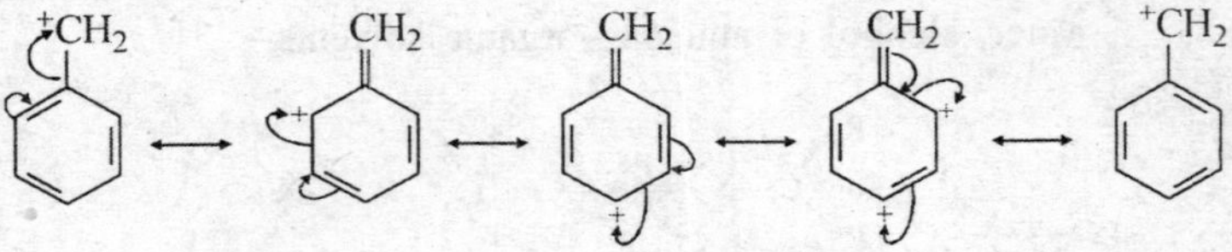

$^{+}CH_2$ ⟷ $CH_2$ ⟷ $CH_2$ ⟷ $CH_2$ ⟷ $^{+}CH_2$

**resonance in benzyl carbocation**

**Fig. 3.10**

4. *Effect of Substituent on the Nucleophilic Substitution in Halobenzenes :* Haloarenes are not reactive in nucleophilic substitution reactions. However, haloarenes (in particular halobenzenes) become reactive to nucleophilic substitution when electron-withdrawing groups such as, – $NO_2$, – CN etc., are present at the *ortho* and *para* positions. The reactivity increases progressively with the number of such groups.

   The increase in the reactivity gets reflected in the progressively milder conditions required for the nucleophilic reaction to take place. For example, the effect of – $NO_2$ group on the hydrolysis of chlorobenzene can be seen in the reaction conditions.

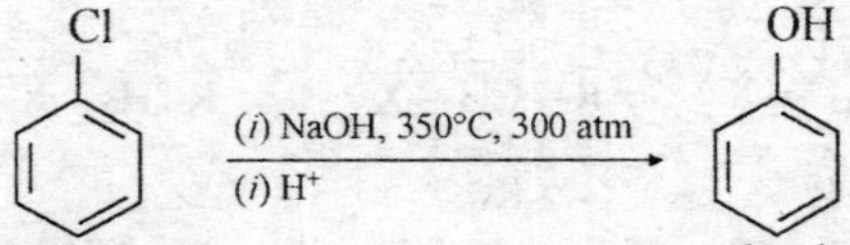

Cl → OH: (*i*) NaOH, 350°C, 300 atm; (*i*) $H^+$

chlorobenzene → phenol

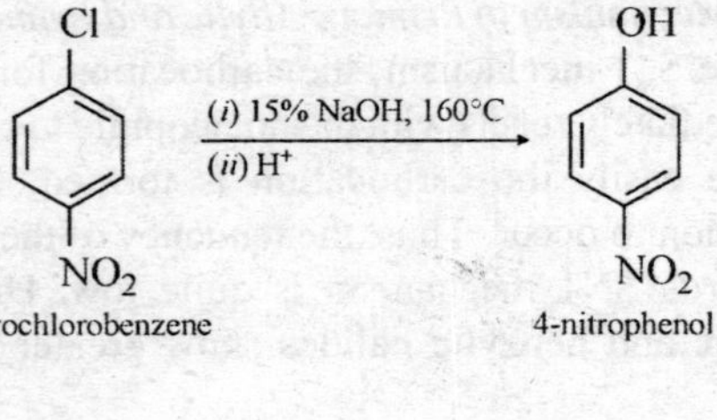

Cl, $NO_2$ → OH, $NO_2$: (*i*) 15% NaOH, 160°C; (*ii*) $H^+$

4-nitrochlorobenzene → 4-nitrophenol

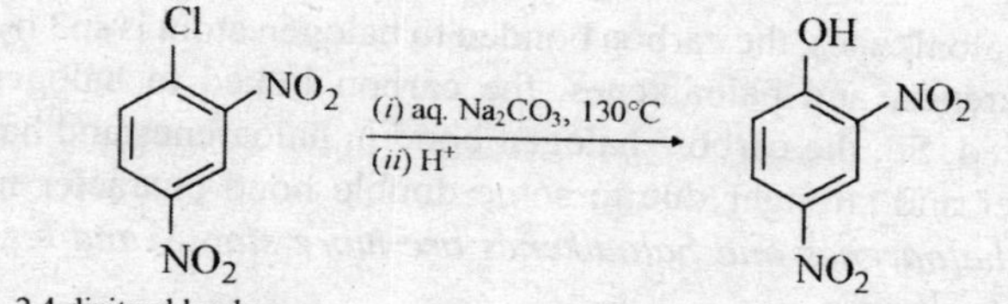

$$\text{2,4-dinitrochlorobenzene} \xrightarrow[(ii)\ H^+]{(i)\ \text{aq. } Na_2CO_3,\ 130°C} \text{2,4-dinitrophenol}$$

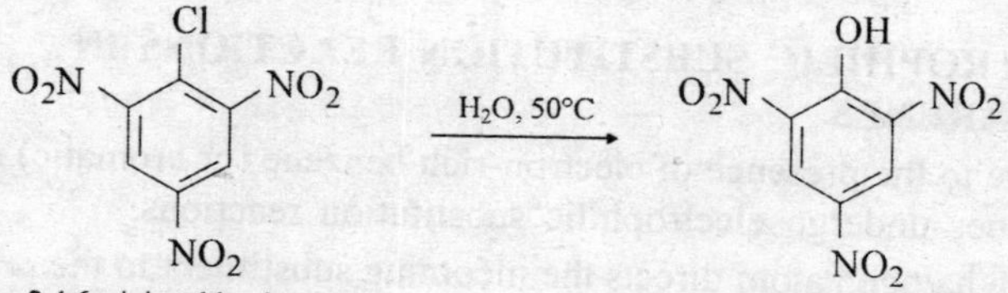

$$\text{2,4,6-trinitrochlorobenzene} \xrightarrow{H_2O,\ 50°C} \text{2,4,6-trinitrophenol (pricric acid)}$$

5. *Nucleophilic Substitution in Haloalkenes and Haloarenes:* Haloalkenes (vinylic halides) and haloarenes (aryl halides) show no reactivity in nucleophilic substitution reactions under the conditions haloalkanes react.

The lower reactivities of haloarenes, and haloalkenes towards nucleophilic substitution is due to the delocalization of electrons by resonance imparting a partial double bond character in the carbon–halogen (C – X) bond.

*This shortens the C – X bond as compared to C – X single bond. As a result C – X bond in haloarenes and haloethenes becomes less reactive as compared to* $C_{sp^3}$*—X bond in haloalkanes.*

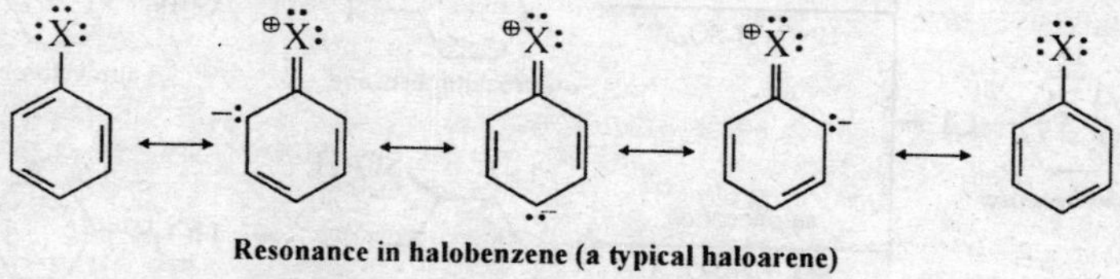

**Resonance in halobenzene (a typical haloarene)**

$$CH_2 = CH—\ddot{\underset{..}{X}}: \longleftrightarrow {}^-CH_2—CH={}^+\underset{..}{X}:$$

**Resonance in haloethene (a typical haloalkene or vinylic halide)**

**Fig. 3.11**

The lower reactivity of haloarenes and haloalkenes as compared to haloalkanes can also be explained as follows.

In haloalkanes, the carbon bonded to halogen atom is sp3 hybridized. In haloarenes, and haloalkenes, the carbon linked to halogen is sp2 hybridized. So, the carbon–halogen bond in haloarenes and haloalkenes is shorter and stronger due to some double bond character in it. As a result, *haloarenes and haloalkenes are more stable, and less reactive than haloalkanes*

## ELECTROPHILIC SUBSTITUTION REACTIONS IN HALOARENES

Due to the presence of electron-rich benzene (or aromatic) ring, the haloarenes undergo electrophilic substitution reactions.

The halogen atom directs the incoming substituent to the *ortho,* and *para* position relative to the halogen atom. Due to – I effect, halogen atom deactivates the benzene ring. As a result, the electrophilic substitution reactions of haloarenes occur slowly and under drastic conditions as compared to the benzene.

Any chemical species seeking electrons is called an *electrophile.* Electrophiles attack a centre of increased electron-density in any molecule, such as benzene ring.

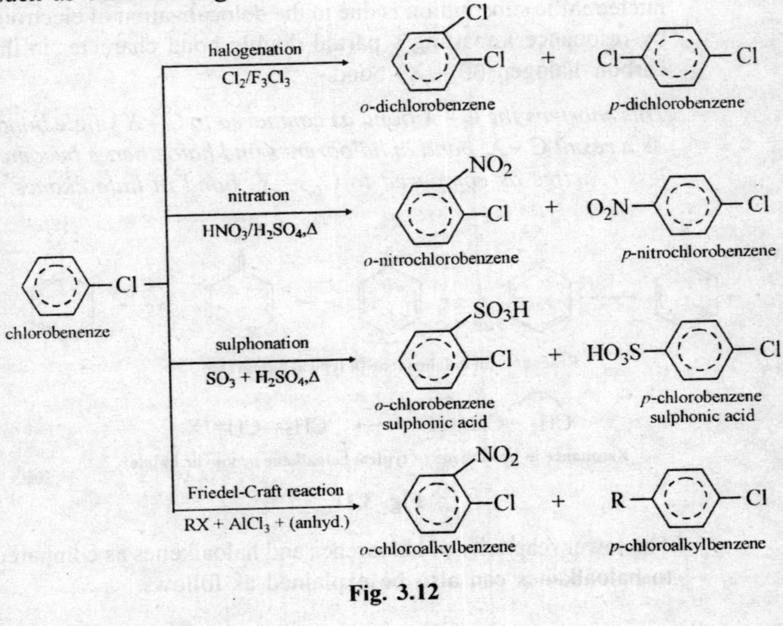

chlorobenenze

halogenation $Cl_2/F_3Cl_3$ → *o*-dichlorobenzene + *p*-dichlorobenzene

nitration $HNO_3/H_2SO_4,\Delta$ → *o*-nitrochlorobenzene + *p*-nitrochlorobenzene

sulphonation $SO_3 + H_2SO_4,\Delta$ → *o*-chlorobenzene sulphonic acid + *p*-chlorobenzene sulphonic acid

Friedel-Craft reaction RX + $AlCl_3$ + (anhyd.) → *o*-chloroalkylbenzene + *p*-chloroalkylbenzene

**Fig. 3.12**

Generally, such reactions are believed to be bimolecular, involving the formation of an intermediate. This step is a slow step, and therefore, rate-determining. The mechanism of electrophilic substitution reaction can be represented by the reaction given above.

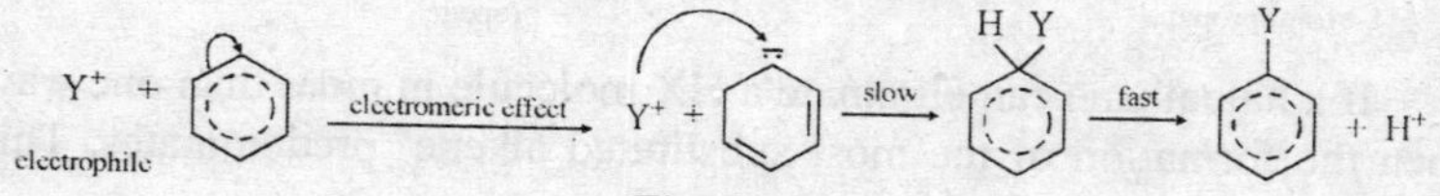

$$Y^+ + C_6H_6 \xrightarrow{\text{electromeric effect}} Y^+ + C_6H_6 \xrightarrow{\text{slow}} [C_6H_6Y]^+ \xrightarrow{\text{fast}} C_6H_5Y + H^+$$

electrophile

**Fig. 3.13**

## ELIMINATION REACTIONS

An elimination reaction involves the removal or one molecule of HX from the molecule of a haloalkane. The elimination reactions, therefore, are also called *dehydrohalogenation reactions.*

In elimination reactions the halogen atom of the haloalkane, and a hydrogen atom form the adjacent carbon atom (β-carbon) are involved. That is why, the elimination reactions of haloalkanes are called β-*elimination reactions.* The order of reactivity in elimination reactions is

Chloro < Bromo < Iodo

Primary < Secondary < Tertiary

The reactivity of elimination reaction also increases with increasing number of substituents at the β-carbon.

Haloalkane when boiled with alcoholic potassium hydroxide form alkenes by the elimination of one molecule of HX.

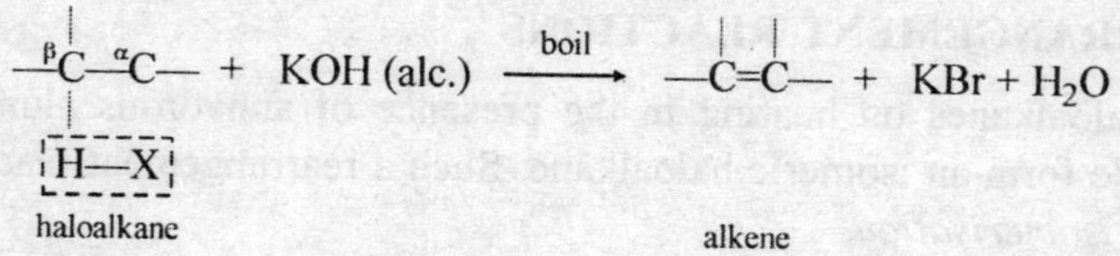

$$-\overset{|}{\underset{|}{{}^{\beta}C}}(H)-{}^{\alpha}C(X)- + KOH\,(alc.) \xrightarrow{\text{boil}} -C{=}C- + KBr + H_2O$$

haloalkane $\qquad$ alkene

***Example:***

Bromomethane on boiling with alcoholic KOH, gives ethene.

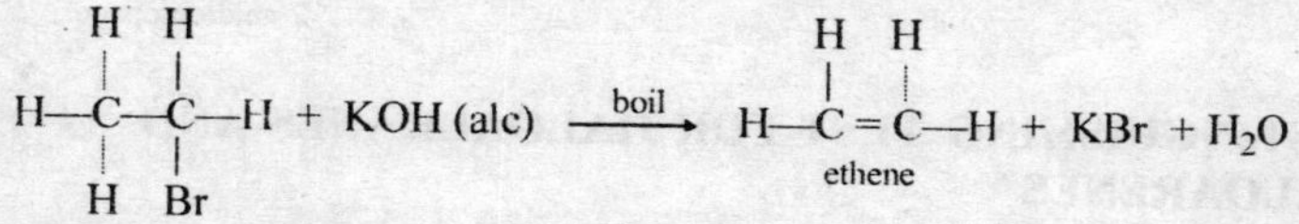

$$H-\underset{H}{\overset{H}{C}}-\underset{Br}{\overset{H}{C}}-H + KOH\,(alc) \xrightarrow{\text{boil}} H-\overset{H}{C}=\overset{H}{C}-H + KBr + H_2O$$

ethene

(b) 1–bromopropane with KOH (alc), gives propene.

$$\underset{\text{1-bromopropane}}{\text{H}-\overset{\text{H}}{\underset{\text{H}}{\text{C}}}-\overset{\text{H}}{\underset{\text{H}}{\text{C}}}-\overset{\text{H}}{\underset{\text{Br}}{\text{C}}}-\text{H}} + \text{KOH (alc)} \xrightarrow{\text{boil}} \underset{\text{propene}}{\text{H}-\overset{\text{H}}{\underset{\text{H}}{\text{C}}}=\overset{\text{H}}{\text{C}}-\overset{\text{H}}{\text{C}}-\text{H}} + KBr + H_2O$$

If a haloalkane can eliminate a HX molecule in more than one way, then the formation of the most substituted alkene' predominates. This is called *Satzeff's rule.*

*Saytzeff rule : In dehydrohalogenation reactions, the preferred product is that alkene which has greater number of alkyl groups attached to the doubly bonded carbon atoms. For example 2*-bromobutane (*see* butyl bromide) can eliminate one molecule of HBr in the following two ways.

$$CH_3-{}^{\beta}CH_2-{}^{\alpha}\underset{\text{Br}}{CH}-{}^{\beta}CH_3 \xrightarrow{\text{KOH (alc)}} CH_3-CH=CH-CH_3 + H_2O + KBr$$

but-2-ene (80%)
(preferred mode of elimination)
(here, a disubsituted alkene is formed)

$$\xrightarrow{\text{KOH (alc)}} CH_3-CH_2=CH-CH_2 + KBr + H_2O$$

but-1-ene (20%)
(here, a monosubstituted alkene is formed)

In this reactions, but-2-ene is a more substituted alkene than but-1-ene. Therefore, but-2-ene is the major product.

## REARRANGEMENT REACTIONS

Haloalkanes on heating in the presence of anhydrous aluminium chloride form an isomeric haloalkane. Such a rearrangement reaction is called *isomerisation.*

$$\underset{\text{1-chloropropane}}{CH_3-CH_2-CH_2Cl} \xrightarrow[\text{3000°C}]{\text{AlCl}_3\text{ (anhyd.)}} \underset{\text{2-chloropropane}}{CH_3-\underset{\text{Cl}}{CH}-CH_3}$$

## DISTINGUISHING TEST FOR HALOALKANES AND HALOARENES

Haloalkanes and haloarenes contain halogen atom in their molecules. To distinguish between the two *silver nitrate test* is conducted.

*Test* : Take about 1 mL of the given compound in a test tube. Add to it about 2 mL of aqueous KOH, and heat to boiling. Cool the solution and acidified with dil. $HNO_3$. To this, add a few drops of $AgNO_3$ solution, and observe for any white precipitate.

(a) Haloalkanes give white precipitate.

(b) Haloarenes and haloalkenes do not give white precipitate.

**Yses of Haloderivatives of Hydrocarbons**

Haloalkanes are very reactive compounds. Therefore, haloalkanes are used as important starting materials for synthesizing a large number of organic compounds.

Haloarenes are used as the starting material for the manufacture of compounds, such as, phenol, aniline, D.D.T. etc.

## GENERAL PROPERTIES OF THE ALKYL HALIDES

The lower members methyl chloride, methyl bromide and ethyl chloride are gases, methyl iodide and the majority of the higher members are sweet-smelling liquids. The order of values of the poiling points (and densities) of the alkyl halides is iodide > bromide > chloride > fluoride. In a group of isomeric alkyl halides, the order of the boiling points is primary > secondary > tertiary. Many of the alkyl halides burn with a green-edged flame.

The chemical reactions of the alkyl halides are similar, but they are not equally reactive, the order of reactivity being iodide > bromide > chloride, the reactivity of alkyl fluorides depends on the nature of the fluoride.

Alkyl iodides are sufficiently reactive to be decomposed by light, the iodide darkening due to the liberation of iodine.

$$2RI \longrightarrow R—R + I_2$$

Alkyl halides (and the polyhalides) are covalent compounds, insoluble in water, with which they cannot form a hydrogen bond, but soluble in organic solvents.

The alkyl halides are classified as primary, secondary and tertiary, according as the halogen atom is present in the respective groups — $CH_2X$ =CHX and ≡CX, *i.e.*, according as the given atom is joined to a primary, secondary or tertiary carbon atom.

The absorption region of alkyl halides (C—X stretch) depends on the nature of the halogen atom, C—F, 1000 $cm^{-1}$ (s), C—Cl, 750 –700 $cm^{-1}$ (s), C—Br, 600–500 $cm^{-1}$ (s), C—I, –500 $cm^{-1}$ (s)

NMR spectroscopy of alkyl halides has been previously discussed. Here we shall mention the problem of fluorine. Apart from being the most electronegative atom of the halogen group, fluorine has nuclear spin, with I = + 1/2 and –1/2 (as for hydrogen). It can there for spin-spin couple with protons and so give rise to splitting, *e.g.*, $CH_3F$ gives a doublet for the methyl group ($\tau$ 5.70), whereas $CH_3Cl$ gives singlet ($\tau$ 6.98) and $CH_3Br$ also a singlet ($\tau$ 7.35). The signal for fluorine is not observed under the conditions used for proton resonance, but with the conditions fulfilled, the fluorine spectrum will be a quartet.

Mass spectra. Loss of one electron from a lone pair on the halogen atom occurs most readily to form the molecular ion, and this then undergoes fragmentation in a way which depends on the nature of the halogen atom and on the nature of the alkyl group. Since Cl and Br exist as isotopes, the molecular ions appear as doublets of mass M and M + 2. However, because the abundance ratios are quite different, alkyl chlorides and bromides may be readily distinguished. Some of the fragmentation patterns are:

$$RCH_2CH_2^+ + X\cdot \xleftarrow{\text{Br. 1}} [RCH_2CH_2{-}X]^+ \xrightarrow{\text{FeCl}} [RCH{=}CH_2]^+ + HX$$

$$\downarrow$$

$$RCH_2^+ + CH_2{=}X^+$$

These are the predominant paths and the intensity of $M^+$ is in the order I > Cl > F, and decreases with increases in the size of the alkyl group. The $RCH_2CH_2^+$ and $[RCH{=}CH_3]^+$ fragments undergo further fragmentation typical of alkanes and alkenes, respectively. Alkyl halide (predominantly for Cl and Br) containing six or more carbon atoms in a straight chain also fragment to give the ion $C_4H_4X^+$.

$$\begin{array}{c} R \curvearrowright H^+ \\ CH_2 \quad CH_2 \\ | \qquad\quad | \\ CH_2 - CH_2 \end{array} \longrightarrow R\cdot + \begin{array}{c} X^+ \\ CH_2 \quad CH_2 \\ | \qquad\quad | \\ CH_2 - CH_2 \end{array}$$

It has already been pointed out that alkyl groups have a + I effect. Before considering the reactions of the alkyl halides, it is important to explain why an alkyl group is electron-repelling. Several explanations have been offered the following is a highly favoured one. As we have seen methane and ethane are non-polar. Thus the methyl group has a zero inductive effect, and this is true for all alkyl groups since all alkanes, whether straight or branched-chain, have a into dipole moment. When, however, one ]hydrogen atom in an alkane is replaced by some polar atom (or group), the alkyl group now exerts a polar effect which is produced by the presence of the polar atom.

Thus, in an alkyl halide, the alkyl group possesses an inductive effect, but it is one which is produced mainly by the mechanism of *interaction polarisation.* It therefore follows that the alkyl group is more polarisable than a hydrogen atom, and since most of the groups attached to the alkyl group are electron-attracting groups, the alkyl groups thus usually becomes an electron-repelling group, *i.e.*, alkyl groups normally here a + 1 effect.

*Optical isomerism : Optical activity* is the name given to the phenomenon exhibited by compounds which, when placed in the path of a beam of plane-polarised. Light, are capable of getting the plane of polarisation to the left or right. Such compounds are said to be *optically* atoms, and are dextrorotatory if the substance rotates the plane of polarisation to the right and levorotatory if to the left. The symbols (+) and (–) are used to indicate the sign of rotation, *e.g.*, (+)-glucose (dextrorotatory). (–)-fructose (levorotatory).

Optical activity may be due to the nature of the crystalline structure of the compound to, but for the present brief discussion we shall consider optical activity that is due to nature of the *molecular* structure of the compound.

When a molecular structure is not supper impossible on its mirror image, that molecule is said to posses *molecular asymmetry,* and as result, exits in two forms, one being the (+)-from and the other the (–)-form, the rotation being numerically equal.

These two forms constitute a pair of enantiomorphs or enantion and a mixture of a pair of enantiomers in equimolecular amounts is called a racemic modification. This is optically inactive (by external compensation) and is represented as DL- or (±)-.

From the chemical point of view it is less important to know the signs of rotation than know *stereochemical relationships i.e.* the relationships between the configurations of various optically active compounds.

These relationships are indicated by the symbols D and I, D (+)-glucose and D(–) -fructose are stereochemically related, *i.e.,* their configurations are related (both are members of the D-series) even through their rotations happen to be optical sign.

It should be noted here that may molecule that contains an asymmetric carbon atom. *i.e.*, a carbon atom joined to for *different* groups, possesses molecular asymmetry.

Furthermore, substitution reactions occurring at an asymmetric carbon atom may be accompanied by *optical inversion (Walden investing i.e.,* a D-substrate, on undergoing reaction, produces the L-product, or *vice versa.*

## GENERAL PROPERTIES AND REACTIONS OF THE DIHALIDES

The dihalides are sweet-smelling, colourless liquids.

1,1,-Dihalides are hydrolysed by aqueous alkalis to the corresponding carbonyl compound (aldehyde or ketone), *e.g.*, isopropylidene dichloride gives acetone.

$$CH_2CCl_2CH_3 + 2KOH \longrightarrow 2KCl + [CH_3C(OH)_2CH_3]$$
$$\longrightarrow CH_3COCH_3 + H_2O$$

It has generally been found that a compound containing two (or more) hydroxyl groups attached to the *same* carbon atom is unstable, and readily eliminates a molecule of water,

1,1,-Dihalides gives alkenes when treated with zinc dust and methanol, and alkynes when treated with ethanolic potassium hydroxide.

1,2- and αω–Dihalides are just as reactive as the alkyl halides. When heated with zinc and methanol, 1,2-dihalides give alkenes, but α,ω-dihalides in which the two halogen atoms are in the 1.3, to the 1,6 positions given cyclic compounds, *e.g.*, trimethylene dibromide gives cyclopropane.

$$CH_2\begin{matrix} \diagup CH_2Br \\ \diagdown CH_2Br \end{matrix} + Zn \longrightarrow CH_2\begin{matrix} \diagup CH_2 \\ | \\ \diagdown CH_2 \end{matrix} + ZnBr_2$$

1,2-Di-iodides tend to eliminate iodine, particularly at raised temperatures, to form alkenes *e.g.*, propene di-iodide gives propene.

$$CH_3CHICH_2I \xrightarrow{heat} CH_3CH=CH_2 + I_2$$

Methylene dichloride, b.p., 40°C, is prepared industrially by the direct chlorination of methane (see methyl chloride). It is used as an industrial solvent. Methylene dibromide, b.p., 97°C, and methylene di-iodide, b.p. 181°C, are prepared by the partial reduction of bromoform and iodoform, respectively, with sodium arsenite in alkaline solution (the yield of $CH_2Br_2$ is 88–90 per cent, $CH_2I_2$, 90–97 per cent), *e.g.*,

$$CHI_3 + Na_3AsO_3 + NaOH \longrightarrow CH_2I_2 + NaI + Na_2AsO_4$$

All the methylene dihalides are used in organic syntheses.

Ethylene dichloride, b.p. 84°C, and ethylidene dichloride, b.p. 57°C, are isomers, the former is prepared from ethylene and chlorine, and the latter by the action of phosphorus pentachloride on acetaldehyde (q.v.).

Ethylene dibromide, b.p. 131°C, is made industrially by passing ethylene into cooled bromine, it is used as a component in anti-knock fluids.

## TRIHALOGEN DERIVATIVES

The most important derivatives are those of methane, and they are usually know by their trivial names, chloroform, $CHCl_3$, bromoform, $CHBr_3$ and iodoform, $CHI_3$.

Coloform may be prepared in the laboratory of industrially by heating ethanol or acetone with bleaching power, or with chlorine and alkali (yield about 40 per cent). The reactions is externally complicated, and the mechanism is obscure.

The equations usually given for the action of bleaching power on ethanol are, (i) oxidation of ethanol to acetaldehyde, (ii) chlorination of acetaldehyde to trichloroacetaldehyde, (iii) decomposition of trichloroacetaldehyde (choral) by free calcium hydroxide (present, in the bleaching power) into chloroform and formic acid.

(i) $CH_3CH_2OH + Cl_2 \longrightarrow CH_3CHO + 2HCl$

(ii) $CH_3CHO + 3Cl_2 \longrightarrow CCl3CHO + 3HCl$

(ii) $2CCl_3CHO + Ca(OH)_2 \longrightarrow 2CHCl_3 + (HCO_2)_2Ca$

When acetone is used, the first product given is trichloroacetone, which is then decomposed by the calcium hydroxide into chloroform and acetic acid.

(i) $CH_3COCH_3 + Cl_2 \longrightarrow CCl_3COCH_3 + 3HCl$

(ii) $2CCl_3COCH_3 + Ca(OH)_2 \longrightarrow 2CHCl_3 + (CH_3CO_2)_2Ca$

Chloroform is also prepared industrially by the chlorination of methane (see methyl chloride).

Chloroform is a sickly, sweet-smelling, colourless liquid, b.p. 61°C. It is sparingly soluble is water but readily soluble in ethanol and ether. It does not burn in air under usual conditions but its vapour may be ignited, when it burns with a green-edged flame. According to Hine (1953 1954), chloroform (and other haloforms) undergoes alkaline hydrolysis to produce the formation and carbon monoxide by what Hine calls the *alpha-elimination mechanism.* This involves the removal of hydrogen and chloride ions form the *same* carbon atom.

The mechanism proposed is an follows, involving the intermediate formation of dichloromethylene

$$CHCl_3 + OH^- \rightleftharpoons :CCl_3^- + H_2O$$

$$:CCl_3^- \longrightarrow :CCl_2 + Cl^-$$

$$:CCl_2 \xrightarrow[H_2O]{OH^-} CO + HCO_2^- + 3Cl^-$$

The removal of a proton is facilitated by the – 1 effect of the chlorine atoms. However, the cannot be the complete story since, although fluorine has a much stronger – 1 effect than chlorine, $CHClF_2$ behaves differently. Hine *et al.* (1954 – 57) have shown that $CHCl_3$, $CHBr_2$ $CHBr_2F$, etc, readily undergo deuterium exchange in alkaline solution.

$$CHCl_3 + OH^- \rightleftharpoons H_2O + :\bar{C}Cl_3 \xrightleftharpoons{D_2O} CDCl_3 + OD^-$$

$CHClF_2$ was found to react more slowly. To account for this, it has been suggested that the $CCl_3^-$ ion is stabilised by resonance. Cl can expand its shell to a decet because it has *d*-orbitals but F has no*d*-orbitals and so this expansion cannot take place.

$$:\ddot{Cl}-\overset{\overset{:\ddot{Cl}:}{|}}{\underset{\underset{:\ddot{Cl}:}{|}}{C}}:^- \longleftrightarrow :\ddot{Cl}-\overset{\overset{:\ddot{Cl}:^-}{\|}}{\underset{\underset{:\ddot{Cl}:}{|}}{C}} \longleftrightarrow \cdots$$

This type or resonance, which requires expansion of an octet, is known as *d orbital resonance.*

When chloroform is treated with zinc and hydrogen chloride in ethanolic solution, methylene dichloride (q.v) is obtained, when treated with zinc and water, methane is obtained.

$$CHCl_3 \xrightarrow{Zn/H_2O} CH_4 + 3HCl$$

When chloroform is warmed with silver power, acetylene is obtained.

$$2CHCl_3 + 6Ag \longrightarrow C_2H_2 + 6AgCl$$

When treated with concentrated nitric acid, chloroform forms *chloropicrin.*

$$CHCl_3 + HNO_3 \longrightarrow CCl_3NO_2 + H_2O$$

Chloropicrin or nitrochloroform (liquid, b.p., 112°C) is used as an insecticide, and has been used as a war-gas. Chloroform adds on to the carbonyl group of ketones in the presence of potassium hydroxide, *e.g.*, with acetone it forms *chloretone* (colourless needles, m.p. 97°C), which is used as a drug. The mechanism is possibly.

$$(CH_3)_2C{=}O + :\bar{C}Hl_3 \longrightarrow (CH_3)_2C(O^-)CCl_3 \xrightarrow{H_2O} (CH_3)_2C(OH)CCl_3$$

## POLYHALOGEN COMPOUNDS

The organic compounds containing more than one halogen atoms in their molecules are termed *compounds.* These compounds are designated as follows.

The compounds containing two halogen atoms in their molecules are called *dihalo*-compounds. Those containing three halogen atoms in their molecules are called *trihalo-* compounds. The compounds containing four halogen atoms in their molecules are called *tetrahalo*-compounds.

Some of the polyhalo compounds are commerically very important. Some common polyhalogen compounds are listed in Table 17.3 given on the next page.

## DICHLOROMETHANE (METHYLENE CHLORIDE, $CH_2Cl_2$)

Dihalogen derivatives of alkanes are obtained by replacing its two hydrogen atoms by two halogen atoms. Dichloromethane is thus a derivative of methane in which two H-atoms of methane molecule are replaced by two Cl atoms.

$$\begin{array}{c} H \\ | \\ C \\ H \diagup \quad \diagdown Cl \\ \quad Cl \end{array}$$

Dichloromethane

## Industrial Preparation

Dichloromethane is prepared on industrial scale by direct chlorination of methane (or natural gas).

$$\underset{\text{methane}}{CH_4} \xrightarrow[-\,HCl]{+\,Cl_2,\,h\nu} \underset{\text{chloromethane}}{CH_3Cl} \xrightarrow[-\,HCl]{+\,Cl_2,\,h\nu} \underset{\text{dichloromethane}}{CH_2Cl_2}$$

$$\xrightarrow[-\,HCl]{+\,Cl_2,\,h\nu} \underset{\text{trichloromethane}}{CHCl_3} \xrightarrow[-\,HCl]{+\,Cl_2,\,h\nu} \underset{\text{tetrachloromethane}}{CCl_4}$$

In this reaction, a mixture of different chloro derivatives is obtained. Dichloromethane is separated from this mixture by fractional distillation.

*Properties* :

(i) Dichloromethane is a colourless, sweet-smelling, volatile liquid (b.p. 40°C).

(ii) It has low inflammability.

*Use* : Dichloromethane is used

(i) as an effective extraction solvent in pharmaceutical and food industries.

(ii) as a paint remover.

## TRICHLOROMETHANE, (CHLOROFORM, $CHCL_3$)

Chloroform is the simplest trihalogen compound in which three hydrogen atoms of methane are substituted by three chlorine atoms.

### Preparation of Chloroform

Chloroform is prepared in the laboratory as well as on the industrial scale by heating ethyl alcohol or acetone with bleaching power or chlorine and alkali.

**From Ethanol**

The formation of chloroform form ethanol takes place as follows

(i) Oxidation of ethyl alcohol by chlorine

$$\underset{\text{ethanol}}{CH_3CH_2OH} + Cl_2 \longrightarrow \underset{\text{ethanal ( acetaldehyde )}}{CH_3CHO} + 2HCl$$

(ii) Formation of chloral by the action of chlorine on acetaldehyde so formed.

$$CH_3CHO + 3Cl_2 \longrightarrow \underset{\text{chloral}}{CCl_3CHO} + 3HCl$$

(iii) Hydrolysis of chloral of chloroform by calcium hydroxide of alkali (NaOH).

$$\underset{\substack{\text{slaked lime}\\\text{(from bleaching powder)}}}{Ca\begin{matrix}\diagup OH\\ \diagdown OH\end{matrix}} + \underset{\text{chloral}}{\begin{matrix}CCl_3CHO\\ CCl_3CHO\end{matrix}} \longrightarrow \underset{\text{chloroform}}{2CHCl_3} + \underset{\text{calcium formate}}{\begin{matrix}HCOO\diagdown\\ HCOO\diagup\end{matrix}Ca}$$

or $$\underset{\text{chloral}}{CCl_3CHO} + NaOH \longrightarrow CHCl_3 + \underset{\text{sod, formate}}{HCOONa}$$

*The net reaction when bleaching power is used is,*

$$\underset{\text{ethanol}}{2CH_3CH_2OH} + 8Cl_2 + \underset{\text{from bleaching powder}}{Ca(OH)_2} \longrightarrow \underset{\text{chloroform}}{2CHCl_3}$$

$$+ \underset{\text{calcium formate}}{(HCOO)_2Ca} + 10HCl$$

**From Acetone (from Propanone)**

When acetone is heated with bleaching power, or with chlorine and alkali, the following reactions take place.

(i) Reaction of chlorine with acetone producing trichloroacetone.

$$\underset{\text{acetone}}{CH_3 \cdot CO \cdot CH_3} + \underset{\text{from bleaching powder}}{3Cl_2} \longrightarrow \underset{\text{trichloroacetone}}{CCl_3 \cdot CO \cdot CH_3} + 3HCl$$

(ii) Trichloroacetone is decomposed by either calcium hydroxide or an alkali to give chloroform.

$$CCl_3 \cdot CO \cdot CH_3 + NaOH \longrightarrow CHCl_3 + \underset{\text{sodium acetate}}{CH_3COONa}$$

$$2CCl_3 \cdot CO \cdot CH_3 + \underset{\text{from bleaching powder}}{Ca(OH)_2} \longrightarrow \underset{\text{chloroform}}{2CHCl_3} + \underset{\text{calcium acetate}}{(CH_3COO)_2Ca}$$

*The net reaction, when acetone is heated with bleaching powder is,*

$$\underset{\text{acetone}}{2CH_3 \cdot CO.CH_3} + 6Cl_2 + \underset{\text{from bleaching powder}}{Ca(OH)_2} \longrightarrow \underset{\text{chloroform}}{2CHCl_3}$$

$$+ \underset{\text{calcium acetate}}{(CH_3COO)_2Ca} + 6HCl$$

**Preparation of Pure Chloroform**

Pure chloroform needed for medicinal use is obtained by heating chloral hydrate with concentrated solution of NaOH.

$$\underset{\text{chloral hydrate}}{CCl_3CHO \cdot H_2O} + NaOH \xrightarrow{\Delta} \underset{\text{chloroform}}{CHCl_3} + \underset{\text{sod. formate}}{HCOONa} + H_2O.$$

## MANUFACTURE OF CHLOROFORM

Chloroform is generally manufactured by the following methods.

(i) **From methane.** Chloroform can be manufactured by chlorinating methane. A mixture of $CH_4 : Cl_2 : N_2$ in the ration 8 : 3 : 80 is passed over a partially reduced sample of cupric chloride chloroform is recovered by distillation.

$$CH_4 + 3Cl_2 \longrightarrow CHCl_3 + 3HCl$$

(ii) **From carbon tetrachloride.** Chloroform can also be prepared on a large scale by partial reduction of carbon tetrachloride using iron filings and water.

$$\underset{\text{carbon tetrachloride}}{CCl_4} + 2H \xrightarrow{Fe/H_2O} CHCl_3 + HCl$$

## PHYSICAL PROPERTIES OF CHLOROFORM

Some important physical properties of chloroform ($CHCl_3$) are.

(i) Chloroform is a heavy, colourless, sweet-smelling liquid, (b.p. = 334 K, (61°C) f.P. = 210 K, (– 63°C). Chloroform dissolves substances like oils, fats, and waxes etc. So, chloroform is a good solvent for oils, fats and waxes.

(ii) Chloroform is insoluble in water but soluble in organic liquids like ether, alcohol etc.

(iii) It is non-inflammable.

(iv) Its vapour, if inhaled causes temporary unconsciousness. Therefore, it is used as a general anaesthetic.

## CHEMICAL PROPERTIES OF CHLOROFORM

Some important chemical properties of chloroform are described below :

1. **Oxidation :** It undergoes oxidation in the presence of light to give highly poisonous substance called phosgene, (carbonyl chloride).

$$2CHCl_3 + O_2 \xrightarrow[\text{air}]{\text{sunlight}} \underset{\substack{\text{phosgene} \\ \text{(carbonyl chloride)}}}{2COCl_2} + 2HCl$$

*It is the presence of phosgene in long-stored chloroform which makes it unfit to be used as anaesthetic. It is for this reason that the chloroform is stored in coloured bottles filled upto the stopper*

*To prevent oxidation of chloroform :* It is very important of prevent oxidation of chloroform. This can be done as follows.

(a) Chloroform should be stored in dark-coloured bottles to cut of light.

(b) The bottle containing chloroform should be filled upto the stopper, and stoppered tightly to avoid the presence of air inside the bottle.

(c) A small quantity of ethanol (1%) should be added to chloroform at the time packaging. Ethanol converts phosgene (if, formed) to non-poisonous, and non-volatile neutral ethyl carbonate.

$$\underset{\text{phosgene}}{COCl_2} + \underset{\text{ethanol}}{2C_2H_5OH} \longrightarrow \underset{\text{ethyl carbonate}}{O{=}C(OC_2H_5)_2} + 2HCl$$

Thus, the formation and accumulation of phosgene in chloroform is prevented.

2. **Reduction :** Chloroform undergoes reduction with Zn. HCl in the presence of ethyl alcohol to give dichloromethane, (methylene chloride), but when reduced with Zn dust and water, methane is obtained.

$$CHCl_3 + 2[H] \xrightarrow[\text{ethanol}]{\text{Zn / HCl}} CH_2Cl_2 + HCl$$

$$CHCl_3 + 6[H] \xrightarrow{\text{Zn / }H_2O} CH_4 + 3HCl$$

3. **Hydrolysis :** Chloroform gets hydrolysed when boiled with aqueous solution of potassium hydroxide.

$$\underset{\text{chloroform}}{H-C(Cl)_3} + 3KOH \xrightarrow[-3KCl]{\text{boil}} \underset{\text{unstable}}{\left[H-C(OH)_3\right]} \xrightarrow{-H_2O}$$

$$\underset{\text{formic acid}}{H-C(=Cl)-OH} \xrightarrow{\text{KOH}} \underset{\text{pot. formate}}{H-C(=Cl)-OK} + H_2O$$

4. **With Primary Amines:** *Carbylamine Reaction* : When chloroform is warmed with a primary amine, and alcoholic potassium hydroxide, an offensive smelling compound called carbylamine (isocyanide) is formed. **Carbylamine reaction is used as a test for chloroform.**

$$\underset{\text{primary amine}}{RNH_2} + CHCl_3 + 3KOH\,(alc) \xrightarrow{\text{warm}} \underset{\substack{\text{carbylamine (isocyanide)} \\ \text{offensive smelling}}}{R\cdot N \equiv C} + 3KCl + 3H_2O$$

***Example:***

$$\underset{\text{aniline}}{C_6H_5NH_2} + CHCl_3 + 3KOH\,(alc) \longrightarrow \underset{\text{phenyl carbylamine}}{C_6H_5N \equiv C} + 3KCl + 3H_2O$$

$$\underset{\text{ethylamine}}{C_2H_5NH_2} + CHCl_3 + 3KOH\,(alc) \xrightarrow{\text{warm}} \underset{\text{ethyne (acetylene)}}{C_2H_5N \overset{\rightarrow}{=} C}$$

$$+ 3KCl + 3H_2O$$

5. **With Silver Powder :** On heating with silver powder at high temperature, ethyne (acetylene) is obtained.

$$\underset{\text{chloroform}}{2CHCl_3} + 6Ag \xrightarrow{\Delta} CH = CH + 6AgCl$$

6. **With Phenol :** When chloroform is heated with phenol at 340 K, and sodium hydroxide followed by hydrolysis, salicyaldehyde, (*o*-hydroxybenzaldehyde) is formed. This reaction is called **Reimer-Tiemen reaction.**

$$\underset{\text{phenol}}{C_6H_5\text{—}OH} + NaOH + CHCl_3 \xrightarrow[\text{hydrolysis}]{340\ K} \underset{\substack{\text{salicylaldehyde}\\ \text{(o-hydroxybenzaldehyde)}}}{C_6H_4(CHO)\text{—}OH} + 3NaCl + H_2O$$

7. **With Acetone :** Chloroform condenses with acetone in the presence of potassium hydroxide to give chloretone. **Chloretone is used as a sleep-inducing *(hypnotic)* drug.**

$$\underset{\text{acetone}}{(CH_3)_2C{=}O} + CHCl_3 \xrightarrow{KOH} \underset{\text{chloretone}}{(CH_3)_2C(OH)(OH)}$$

## Uses of Chloroform

Chloroform is used,

(i) as an anaesthetic. It has now being replaced.

(ii) as a solvent for fats, waxes and resins.

(iii) as a preservative for anatomical specimen.

(iv) for the preparation of chloretone, chloropicrin etc.

## TRIIODOMETHANE (IODOFORM, $CHI_3$)

Iodoform is another simple trihalogen derivative of methane. Iodoform is obtained by substituting three hydrogen atoms of methane by three iodine atoms.

**Preparation of Iodoform**

Iodoform is obtained by the action of iodine on ethanol (ethyl alcohol), of on propanone (acetone) in the presence of an alkali. This reaction is called *haloform reaction,* and is commonly known as *iodoform test.*

**1. From Ethanol :** When ethanol is warmed (60°C) with iodine in the presence of an alkali (NaOH, or $Na_2CO_3$), iodoform is obtained,

**Step. 1.** $NaOH + I_2 \longrightarrow \underset{\text{sod. hypoiodite}}{NaOI} + NaI + H_2O$

**Step. 2.** Oxidation of ethanol to ethanal (acetaldehyde).

$$\underset{\text{ethanol}}{CH_3CH_2OH} + NaOH \longrightarrow \underset{\text{ethanal}}{CH_3CHO} + NaI + H_2O$$

**Step. 3.** Ethanal so formed undergoes iodination.

$$\underset{\text{ethanal}}{CH_3CHO} + 3NaOI \longrightarrow \underset{\text{tri-iodoacetaldehyde}}{CI_3CHO} + 3NaOH$$

**Step. 4.** Hydrolysis (cleavage) of $CI_3CHO$ by alkali.

$$CI_3CHO + NaOH \longrightarrow \underset{\substack{\text{iodoform}\\ \text{(yellow ppt.)}}}{CHI_3} + HCOONa$$

**Net Reaction**

**2. From Acetone :** Iodoform can also be prepared by warming acetone with iodine in the presence of an alkali. The reactions are

**Step. 1.** $2NaOH + I_2 \longrightarrow NaOI + NaI + H_2O$

**Step. 2.** Iodination of acetone.

$$CH_3 \cdot CO.CH_3 + 3NaOI \longrightarrow \underset{\substack{\text{1,1,1-tri-iodopropane}\\ \text{(triiodoacetone)}}}{CI_3 \cdot CO \cdot CH_3} + 3NaOH$$

**Step. 3.** Hydrolysis (cleavage) of $CI_3 \cdot CO \cdot CH_3$ by alkali.

$$\underset{\text{triiodoacetone}}{CI_3 \cdot CO \cdot CH_3} + NaOH \longrightarrow \underset{\substack{\text{iodoform}\\ \text{(yellow ppt.)}}}{CHI_3} + CH_3COONa$$

**Net Reaction**

$$\underset{\substack{\text{propanone}\\ \text{(acetone)}}}{CH_3 \cdot CO \cdot CH_3} + 3I_2 + 4NaOH \xrightarrow{\text{warm}} \underset{\substack{\text{iodoform}\\ \text{(yellow ppt.)}}}{CHI_3} + 3NaI$$

$$+ CH_3COONa + 3H_2O.$$

The reactions for the preparation of iodoform when $Na_2CO_3$ is used in place of NaOH are as follows.

*(a) when ethanol is used.*

$$\underset{\text{ethanol}}{CH_3CH_2OH} + 4I_2 + 3Na_2CO_3 \xrightarrow{\text{warm}} \underset{\text{iodoform}}{CHI_3}$$

$$HCOONa + 5NaI + 2H_2O + 3CO_2$$

*(b) when propanone (acetone) is used.*

$$\underset{\substack{\text{propanone}\\(\text{acetone})}}{CH_3 \cdot CO \cdot CH_3} + 3I_2 + 2Na_2CO_3 \xrightarrow{\text{warm}} \underset{\text{iodoform}}{CHI_3}$$

$$CH_3COONa + 3NaI + H_2O + 2CO_2$$

## Physical Properties of Iodoform

Iodoform is a yellow crystalline solid (hexagonal plates) with characteristic unpleasant odour. It melts at 392 K, (119°C).

## Chemical Properties of Iodoform

Iodoform shows chemical behaviour similar to that of chloroform, but it is less stable. Some of the typical reactions given by iodoform are.

1. **Stability :** On heating, iodoform decomposes to give iodine vapour. Due to the liberation of free iodine iodoform shows antiseptic action. Moisture, air and light also cause the decomposition of iodoform.

2. **Hydrolysis :** On boiling with alcoholic potash (KOH (alc)), iodoform gives potassium formate.

$$\underset{\text{iodoform}}{H{-}C\begin{cases} I + KOH\,(alc.) \\ I + KOH\,(alc.) \\ I + KOH\,(alc.) \end{cases}} \xrightarrow{\Delta} 3KI + \underset{\text{(unstable)}}{\left[ H{-}C\begin{matrix} OH \\ OH \\ OH \end{matrix} \right]}$$

$$\xrightarrow[-H_2O]{} \underset{\text{formic acid}}{H{-}C\begin{matrix} {/\!\!/}OH \\ \diagdown OH \end{matrix}} \xrightarrow{KOH} \underset{\text{formic acid}}{H{-}C\begin{matrix} {/\!\!/}OH \\ \diagdown OH \end{matrix}} + H_2O$$

3. **Heating with Silver Powder :** When heated with Ag powder, iodoform gives ethyne (acetylene).

$$CHI_3 + 6Ag + I_3CH \longrightarrow \underset{\text{ethyne (acetylene)}}{CH \equiv CH} + 6AgI$$

4. **With Primary Amines:** *Carbylamine Reaction :* Iodoform when heated with a primary amine and alcoholic potash, gives an offensive smelling compound known as carbylamine.

$$\underset{\text{aniline}}{C_6H_5NH_2} + \underset{\text{iodoform}}{I_3HC} + 3KOH\,(alc) \longrightarrow \underset{\text{phenyl carbylamine}}{C_6H_5NC} + 3KI + 3H_2O$$

5. **With Silver Nitrate:** Iodoform gives yellow coloured precipitate of AgI when heated with alcoholic silver nitrate.

**Uses of Iodoform**

Iodoform is used as an antiseptic. Its antiseptic property is due to the liberated iodine. However, due to its unpleasant smell, it has been replaced by other formulations containing iodine.

## IODOFORM TEST

Iodoform test is given by all the compounds which contain a methyl group attached to the carbonyl group *i.e.*, $CH_3 - \overset{|}{C} = O$ group, or the compounds that can produce $CH_3 - \overset{|}{C} = O$ group upon oxidation, *e.g.*, those containing ,

$$CH_3 - \underset{OH}{\overset{|}{\underset{|}{C}}}H = O \text{ group.}$$

Thus, *ethanol propanone (acetone), and 2-propanol (isopropyl alcohol) give iodoform test.*

*Procedure :* Take about 1 g of $Na_2CO_3$ in 5 mL of water in a boiling tube. Add about 0.5 mL of ethanol (or the given compound), and warm the solution to about 60°C on a water bath. Now, add 0.5 g of iodine gradually till a permanent yellow-brown colour is obtained. At this stage a few drops of dil., sodium hydroxide solution are added to discharge thee yellow colour. Yellow crystals of iodoform immediately separate out.

### Usefulness of the Iodoform Test

Iodoform test can be used to distinguish between the following pairs.

| *Pairs of compounds* | *Compound which gives positive iodoform test* |
|---|---|
| 1. Ethanol and Methanol | Ethanol |
| 2. Ethanol and Propan-1-ol | Ethanol |
| 3. Methanol (formaldehyde) and Ethanal (acetaldehyde) | Ethanal (acetaldehyde) |
| 4. Pentan-2one (Methyl propyl ketone), and Pentan-3-one (diethyl ketone) | pentan-2-one |

Chlorine and bromine can replace iodine in iodoform to give chloroform and bromoform respectively.

In this generalised form, the reaction is know as *haloform reaction.*

## TETRACHLOROMETHANE (CARBON TETRACHLORINE, $CCl_4$)

Carbon tetrachloride is the most important tetrahalogen derivatives of alkanes.

### Preparation of Carbon Tetrachloride

Carbon tetrachloride can be prepared on the industrial scale by the following methods.

*(i) From methane:* By the chlorination of methane in the presence of sunlight.

$$CH_4 + 4Cl_2 \xrightarrow{hv} CCl_4 + 4HCl$$

*(ii) From carbon disulphide:* Carbon tetrachloride can be obtained by the action of chlorine on carbon disulphide in the presence of aluminium chloride as a catalyst.

$$\underset{\text{carbon disulphide}}{CS_2} + 3Cl_2 \xrightarrow{AlC_3} CCl_4 + \underset{\text{sulphur monochloride}}{S_2Cl_2}$$

## PHYSICAL PROPERTIES OF CARBON TETRACHLORIDE

(i) Carbon tetrachloride is a colourless heavy liquid (b.p. 350 K, (77°C)).

(ii) It is insoluble in water but soluble in organic solvents.

(iii) It is a good solvent for oils, fats and resins.

(iv) It is non-inflammable.

(v) It is stable at red heat (≈ 500°C)

In past carbon tetrachloride was used as fire extinguishing material, under the name *pyrene.* When a find spray of liquid $CCl_4$ is directed at the fire, it vaporises. The dense vapour of $CCl_4$ cut off the supply of air to the burning article. As a result, the fire gets extinguished.

However, at higher temperature in the presence of steam, carbon tetrachloride gets oxidised to phosgene. So, care should be taken while using carbon tetrachloride in fire fighting operations,

$$CCl_4 + H_2O\,(\text{steam}) \xrightarrow{500°C} \underset{\substack{\text{phosgene} \\ \text{(a poisons gas)}}}{COCl_2} + 2HCl$$

## CHEMICAL PROPERTIES OF CARBON TETRACHLORIDE

Some chemical reactions shown by carbon tetrachloride are :

(i) *Reduction :* Carbon tetrachloride on reduction with moist iron filings gives chloroform.

$$CCl_4 + 2H \xrightarrow{Fe.H_2O} CHCl_3 + HCl$$

(ii) *Hydrolysis :* In the presence of steam, $CCl_4$ gives phosgene.

$$CCl_4 + \underset{\text{(steam)}}{H_2O} \longrightarrow \underset{\substack{\text{phosgene} \\ \text{(a poisonous gas)}}}{COCl_2} + 2HCl$$

## USES OF CARBON TETRACHLORIDE

Carbon tetrachloride is used,

(i) As a solvent for fats, oils and waxes etc,

(ii) In manufacturing freon (dichlorodifluoromethane).

(iii) As a fire extinguishing agent, (with all precautions).

Carbon tetrachloride is highly toxic. Moderate exposure to it causes headache, nausea and dizziness. It harms gastrointestinal tract, central nervous system, liver and lungs. It irritates eyes and dires the skin. Prolonged exposure may cause unconsciousness or even death. It is suspected to be carcinogenic.

## *p*-DICHLOROBENZENE

Preparation. *p*-dichlorobenzene is prepared by chlorination of benzene.

$$\underset{\text{benzene}}{C_6H_6} + Cl_2 \xrightarrow[-HCl]{\text{dark, iron}} \underset{\text{chlorobenzene}}{C_6H_5Cl} \xrightarrow[-HCl]{+Cl_2\ \text{(iron, in dark)}} \underbrace{\underset{o-}{C_6H_4Cl_2} + \underset{p-}{C_6H_4Cl_2}}_{\text{dichlorobenzene}}$$

In the mixture of o-and *p*- dichlorobenzene, the *p*- variety is present in larger amounts. The reaction proceeds through electrophilic substitution.

*Properties and uses :* It is a white, volatile solid (m.p. 52°C). It readily sublimes. It is used as a general insecticide, germicide, soil fumigant, deodorant and moth-repellent.

## FREONS

The chlorofluoro compounds of methane and ethane are collectively called *freons*. Freons are extremely stable, nonreactive, nontoxic, noncorrosive and easily liquefiable gasses.

Dichlorodifluoromethane ($CCl_2F_2$, Freon 12) is one of the most common industrially used freons.

## DICHLORODIFLUOROMETHANE ($CF_2Cl_2$)

Freon is dichlorodifluoromethane. It is a non-toxic gas which can be liquefied very easily. It is non-corrosive, and has a pleasant odour.

*Preparation :* It is made from carbon tetrachloride and dry hydrogen fluoride, or antimony trifluoride ($SbF_3$) using antimony pentachloride as a catalyst. This reaction is called *Swarts reactions*.

$$CCl_4 + 2HF \longrightarrow \underset{\text{freon 12}}{CF_2Cl_2} + 2HCl$$

$$3CCl_4 + 2SbF_3 \xrightarrow{SbCl_5} \underset{\text{freon 12}}{3CCl_2F_2} + 2SbCl_3$$

*Uses :* (i) It is used as a refrigerant in refrigerant and air conditioners.

(ii) It is used as a propellant for aerosols and foams to spray out deodorants, hair sprays, shaving foams etc.

*Harmful effects :* Freon is found to destroy ozone in the upper atmosphere.

The depletion of ozone layer is very harmful to the life on the earth. Efforts are being made to restrict the use of freon, and related compounds.

## BENZENEHEXACHLORIDE (BHC, $C_6H_6CL_6$)

Commercially, benzenehexachloride (BHC) is known as *Lindane* or *Gammaxzne.*

*Preparation :* BHC is obtained by heating benzene with chlorine in presence of sunlight/ultraviolet radiation. It is a mixture of several isomers, but only the gamma isomer exhibits insecticidal/pesticidal activity.

$$\underset{\text{benzene}}{C_6H_6} + 3Cl_2 \xrightarrow[UV]{h\nu} \underset{\text{1,2,3,4,5,6-hexachlorocyclohexane}}{C_6H_6Cl_6}$$

## DICHLORO-DIPHENYL TRICHLOROACETALDEHYDE, (DDT)

The IUPAC name of dichloro-diphenyl trichloroacetaldehyde is 2,2-bis (p-chlorophenyl)- 1, 1, 1 - trichloroacetaldehyde.

*Preparation :* DDT can be prepared by heating a mixture containing chloral (trichloroacetaldehyde) and chlorobenzene (mole ratio 1 > 2) in the presence of conc. $H_2SO_4$.

$$\underset{\text{chlorobenzene}}{C_6H_5\text{-}Cl} + \underset{\text{chloral}}{CCl_3CHO} \xrightarrow{\text{conc. } H_2SO_4} Cl\text{-}C_6H_4\text{-}\underset{\underset{CCl_3}{|}}{CH}\text{-}C_6H_4\text{-}Cl$$

*Properties :*

(i) It is white crystalline sodium (m.p. 382 K).

(ii) It is practically odourless.

(iii) It is insoluble in water, but readily dissolves in benzene, kerosene, alcohol and freon.

(iv) It is remarkably persistent, *i.e.* DDT is non-biodegradable.

*Uses :* It is used as a contact insecticide. DDT is used to kill insects, *e.g.*, mosquitoes. However, its extreme chemical stability makes it potentially dangerous towards humans and animals in its long-term effects.

The use of DDT has been banned in many countries.

## PERFLUOROCARBONS (PFCs)

The fluorocarbons having the general formula $C_nF_{2n+2}$ are called *perfluorocarbons*. Thus perfluorocarbons are the derivatives of alkanes in which all hydrogen atoms of the alkane molecule are replaced by fluorine atoms.

*Preparation :* Perfluorocarbons (PFCs) are obtained by controlled fluorination of vaporized alkanes mixed with nitrogen gas, in the presence of a catalyst.

$$C_7H_{16} + 16F_2 \xrightarrow[\text{CoF}_2\text{ (catalyst)}]{\text{vapour phase, N}_2\text{ 300°C}} \underset{\text{perfluorohep tan e}}{C_7F_{16}} + 16\,HF$$

*Properties :* Perfluorocarbons are,

(i) colourless, odourless, non-inflammable, noncorrosive and unreactive gases, liquids or solids.

(ii) nonpolar and extremely stable, stable even to ultraviolet and other high energy radiations, and therefore, do not deplete ozone layer.

(iii) good electrical insulators

*Use :* Perfluorocarbons are used :

(i) as lubricants, surface coatings, dielectries and heat transfer media in high voltage electrical equipments.

(ii) for vapour phase soldering, leak detection of sealed microchips etc., in electronic industry.

(iii) fin skin care cosmetics, wound healing, retina rearrachment, carbon monoxide poisoning and medical diagnosis.

Chloroform has been used in surgery as an anesthetic, and for this purpose it had to be pure. In the presence of air and light, chloroform slowly forms carbonyl chloride, which is extremely poisonous.

$$CHCl_3 + \frac{1}{2}O_2 \longrightarrow COCl_2 + HCl$$

Chlorine, water and carbon dioxide are also produced. Anesthetic chloroform was therefore kept in well-stopped dark-brown or blue bottles. Ethanol was also added (1 per cent), but its functional is not quite clear. According to some authors, it *retards* the decomposition of the chloroform. This is supported by the fact that infra-red measurements of such mixtures show the absence of the carbonyl frequency.

A dedicate test for chloroform is the isocyanide test. This is carried out by heating chloroform with ethanolic potassium hydroxide and aniline, where by phenyl isocyanide is formed and is readily detected by its nauseating odour.

$$CHCl_3 + 3KOH + C_6H_5NH_2 \longrightarrow C_6H_5NC + 3KCl + 3H_2O$$

Chloroform is widely used in industry as a solvent for fats, waxes, resins, number, etc.

Bromoform may be prepared by methods similar to those used for chloroform, but it is prepared industrially by the electrolysis of an aqueous solution of acetone or ethanol containing sodium carbonate and potassium, bromide (acetone gives a better yield than ethanol). The solution is maintained at about 20°C, and hydrobromic acid is run in to neutralise the sodium hydroxide produced during the electrolysis .Bromine is set free at the anode, and probably reacts in the same way as does chlorine in the preparation of chloroform.

Bromoform is a liquid, b.p. 149.5°C, and smells like chloroform, which it closely resembles chemically.

Iodoform is prepared industrially by the electrolysis of an aqueous solution of ethanol or acetone containing sodium carbonate and potassium iodide (ethanol gives a better yield than acetone). The solution is maintained at 60–70°C, and a current of carbon dioxide is passed through the solution to neutralise the sodium hydroxide formed.

Iodoform crystallises in yellow hexagonal plates, m.p. 119°C. It is insoluble in water, but is readily soluble in ethanol and ether. It is used as an antiseptic, but its antiseptic properties are due to the liberation of free iodine, and not to iodoform itself. Iodoform chemically resembles chloroform and bromoform.

## POLYHALOGEN DERIVATIVES

*Carbon tetrachloride* : $CCl_4$, is prepared industrially in several ways.

(i) By the action of chlorine on carbondisulphide in the presence of aluminium chloride as catalyst.

$$CS_2 + 3Cl_2 \xrightarrow{AlCl_3} CCl_4 + S_2Cl_2$$

The sulphur monochloride is removed by fractional distillation, and the carbon tetrachloride a then shaken with sodium hydroxide, and distilled.

(ii) By the chlorination of methane (see methyl chloride).

(iii) By *chlorinolysis.* This term described the process of chlorinating an organic compound under conditions which rupture the carbon-carbon bond to yield chloro-compounds with fewer carbon atoms than the original compound.

Chlorinolysis may be effected with or without a catalyst, *e.g.*, the hydrocarbon and chlorine are heated at high temperature (300–40°C) and under high pressure (about 70 kg/cm$^2$). The product is usually a mixture, *e.g.*, propane gives both carbon tetrachloride and hexachloroethane.

$$C_3H_8 \xrightarrow{Cl_2} CCl_4 + C_2Cl_6$$

Carbon tetrachloride is a colourless liquid b.p. 77°C, which has a sickly smell. It is insoluble in water but readily soluble in ethanol and ether. Since its vapour is non-inflammable, carbon tetrachloride is widely used as an industrial solvent (for fats, oils, resins, lacquers, etc). It is also used as a fire-extinguished under the name of *Pyrene.*

Carbon tetrachloride is stable at red heat (about 5000°C), but when its vapour comes into consist with water vapour at this temperature, some carbonyl chloride is formed.

$$CCl_4 + H_2O \longrightarrow COCl_2 + 2HCl$$

Hence after using Pyrene to extinguish a fire, the room should be well ventilated.

Carbon tetrachloride is reduced by moist iron filings to chloroform (q.v). The alkaline hydrolysis of carbon tetrachloride gives the same products (formate and carbon monoxide) chloroform, but the rate of reaction is slower (Hing, 1954).

1,1,2,2-Tetrachloroethane (acetylene tetrachloride b.p. 146°C, is manufactured by the reaction between acetylene and chlorine in the presence of ferric chloride at 80°C. It is non-inflammable, and hence is widely used as solvent for oils, fats, paints, varnishes, rubber, etc.

Tetrachloroethane is the striating material for the industrial preparation of trichloroethylene (by 88°C) and tetrachloroethylene (b.p. 121°C), both are used as solvents.

$$CHCl_2CHCl_2 \xrightarrow[\text{cal.}]{300\text{—}500^\circ C} CCl_2 = CHCl \xrightarrow[\text{light}]{Cl_2} CCl_3CHCl_2$$

$$\xrightarrow[\text{cal.}]{300^\circ C} CCl_2 = CCl_2$$

1,2-Dichloroethylene may be prepared by the action of zinc dust on 1.1.2.2-tetrachloroethane.

$$CHCl_2CHCl_2 \xrightarrow{Zn} CHCl = CHCl$$

It is manufactured by the action of iron brings on hexachloroethane. It is a liquid, and exists as two geometrical isomers, *cis,* b.p. 60°C, *trans,* b.p. 48°C, and is used as a rubber solvent, etc.

Hexachloroethane *(perchloreoethane)*. m.p. 187°C, is manufactured by passing a mixture of s-tetrachlroethane and chlorine over aluminium chloride, and by the chlorinolysis of propane (see carbon tetrachloride, above). It is used as a moth repellent.

## FLUORINE DERIVATIVES OF THE ALKANES

Most organic compounds burn or explode when treated with fluorine gas. This great reactivity a partly due to the low dissociation energy of the fluorine molecule (150-6 kJ $mol^{-1}$). Also, one of the termination steps is $R\cdot + F\cdot \rightarrow RF$. This is very strongly exothermic (447.7kJ), and the energy liberal is larger than that required to break a C—C bond (347.3 kJ), thereby resulting in fission of the organic molecule.

Aliphatic fluorine compounds may be obtained in several ways.

(i) Direct fluorination of hydrocarbons may be carried out successfully by diluting the fluorine was nitrogen and carrying out the reaction in a metal tube packed with copper gauze at a temperature 150–350°C.

It is very difficult to control the fluorination, and the product is usually a complex mixture *e.g.*, methane gives $CH_3F$, $CH_2F_2$, $CHF_3$, $CHF_3$ $CF_4$, $C_2F_6$ and $C_3F_8$, ethane gives $CF_4$, $C_2F_6$, $CH_3CHF_4CH_2FCHF_2$, no mono- or s-difluorethane is obtained.

When catalysts other than copper (actually $CuF_2$) are used, *e.g.*, AgF, $CoF_2$, $MnF_2$, prefluorinationn compounds are obtained, *e.g.*, n- heptane gives perfluoroheptane.

$$2AgF + F_2 \longrightarrow 2AgF_2$$

$$C_7H_{16} \xrightarrow{AgF_2} C_7F_{16}$$

It appears that when a catalyst is used, the perfluoro-compound obtained usually has the same number of carbon atoms as the original compound, if no catalyst is used, fluoro-compounds with fewer carbon atom are usually obtained.

The mechanism of direct fluorination is still obscure, but it appears that the first step is the conversion of the catalyst into a higher fluoride, *e.g.*, AgF in to $AgF_2$ into $CoF_3$, etc. Some of these higher fluorides have been isolated, *e.g.*, $AgF_2$.

(ii) Alkenes and alkynes add on hydrogen fluoride under pressure to form fluoro under pressure to form fluoro-derivatives of the alkanes, *e.g.*,

$$C_2H_4 + HF \longrightarrow C_2H_5F \text{ (f.g. g)}$$

$$CH_3C \equiv CH + 2HF \longrightarrow CH_3CH_2CH_3 \text{ (f.g.g)}$$

If the unsaturated compound contains a halogen atom (other than fluorine), this atom may be replaced by fluorine, *e.g.*,

$$RCCl{=\!=}CH_2 + HF \longrightarrow RCFClCH_3$$

$$RCCl{=\!=}CH_2 + 2HF \longrightarrow RCF_2CH_3 + HCl$$

Lead tetrafluoride (from lead dioxide and hydrogen fluoride) is particularly useful for the addition of two fluorine atoms to an alkene containing chlorine, *e.g.*,

$$CCCl_2 = CCl_2 + PbF_4 \xrightarrow{25°C} CFCl_2CFCl_2 + PbF_2.$$

It is also possible to add fluorine directly without a catalyst to highly halogenated alkenes ,*e.g.*,

$$CFCl{=}CFCl + F_2 \longrightarrow CF_2ClCF_2Cl$$

(iii) Fluorine compounds may be prepared indirectly by heating organic halides with inorganic fluorides such as $AsF_3$, AgF, $Hg2F_2$, etc, *e.g.*,

$$C_2H_5Cl + AgF \longrightarrow C_2H_5F + AgCl \text{ (ex)}$$

This method was first used by Swarts (1898), and so is known as the Swarts reactions.

When the organic halide contains two or three halogen atoms attached to the same carbon atom, the best yield of fluoride is obtained when $CoF_3$ is used, but $SbF_3$ gives yield almost as good (and is more accessible). *e.g.*,.

$$3CH_2CCl_2 + 2SbF_3 \longrightarrow 3CH_3CF_2CH_3 + 2SbCl_3 \text{ (v.g.,)}$$

Alternatively, hydrogen fluoride may be used under pressure in the presence of catalyst, *e.g.*,

$$CCl_4 \xrightarrow[(C,\, FeCl_3)]{HF,300^\circ C} CF_2Cl_2 + CFCl_3$$

(iv) Another method of fluorination is the direct electrolytic method. Nickel electrodes are used, and electrochemical fluorination takes place at the anode, the reaction being carried out by the electrolysis of a solution of the organic compound in anhydrous hydrogen fluoride, *e.g.*,

$$CH_3CO_2H \longrightarrow CF_3COF$$

$$C_2H_5OC_2H_5 \longrightarrow C_2H_5OC_2F_5$$

The particular merit of this method is that is usually leaves untouched many types of functional groups. Sulphur tetrafluoride (Smith, 1962) and phenylsulphur trifluoride, $PhSF_3$ (Sheppard, 1962) are capable of replacing *oxygen* atoms by fluorine, *e.g.*,

$$RCO_2H \xrightarrow{SF_4} RCF_3\,;\ R_2CO \longrightarrow R_2CF_2;$$

$$ROH \longrightarrow RF$$

Bromine trifluoride is also useful for replacing oxygen atoms (Stevens, 1961).

The lower n-alkyl fluorides are gases. The first four members are stable, and the higher members tend to decompose spontaneously into alkene and hydrogen fluoride, *e.g.*,

$$CH_3CH_2CH_2CH_2CH_2F \longrightarrow CH_3CH_2CH_2CH{=\!=}CH_2 + HF$$

Secondary and tertiary alkyl fluorides are so unstable that it is impossible to prepare them free from alkene.

1,2-Dichlorides are also usually unstable, *e.g.*, ethylene difluoride decomposes spontaneouly 0°C into hydrogen fluoride and butadiene.

$$2CH_2FCH_2F \longrightarrow 4HF + CH_2{=}CH{-}CH{=}CH_2$$

Alkyl fluorides are readily hydrolysed by strong acids to the corresponding alcohols, alkalis not effect. On the other hand, ethylene difluoride is immediately hydrolysed by water to glycol

$$CH_2FCH_2F + 2H_2O \longrightarrow HOCH_2CH_2OH + 2HF$$

Fluorides with or three fluorine atoms on the same carbon atom are stable to water and acids, *e.g.*, $CHF_3$. $CHF_2CHF_2$.etc.

Alkyl fluorides do not react with sodium, *i.e.*, do not underego the Wurtz reaction, and do not for Grignard reagents. An intersecting compound is trifluomethyl iodide, $CH_3I$. It is converted fluoroform. $CHF_3$, by potassium hydroxide, and it combines directly with many non-metals such As, Sb. S. Se to give,, *e.g.*, with phosphorus, $(CF_3)_3P$, $(CF_3)_2PI$ and $CF_3PI_2$.

Chlorofluoro-derivatives of methane and ethane are used as refrigerants and for air conditions under the name of *Freons.* Which are prepared by the action of hydrogen fluoride on carbon in chloride, chloroform and hexachloroethane.

Tetrafluoroethylene, b.p. – 76, is prepared by the action of antimony trifluoride and hydrogen fluoride on chloroform, and then heating the chlorodifluoroformethane so produced at 800°C.

$$CHCl_3 \xrightarrow[HF]{SbF_3} CHF_2Cl \xrightarrow{\text{Sod C}} C_2F_4 + 2HCl \text{—other products.}$$

When tetrafluoroethylene is polymerised, the plastic. Teflon is produced. Teflon is difficult to work is inert to chemical reagents, even to boiling aqua regia.

Polychlorofluoroethylenes are valuable as oils and greases. Perfluoroheptane is useful in product for the separation of uranium isotopes by gaseous diffusion.

The C—F bond is much stronger than nay other C—X bond and increases in strength as the number of fluorine atoms attached to the carbon atom increases ($CH_3F$. $CF_4$, 485 kJ). It is largely because of this that *perfluoro* compounds are stable to heat and very reason to attack by chemical reagents.

## METAL ALKYLS

Alkyl derivatives of practically all the metals have been prepared; they are named as the alkyl metal, *e.g.*, $(CH_3)_2Hg$, dimethylmercury, $(C_2H_5)_4Pb$, tetraethyl-lead.

Since metal displacement reactions depend on the standard electrode potentials of the metals concerned, it is useful to refer to the electrochemical series *(i.e.*, the arrangement of metals in order of their standard electrode potentials).

Li > K > Na > Mg > Al > Zn > Cd > $Sn^{2+}$ ($\longrightarrow$ Sn) > $Pb^{2+}$ ($\longrightarrow$ Pb) > H > $Sn^{4+}$ ($\longrightarrow$ $Sn^{2+}$) > Bi > Cu > $Ag^+$ ($\longrightarrow$ Ag) > Hg > $Pb^{4+}$ ($\longrightarrow$ $Pb^{2+}$)

### General Methods of Preparation

1. *Direct displacement.* This is the reaction between an organic halide and a metal.

$$RX + 2M \longrightarrow RM + MX$$

The order of reactivity of the alkyl halides is RI > RBr > RCl, and the higher the metal is in the electrochemical series, the more reactive it is. With the less reactive metals, use of an alloy (usually with sodium) is very convenient.

2. *Exchange between organometallic compounds and metal halides.* This is the most widely used method; it takes place when $M^1$ is higher in the elect-o-chemical series than $M^2$. Thus, with $M^1$

$$RM^1 + M^2X \longrightarrow RM^2 + M^1X$$

as magnesium, *i.e.*, using a Grignard reagent, only a few metals (those which occur above Mg in the electrochemical series) fail to react.

3. *Halogen-metal exchange.* This is the reaction between an organic halide and an organo.

$$R^1X + R^2M \longrightarrow R^1M + R^2X$$

metallic compound. This method is particularly useful for preparing orgaholithium compounds.

4. *Metal exchange (transmetalation).* This is the reaction between a metal which is higher in the electrochemical series and, an organo-compound of a metal which is lower in the series.

$$M^1 + RM^2 \longrightarrow RM^1 + M^2$$

**Group IA Metals**

The alkyl derivatives of these metals are generally best prepared by the metal exchange reaction (method 4, above) using a dialkylmercury.

$$2Na + Me_2Hg \longrightarrow 2Me^- Na^+ + Hg$$

Primary alkylsodium compounds are best prepared by reaction between an alkyl-lithium and sodium t-butoxide (Lochmann *et al.,*, 1966).

$$Bu^tONa + RLi \longrightarrow Bu^tOLi + R^- Na^+$$

Alkyl-lithium compounds may be prepared by direct displacement (method 1, above), the chlorides give best yields, *e.g.,*

$$BuCl + 2Li \xrightarrow[\text{hexane}]{N_2} BuLi + LiCl$$

When the lithium compounds cannot be prepared this way, then the halogen-metal exchange reaction is used (method 3, above), n-butyl-lithium is normally used, *e.g.,*

$$BuLi + RX \longrightarrow RLi + BuX$$

When alkyl-lithium compounds are prepared in solution, hydrocarbon solvents are generally used. The use of ether encourages the Wurtz reaction, and, more important, lithium alkyls attack ethers. This reaction probably proceeds via metalation.

$$CH_3CH_2OC_2H_5 + BuLi \longrightarrow BuH + CH_3CHLiOC_2H_5 \longrightarrow$$
$$CH_2 = CH_2 + C_2H_5OLi$$

The sodium and potassium alkyls are insoluble, colourless, highly reactive solids. They are electrovalent compounds, *i.e.,* salts. The lithium compounds are colourless liquids or readily fusible solids, and are covalent, *e.g.*, $CH_3Li$, but, according to the nature of the alkyl group, they may have partial ionic character, *e.g.*, n-butyl-lithium. The large covalent character of the C – Li bond is shown by the fact that when R is an optically active group, RLi racemises slowly in non-polar solvents.

Lithium alkyls are associated in the liquid state and some are associated in the vapour state. These 'polymers' are believed to be electron-deficient molecules.

From the practical point of view, lithium compounds, are the* most satisfactory to use in organic synthesis, since they are very easily prepared and dissolve in organic solvents. Generally speaking the lithium compounds (and those of sodium and potassium) behave like the Grignard reagents, but the lithium compounds are usually more reactive and the yield of product is often better. Because of their sensitivity to oxygen and to water, reactions with lithium alkyls are best carried out in an atmosphere of dry nitrogen. Some reactions of the lithium compounds are as follows. With carbon dioxide a good yield of ketone is obtained.

$$RLi + CO_2 \longrightarrow RC(=O)OLi \xrightarrow{RLi} R_2C(OLi)O \xrightarrow{H_2O} R_2CO$$

A ketone is also the main product when a carboxylic acid is used as starting material.

$$R^1C(=O)OH \xrightarrow{R^2Li} R^1C(=O)OLi \xrightarrow{R^2Li} R^1C(OLi)_2{-}R^2 \xrightarrow{R_2O} R^1COR^2$$

Lithium compounds also react with cyanides to give ketones, and with aldehydes and ketones to give alcohols in better yields than with Grignard reagents.

It has been pointed out on that highly sterically hindered t-alcohols cannot be prepared by the Grignard reaction. On the other hand, many of these alcohols can be prepared by means of lithium alkyls, *e.g.*,

$$(Me_2CH)_2CO + Me_2CHLi \longrightarrow (Me_2CH)_3COLi$$

$$\xrightarrow{H_3O^+} (Me_2CH)_3COH \quad (53\%)$$

Thus, orgaholithium compounds are less susceptible to steric hindrance than Grignard reagents. This is also illustrated by the fact that addition of orgaholithium compounds to α, β-unsaturated carbonyl compounds is usually almost exclusively 1,2. The main difference between the two reagents is that lithium compounds add to an ethylenic bond, whereas Grignard reagents do not, *e.g.,*

$$Bu^tLi + CH_2H_4 \longrightarrow ButCH_2Li$$

Apart from the synthetic uses described above, organolithium compounds are used in the preparation of carbenes, Wittig reagents, arynes, etc. The lithium alkyls, particularly the n-butyl compound, mixed with titanium chloride, are used as catalysts for the polymerisation of propene and isoprene (Ziegler-Natta catalyst).

**Group IIB Metals**

*Organozinc compounds.* Zinc alkyls may be prepared in several ways, *e.g.,*

(i) By heating an alkyl iodide with zinc, or better, a zinc-copper couple. Alkylzinc iodide is formed first, and this then disproportionates on distillation.

$$2RI + 2Zn \longrightarrow 2RZnI \longrightarrow P_2Zn + ZnI_2$$

(ii) By reaction between zinc chloride and an aluminium alkyl,

$$e.g.,\ ZnCI_2 + 2Et_3\ Al \longrightarrow Et_2Zn + 2Et_2A1Cl$$

The zinc dialkyls (collinear molecules) are volatile liquids, spontaneously inflammable in air and readily attacked by water. Their reactions are similar to those of the Grignard reagents, but the zinc compounds are less reactive; they do not react with carbon dioxide and react only slowly with cyanides, ketones and esters (see also below).

*Alkylzinc halides.* These may be intermediates in the preparation of zinc dialkyls (see above) and can be obtained as polymers as follows.

$$2ZnCl_2 + 2Et_2Zn \longrightarrow [EtZnCl]_4$$

lodomethylzinc iodide, $ICH_2ZnI$, is an alkylzinc halide, and reacts with alkenes to form cyclo-propanes.

*The Reformatsky reaction* (1887). The Reformatsky reaction is the reaction between an α-bromoacid ester and a carbonyl compound (aldehyde, ketone) in the presence of zinc to form a β-hydroxy-ester.

$$R^1CHBrCO_2Et + Zn \longrightarrow BrZnCHR^1CC_2Et \xrightarrow{R_2^2CO}$$

$$R_2^2C\,H(OZnBr)CHR^1CO_2Et \xrightarrow{H_3O^+} R_2^2C\,(OH)CHR^1CO_2Et$$

The intermediate organozinc bromides have been isolated (Cure *et* al., 1966).

The use of zinc instead of magnesium in the Reformatsky reaction is based on the fact that the Grignard reagent formed immediately attacks the ester group of a second molecule.

A difficulty with the Reformatsky reaction is the conversion of the /β-hydroxy-ester into the corresponding acid; in general, hydrolysis leads to the α,β-unsaturated ester and acid. However, Read *et al.,* (1969) have shown that by refluxing the adduct (Zn + carbonyl compound + ester) in benzene, and then followed by acid, good yields of hydroxy-acid are obtained.

*Organocadmium compounds.* Cadmium dialkyls are readily prepared by method 2

$$2RMgX + CdCl_2 \longrightarrow R_2Cd + 2MgClX$$

Cadmium dialkyls (collinear molecules) are volatile liquids; they are chemically similar to the zinc compounds, but are less reactive.

Dialkylcadmium compounds are used to prepare ketones from acid chlorides, they are prepared *in situ:*

$$2R^1COCl + R^2_2Cd \longrightarrow 2R^1COR^2 + CdCl_2$$

This ketone synthesis may be used to prepare long-chain monocarboxylic acids as follows.

$$R(CH_2)_xOH \xrightarrow[\text{(ii) Mg} \atop \text{(iii) }CdCl_2]{\text{(i) HBr}} [R(CH_2)_x]_2Cd$$

$$\xrightarrow{ClOC(CH_2)_yCO_2C_2H_5} R(CH_2)_xCO(CH_2)_yCO_2C_2H_5$$

$$\xrightarrow[HCl]{Zn/Hg} R(CH_2)_xCH_2(CH_2)_yCO_2H$$

An interesting point in connection with syntheses involving cadmium dialkyls is that Kollonitsch (1960) has shown that *the pure* compounds react very slowly (if at all) with acid chlorides. In the presence of magnesium (and other ether-soluble halides), reaction is fast.

*Organomercury compounds.* Mercury dialkyls are best prepared by the same method used for the cadmium compounds.

$$2RMgX + HgCl_2 \longrightarrow R_2Hg + 2MgClX$$

When X = Br or I, the main product is an alkylmercury bromide or iodide, *e.g.,*

$$RMgBr + HgCl_2 \longrightarrow RHgBr + MgBrCl$$

The mercury dialkyls (collinear molecules) are poisonous liquids, not spontaneously inflammable in air; they are not decomposed by water but are by dilute acids. With halogen acids, the alkylmercury halide is obtained.

$$R_2Hg + HCl \longrightarrow RHgCl + RH$$

Trihalogenomethyl derivatives of mercury have also been prepared, *e.g.,*

$$PhHgCl + CHBrCl_2 + Bu^tOK \longrightarrow PhHgCBrCl_2 + KCl + Bu^tOH$$

These compounds are a useful source of dihalogenomethylenes, *e.g.,*

$$PhMgCBrCl_2 \longrightarrow PhHgBr + [CCl_2]$$

*Mercuration. i.e.,* the introduction of a mercuri-acid group, has important applications in aromatic chemistry. The commonest two groups are *chloromercuri* (– HgCl) and *acetoxymercuri* (– $HgOOCCH_3$). These are discussed elsewhere, *oxymercuration* has been discussed on.

## Group IIIB Metals

*Organoaluminium compounds. Aluminium alkyls.* A number of trialkyls, are prepared commercially by heating aluminium powder, hydrogen, a 1-alkene, and an aluminium trialkyl under pressure, *e.g.,*

$$Al + 3/2H_2 + 2Et_3Al \longrightarrow 3Et_2 \xrightarrow{3C_2H_4} 3Et_3Al$$

Dialkylaluminium hydrides are also prepared by the above method. On the other hand, mono-, di- and tri-alkylaluminium compounds may be prepared by reaction between aluminium chloride and the appropriate amount of Grignard reagent.

The aluminium n-trialkyls are dimers, but branched-chain compounds are monomers (due to the steric effect), *e.g.,* $Pr_3^i Al$. The dimers are electron-deficient molecules and have alkyl-bridges *(cf.* alkyl-boranes), *e.g.,* trimethylaluminium dimer is (I) and the bridging carbon atoms are five-co-ordinate. These dimers dissociate to a small extent in solution; in the gaseous phase dissociation is much greater, *e.g.,* $Me_6Al_2$ at 215°C and under low pressure is 100 per cent monomer (Haaland *et al.,* 1969).

The N.M.FiL spectroscopy of $Me_6Al_2$ (at room temperature) has shown that all the methyl groups are equivalent. This has been explained on the basis that the terminal and bridging methyl groups in (I) are exchanging rapidly (Muller *et al.,* 1960). Molecules such as this, which exhibit site exchange, are called *fluxional molecules.*

$$\begin{array}{c} Me \quad\; Me \quad\; Me \\ \diagdown \; \diagup \; \diagdown \; \diagup \\ Al \qquad Al \\ \diagup \; \diagdown \; \diagup \; \diagdown \\ Me \quad\; Me \quad\; Me \end{array}$$

Dialkylaluminium hydrides (which exist as trimers) are formed when trialkyls are heated, alkene is eliminated, *e.g.,*

$$Bu^t_3Al \xrightarrow{150°C} Bu^t_2AlH + Me_2C = CH_2$$

This reaction is *reversible,* and because of the ease with which isobutene is lost, tri-t-butylaluminium is used to prepare other aluminium alkyls by exchange with alkenes.

$$Bu^t_3Al + 3RCH = CH_2 \longrightarrow (RCH_2CH_2)_3Al + 3Me_2C = CH_2$$

Since aluminium trialkyls are decomposed by water, the overall reaction is reduction of an alkene.

$$(RCH_2CH_2)_3Al \xrightarrow{H_2O} 3RCH_2CH_3 + Al(OH)_3$$

Di-isobutylaluminium hydride is a very useful reducing reagent for various functional groups, *e.g.,* aldehydes and ketones give alcohols; esters give alcohols, but at low temperatures, aldehydes are obtained; cyanides give aldehydes or amines. The mechanism is similar to that with lithium aluminium hydride, *i.e.,* involves hydride ion transfer.

$$R_2CO + Bu^t_2AlH \longrightarrow R_2CHOAl\,Bu^i_2 \xrightarrow{H_2O} R_2CHOH$$

There is, however, an added complication; if only the $\diagdown\!\!\!\diagup Al—H$ bond were involved, then one molecule of the hydride will supply one hydride ion. In practice, particularly with aldehydes, *both* isobutyl groups are also involved, *three* molecules of aldehyde are reduced by *one* molecule of $Bu^i_2AlH$, and two molecules of isobutene are formed. This could be explained by direct β-hydride ion transfer from the isobutyl groups.

Many other aluminium compounds are used as reducing reagents, particularly LAH.

Aluminium trialkyls, like the alkylboranes, form co-ordination compounds, *e.g.,*

$$Me_6Al_2 + 2Me_3N \longrightarrow 2Me_3Al - NMe_3$$

Triethylaluminium is the starting point for the commercial preparation strong straight-chain alcohols. Long chains are produced by insertion of ethylene when heated with this under pressure at about 100°C. The length of each individual chain, is usually different, for simplicity, the reaction is written as follows.

$$AlEt_3 \xrightarrow{3nC_2H_4} Al\ [(C_2H_4)_nEt]_3 \xrightarrow{O_2} Al[O(C_2H_4)_nEt]_3$$

$$\xrightarrow{HO_2} 3Et(C_2H_4)_nOH + Al(OH)_3$$

The most important use of aluminium trialkyls is as a catalyst for the low temperature polymerisation of alkenes.

**Group IVB Metals**

*Organotin compounds. Tetra-alkylstannanes* are conveniently prepared as follows (methods 1, 2, and 3,).

$$4RX + 4Na/Sn \longrightarrow R_4Sn + 3Sn + 4NaX$$

$$SnCl_4 + 4RMgX \longrightarrow R_4Sn + 4MgClX$$

$$3SnCl_4 + 4R_3Al + 4NaCl \longrightarrow 3R_4Sn + 4NaAlCl_4$$

The tetra-alkylstannanes are stable liquids, they are unaffected by air or water. They react with halogens to form alkylt in halides.

$$R_3Sn \xrightarrow{I_2} R_3SnI + RI$$

*Tetravinyltin,* b.p. 160°C, may be prepared as follows.

$$CH_2 = CHBr + Mg \xrightarrow{THF} CH_2 = CHMgBr \xrightarrow{SnCl_4}$$

$$(CH_2 = CH)_4Sn$$

This compound is of particular interest in that the vinyl group is very labile, and hence is used to prepare vinyl derivatives of other metals, *e.g.*,

$$(CH_2 = CH)_4Sn + 4PhLi \longrightarrow 4CH_2 = CHLi + Ph_4Sn$$

*Organotin halides* may be prepared, *e.g.*,

$$R_4Sn + SnCl_4 \xrightarrow{20°C} R_3SnCl + RSnCl_3$$

$$\xrightarrow{180°C} 2R_2SnCl_2$$

They are reduced by LAH to the corresponding tin hydride, *e.g.*,

$$R_2SnCl \xrightarrow{LiAlH_4} R_2SnH_2$$

These hydrides selectively reduce the carbonyl group in α,β-unsaturated aldehydes and ketones.

Alkenes (and alkynes) undergo *hydrostannation* when heated with organotin hydrides, *e.g.*,

$$R^1_3SnH \longrightarrow H.\ R^1_3Sn\ . \xrightarrow{R^2CH=CH_2}$$

$$R^1_3SnCH_2\dot{C}HR^2 \xrightarrow{R^1_3SnH} R^1_3SnCH_2CH_2R^2 + R^1_3Sn\text{ , etc.,}$$

Many organotin compounds are used as polymerisation catalyst, bacteriocides, etc.,

*Organolead compounds. Lead tetra-alkyls* may be prepared by methods used for the tin compounds (see above). Tetraethyl- and tetraethyl-lead, which are used as anti-knock additives to petrol, are manufactured by the reaction between the alkyl chloride and a lead-sodium alloy.

$$4RtCl + 4Na.Pb \longrightarrow Et_4Pb + 4NaCl$$

Another industrial method is the electrolysis of an ethereal solution of methyl-or ethyl-magnesium chloride using a lead anode.

$$4RMgCl + Pb \longrightarrow R4Pb + 2Mg + 2MgCl_2$$

A point of interest is that lead (II) halides generaly give lead tetra-alkyls, *e.g.*,

$$2PbCl_2 + 4RMgX \longrightarrow R_4Pb + Pb + 2MgX_2 + 2MgCl_2$$

$$2PbCl_2 + 4RLi \longrightarrow R_4Pb + Pb + 4LiCl$$

The lead tetra-alkyls are stable liquids, they are very toxic, and are unaffected by air or water.

*Alkyl-lead hydrides* and halides are known, trialkyl-lead hydrides are sufficiently reactive to add to alkenes (or alkynes) at 0°C in the absence of a catalyst.

## Group VB Metals

*Organoarsenic compounds.* In many ways, the chemistry of the organic compounds of arsenic (a *metalloid)* is similar to that of the phosphorus compounds; the methods of nomenclature of the two series are also similar (see p. 408). *Trialkylarisines* may be prepared as follows.

$$3RMgX + AsBr_3 \longrightarrow R_3As + 3MgBrX$$

*Alkylarsenic hydrides* and *halides* can be prepared in various ways. One of the starting compounds is

(i) an alkylarsonic acid, or

(ii) a dialkylarsinic acid.

(i) $$RX + Na_3AsO_3 \longrightarrow NaX + RAsO(ONa)_2$$

$$\xrightarrow{H_3O^+} RAsO(OH)_2 \xrightarrow{Zn/HCl} RAsH_2$$

(ii) $$RAsO(OH)_2 \xrightarrow[\text{(ii) } H_3O^+]{\text{(i) } RX - NaOH}$$

$$R_2AsO(OH) \xrightarrow[HCl]{Zn} R_2AsH$$

The arsines are oxidised by air; $RAsH_2$ gives an alkylarsonic acid, $R_2AsH$ a dialkylarsinic acid, and $R_3As$ an arsine oxide. Arsonium salts may be prepared by reaction between an alkyl halide and a trialkylarsine or arsenic.

$$2As + 4MeI \longrightarrow Me_4As^+I^- + AsI_3$$

Arsonium salts readıly form *alkylidenearsoranes (arsenic ylids).*

$$Ph_3AsMe^+Br^- \xrightarrow{PhLi} Ph_3As = CH_2$$

With moist silver oxide, they give arsonium hydroxides, $R_4As^+OH^-$ which are strong bases.

Pentaphenylarsenic has been prepared as a crystalline solid, and is isomorphous with the corresponding phosphorus compound. Pentavalent arsenic compounds of the type $R_3AsX_2$ are also known (prepared by the addition of $X_2$ to $R_3As$). Lewisite is $ClCH = CH — AsCl_2$

*Organoantimony compounds. Trialkylstibines* have been prepared, *e.g.*,

$$3MeMgI + SbCl_3 \longrightarrow Me_3Sb + 3MgClI$$

$$SbF_3 + 3Et_3Al \longrightarrow Et_3Sb + 3Et_2AlF$$

They are stable liquids, but the alkylstibine hydrides are unstable volatile liquids. Trimethylstibine combines with methyl iodide to form tetramethylstibonium iodide, $Me_4Sb^+I^-$, which, with moist silver oxide, gives the corresponding hydroxide, a strong base. Pentaphenylantimony has been prepared, and its structure has been shown to be a square pyramid, with the antimony atom lying inside the pyramid (cf. the analogous phosphorus compound). Me,Sb is also known, and pentavalent compounds of the type $R_3SbX_2$ have been prepared.

*Organobismuth compounds. Trialkybismuthines* are prepared as follows.

$$3RMgX + BiCl_3 \longrightarrow R_3Bi + 3MgClX$$

These compounds do *not* add a molecule of alkyl halide to form bismuthonium salts, of which only a few are known, *e.g.*, bismuthonium tetraphenylborate.

$$Ph_5Bi + Ph_3B \longrightarrow Ph_4Bi + Ph_4B^-$$

Pentaphenylbismuth is an unstable solid, other pentacovalent compounds of the type $Ph_3BiX_2$ (X = C1, Br) are also known.

## Alkyl free radicals

A characteristic property of many metal alkyls is the generation of free radicals when they are subjected to pyrolysis or photolysis, *e.g.*,

$$Et_4Pb \xrightarrow{\text{Heat}} 4Et. + Pb$$

$$Me_2Hg \xrightarrow{hv} 2Me. + Hg$$

Because of the presence of an unpaired. electron, free radicals may be detected by means of paramagnetic susceptibility measurements and electron spin resonance spectroscopy. Chemical methods have also been used to detect free radicals.

Alkyl free radicals are formed as reactive intermediates in many types of reactions, land a common method of generating them in solution is to use acyl peroxides, but better : reagents are azonitriles.

The configuration of alkyl free radicals is uncertain. The methyl free radical has been shown, from ultraviolet and Raman spectra studies (Herzberg *et al.,* 1956), to have high symmetry. Thus, this radical is planar. On the other hand, infra-red and ESR studies indicate that $CF_3$ is tetrahedral.

In general, reactions involving free radicals which are asymmetric (at the free-radical centre) result in racemised product. However, in certain trapping experiments, optically active products have been obtained (*inter alia,* Jacobus *et al,,* 1969). Thus, there is chemical evidence to show that *some* free radicals are transiently non-planar.

## EXERCISES

1. Comment on the following equations.

$$HOCH_2CH_2COCH_2CH_2CO_2Et \xrightarrow{1RMgX} ? \xrightarrow{1RMgX} ?$$
$$\xrightarrow{2RMgX}$$

2. Complete the following equations and comment where necessary

(i) $n\text{-}C_4H_9MgBr + \xrightarrow[(ii)\,H^+]{(i)\,Et_2O} n\text{-}C_7H_{15}OH$

(ii) $3EtMgBr + (EtO)_2CO \xrightarrow[(ii)\,aq.NH_4Cl]{(i)\,Et_2O} ?$

(iii) $PhMgBr + MeOCH_2CN \xrightarrow[(ii)\,H^+]{(i)\,Et_2O} ?$

(iv) $C_6H_{11}\ MgBr + BrCH_2CBr = CH_2 \xrightarrow{Et_2O} ?$

(v) $CH_2 = CO + EtMgBr \xrightarrow[(ii)\,H^+]{(i)\,Et_2O} ?$

(vi) $MeCH\overset{O}{—}CH_2 + Me_2CHMgBr \xrightarrow[(ii)\,H^+]{(i)\,Et_2O} ?$

(vii) $EtCH = CHCOEt \xrightarrow[(ii)\,H^+]{(i)\,EtMgBr + CuBr_2O} ?$

(viii) $PrMgBr + MeC(OMe)_3 \xrightarrow[(ii)\,H^+]{(i)\,C_6H_6} ?$

(ix) $ClCH_2CO_2Et + 2RMgX \xrightarrow[(ii)\,H^+]{(i)\,Et_2O} ?$

(x) $MeCH\overset{O}{—}CH_2 + EtMgBr \xrightarrow[(ii)\,H^+]{(i)\,Et_2O} ?$

(xi) $(CH_2)_5Br_2 \xrightarrow[Et_2O]{2Mg} ? \xrightarrow[(ii)\,H^+]{(i)\,2MeONH_2} ?$

(xii) $MeCOCH_2CO_2Et + Me_2Cd \xrightarrow[(ii)\,H^+]{(i)\,Et_2O} ?$

(xiii) $Et_2Cd + CC1_3CHO \xrightarrow[(ii)\,H^+]{(i)\,Et_2O} ?$

3. Starting with EtOH and n-PrOH, prepare pent-2-ene via a Grignard reagent. Is there more than one way ? If so, which is the best and why ?

4. Convert n-$C_5H_{11}CO_2H$ into n-$C_4H_9CO_2H$ by *two* methods.

5. $R^1CH = CHCOR^2 + R^3MgX \longrightarrow$ 1,2-4-1,4-products

   How would the. ratio of the products change if $R^2$ = Et, isoPr, t-Bu ? Explain.

6. $EtCH_2CHBrMe \xrightarrow[(-HBr)]{EtOK} \underset{A}{I-ene} + \underset{B}{2-ene} + \underset{C}{2-ene}$

   What are A, B and C and which one predominates? (Assume the mechanism is E2).

7. Discuss the reaction : $ClCH_2CHCl_2 \xrightarrow{OH^-} CH_2 = CHI_2$

8. The hydrolysis of n-BuCl in aqueous ethanol is accelerated in the presence of NaI. Explain and draw the energy profiles for the uncatalysed and catalysed reactions.

9. What is the simplest alkane that is optically active?

10. The NMR spectrum of the compound $C_3H_5Cl_2$ showed two signals, one a doublet and the other a quadrat. What is the structure of the compound?

11. State the number of signals and their multiplicity for the NMR spectra of.

    (i) $CH_2F_2$.

    (ii) $Me_2CCH_2Cl$.

    (iii) $Me_3CCl$.

    (iv) $ClCH_2CHClCCl_3$.

    (v) $ClCH_2CCl_2.CH_3$.

12. The table gives some of the peaks in the mass spectra of two isomeric s straight chain alkyl halides. A and B. One is a primary halide. What is this one? Give your reasons.

| *m/e* | 41 | 43 | 57 | 85 | 135 | 164 | 166 |
|---|---|---|---|---|---|---|---|
| (A%) | 38 | 100 | 22 | 49 | 50 | 2.3 | 2.3 |
| B(%) | 38 | 100 | 12 | 57 | 1 | 1 | 1n |

13. Convert $CHD_2I$ into $CHD_2CH_2OH$. What alkenes would you expect to get by the acid-catalysed dehydration of the alcohol? Explain.

14. Carry out the following conversions via Grignard reagents, and use any other compounds you like.

(i) $MeCO_2H \longrightarrow MeCOEt$, (ii) $EtBr \longrightarrow EtCOEt$,

(iii) $Me_2CHOH \longrightarrow Me_2CHNH_2$,

(iv) $MeCOEt \longrightarrow EtCH(Me)CHBrMe$.

15. Complete the following equations.

(i) $4PhLi + (CH_2 = CH)_4Sn \longrightarrow ?$

(ii) $2EtNa + Hg \longrightarrow ?$

(iii) $BuLi + MeONH_2 \longrightarrow ? \xrightarrow{H_2O} ?$

(iv) $Me_2CHLi + C_2H_4 \longrightarrow ? \xrightarrow[(ii)\,H_2O]{(i)\,CO_2} ?$

(v) $Me_3CCl + Me_2Zn \longrightarrow ?$

(vi) $2EtI + Na_2/Hg \longrightarrow ?$

(vii) $Et_2Hg + 2K \longrightarrow ?$

(viii) $4RMgCl + SnCl_4 \longrightarrow ?$

(ix) $Me_2Hg + Zn \longrightarrow ?$

(x) $2CH_2(CN)COCl. + Et_2Cd \longrightarrow ?$

(xi) $3Ph_2Hg + 2Al \xrightarrow{heat} ?$

(xii) $CH_2 = CHCOMe \xrightarrow{Me_2SnH_2} ?$

(xiii) $R_2CO + BrCCH_2CH = CHCO_2Et \xrightarrow[(ii)\,H^+]{(i)\,Zn}$

(xiv) $EtMgX \xrightarrow{PbCl_2} ?$

(xv) $(PhCH_2)_4Sn \xrightarrow{heat} ?$

(xvi) $2MeI + Zn/Cu \longrightarrow ?$

(xvii) $EtLi + Br_2 \longrightarrow ?$

16. Distinguish between ethylene dichloride and ethylidene dichloride :

(i) chemically

(ii) by one physical method.

17. Draw all possible stereoisomers for.

(i) $MeCHBrCH=CH_2$.

(ii) MeCHOHCH=CHMe.

(iii) $MeCHCCO_2H$,

(iv) $BrCH_2CH_2CO$.

(v) $Me_2C(OH)CO_2H$.

18. Explain the following reactions.

(i) n-BuBr + KCN $\xrightarrow{EtOC—H_2O}$ n-BuCN

(ii) t-BuBr + KCN $\xrightarrow{EtOC—H_2O}$ $Me_2C=CH_2$

19. Explain the following reactions.

(i) EtOH + HI $\longrightarrow$ EtI (ii) EtOH + HCN $\longrightarrow$ no reaction

20. Calculate $\Delta H^\theta$ for the reactions.

$$C_2H_6 + Br_2 \longrightarrow C_2H_5Br + HBr$$

Is the reaction thermodynamically favourable? If so, will it proceed in the dark at room temperature. If as why not?

21. Treatment of $Me_3CCH=CH_2$ and $Me_3CCHOHMe$ with concentrated hydrochloric acid gives the term two isomeric alkyl chlorides. What are these two products? Explain.

22. Complete the following equations.

(i) $ClCH_2CH_2CH_2Br + KCN \xrightarrow[heat]{EtOH—H_2O}$ ?

(ii) PbCHO $\xrightarrow{SF_4}$ ?

(iii) $BrCH_2CH = CHCO_2Me \xrightarrow{AgF}$ ?

(iv) isoPrBr $\longrightarrow$ PrH

(v) $Me_2C = CH_2 + HCl \xrightarrow{petroxide}$

23. What are A and B ? Formulate the reaction. Could you use MeMgI instead of $Me_2Zn$ ? Explain.

24. Suggest a mechanism for the reaction.

$$R_2CO + Et_3Al \longrightarrow R_2CHOAlEt_2 + C_2H_4$$

25. Using any compounds you like, convert.

   (i) $Me_2CO \longrightarrow Me_2C(OH)CH_2CO_2H$;

   (ii) $EtOH \longrightarrow Me_2CHCHO$;

   (iii) $Me_3COH\ Me_3CCH_2CH = CH_2$;

   (iv) $MeCHO \longrightarrow EtMeCHCO_2H$;

   (v) $HOCH_2(CH_2)_2CH_2OH\ Me_2C(OH)(CH_2)_4C(OH)Me_2$.

# 4

# DYES AND PHOTOCHEMISTRY

## INTRODUCTION

The dyeing of cloth and the fermenation of sugar to produce alcohol are the two specialized areas of organic chemistry that date back to antiquity.

Fabrics dyed with indigo and maddar have been found in the tombs of predynastic Egypt. A few matters extracted from insects and tropical woods formed the only sources of dyes until the middle of the last centuary.

The first synthetic dye was made in 1856 by William E. Perkin, which he named *mauveine.* He obtained this dye by oxidation of impure aniline.

A few years later, the structures of natural dyes, indigo and alizarin, were determined, and these compounds were prepared by synthesis.

However, the major development in chemistry of synthetic dyes was the discovery of diazotization and azo coupling. Since 1856, tens of thousands of dyes have been synthesized. Well over a thousand are commercially available now.

*Colour.* When white-light (750-400 nm) falls on a substance, the light may be totally reflected or totally absorbed. In the former case, the substance appears white; in the latter, black.

If a certain proportion of the light is absorbed and the rest reflected, the substance has the colour of the *reflected* light.

If only a *single* band is absorbed, the substance has the *complementary* colour (of the absorbed band).

**Table 4.1**

| nm (mm) | Colour absorbed | Visible (complementary) colour |
|---|---|---|
| 400-435 | violet | yellow-green |
| 435-480 | blue | yellow |
| 480-490 | green-blue | orange |
| 490-500 | blue-green | red |
| 500-560 | green | purple |
| 560-580 | yellow-green | violet |
| 580-595 | yellow | blue |
| 595-605 | orange | green-blue |
| 605-750 | red | blue-green |

If a substance absorbs all visible light except one band, which it reflects, the substance will have the colour of that reflected band. Thus, a substance can appear blue because it absorbs the yellow portion of the spectrum only; or because it absorbs *all the visible spectrum except blue.* The shades, however, will be different. Apparently no dye gives a pure shade, *i.e.,* does not reflect only one band of wave-lengths; *e.g.,* malachite green reflects green light, but also, to a small extent, red, blue and violet.

Many substances which appear to be colourless nevertheless have absorption spectra, but in these cases, absorption takes place in the infra-red or ultra-violet, and not in the region of the visible spectrum.

## RELATION BETWEEN COLOUR AND CONSTITUTION

The early theories of colour gave rise to a terminology which, with some modifications, is used in ultra-violet and visible spectroscopy. Witt (1876) was the first to show that colour usually appeared in a compound when it contained a group with multiple bonds. These groups were called *chromophores, e.g.,* $NO_2$, NO, N = N, quinonoid structure, etc. Witt also observed that certain groups which, although not chromophores, deepen colour when introduced into a coloured molecule. These groups were called *auxochromes, e.g.,* OH, $NH_2$, Cl, $CO_2H$. Other terms introduced later in connection with dyes are *bathochromtic* and *hypsochromic groups.* These are, respectively, groups that bring about deepening and lightening of colour. Deepening of colour in *dye chemistry* is usually taken to mean the following changes: yellow → orange → red → purple → blue → green → black. This is effectively the order of the visible *complementary* colour of the colour *absorbed*. Hence, we may say that the colour of one compound is deeper than that of another if the wavelength of maximum absorption of the former is longer than that of the latter. Because visible colour is the complementary colour of the absorbed band, bathochromic groups are said to have a red shift and hypsochromic groups a blue shift.

There are two approaches to the modern theories of colour, the *valence bond* and the *molecular orbital approach*. Both have some aspects in common, and these will be discussed first. Light consists of electromagnetic waves which are composed of a magnetic and an electric component, both oscillating in planes perpendicular to each other and perpendicular to the direction of propagation. It is only the electric component which is usually involved in inter-action with matter,, and this interaction is conveniently described in terms of light behaving as a train of discrete energy packets called *photons* or *quanta*. When light is absorbed by a molecule, the molecule undergoes transition from a state of lower energy ($E_0$) to a state of higher energy $E_1$), and the *change* in energy, ΔE, is given by: $\Delta E = (E_1 - E_0)$. In a monatomic molecule, the energy absorbed can only be used to raise the *electronic* energy level from the lower *(ground state)* to a higher (an *excited state)* level. Since electrons occupy *definite* energy levels, changes from one level to another occur in discrete packets which are therefore not continuous, *i.e.*, the energy levels are *quantised*. Hence, in a monatomic molecule, absorption can occur only if the photon has an energy which is equal to the energy difference $E_1 - E_0$. Since the energy of a photon, E, is given by $E = hv$, it follows that

$$\Delta E = E_1 - E_0 = hv = hc/\lambda$$

where *h* is Planck's constant and c is the velocity of light. By insertion of the appropriate values, the value of A£ is given by

$$\Delta E = 11.97 \times (10^4/\lambda) \text{ kJ mor}^{-1} \text{ } (2.86 \times (10^4/\lambda) \text{ kcal mol}^{-1})$$

where λ is in nm (mμ). The value of ΔE for any particular wavelength is known as an *einstein* (for that given wavelength).

In polyatomic molecules, the total energy state is the sum of the electronic, vibrational, and rotational components. Since electronic transitions are associated with large amounts of energy relative to vibrational and rotational transitions (which are also quantised), $\Delta E$ is large for the former, *i.e.,* the frequency of the photon required to bring about electronic transitions is high (and consequently the wavelength is short). Thus, electronic transitions occur in the visible and ultra-violet parts of the spectrum, whereas vibrational and rotational transitions occur respectively in the near and far infra-red.

If, after absorption of light, only electronic transitions occurred, each transition would be associated with some particular line. However, since an electronic level is associated with a large number of vibrational and rotational levels, a transition from one electronic level to another may be accompanied by vibrational and rotational transitions. This results in the absorption of a range of photons, and consequently the spectrum consists of a family of lines which are usually so closely - spaced that they merge (for low resolution) into a broad band.

Let us now consider this problem in more detail. Figure 4.1 *(b)* represents (diagrammatically) two electronic energy states of a diatomic molecule, $E_0$ being the ground state and $E_1$ *the first* excited state. Excitation from the ground state (bonding M.O.) to the first excited state (anti-bonding M.O.) can occur in two possible ways, (i) with retention of electron spin, *i.e.,* the electron spins are still anti-parallel; this is represented as an $S_0 \rightarrow S_1$ transition, (ii) with spin inversion, *i.e.,* the electron spins are now parallel; this is represented as an $S_0 \rightarrow T_1$ transition.

In the ground state of a diatomic molecule, the M.O. is a $\sigma$-bonding type, and when excited, the M.O. is the $\sigma^*$-anti-bonding type. Their energy levels are represented diagrammatically in Fig. 4.1(a). In the ground state, normal covalent compounds have all their electrons paired, *i.e.,* the molecule is in the *singlet state,* $S_0$. When the molecule is excited, the promoted electron can still be paired to give the first excited *singlet* state, $S_1$ or can be unpaired (diradical) to give the first excited *triplet* state, $T_1$. These two states are said to have different *multiplicities,* and since the energy levels for $S_1$ and $T_1$ are both obtained from the relationship $\Delta E = E_1 - E_0$, they have the *same* levels independent of spin assignment. However, according to *Hund's rule of multiplicity, a system of highest multiplicity has the lowest energy.* Hence, the energy of a triplet state (multiplicity of 3) is lower than that of its corresponding singlet state (multiplicity of 1); *i.e.,* $T_1 < S_1$, $T_2 <$ S2, etc. By taking into account the different spin relationships, it is possible to develop the individual orbital states shown in Fig. 4.1(b) but, in practice, the *Jablonski diagram,* Fig. 4.1(c), which is a simplified version of Fig. 4.1(b), is generally used.

For polyatomic molecules, the orbital level diagram Fig. 4.1(a) is basically the same, but naturally contains far more levels.

Another problem connected with electronic transitions is that theoretical considerations have shown that some of these transitions are *allowed, i.e.,* have high probability, or are *forbidden, i.e.,* have low probability. The essential condition for an electronic transition to take place is that the *transition dipole* (which is related to the actual dipole moment of the molecule) is not zero. Thus, there are *selection rules,* which are the conditions for which the transition dipole is not zero. One transition that would lead to a zero transition dipole is that in which there is electron spin inversion on excitation. This is the basis of the *selection rule of multiplicity:: no change in spin multiplicity occurs during a transition.* Thus, singlet-singlet and triplet-triplet transitions are allowed, but singlet-triplet and triplet-singlet transitions are forbidden. This rule, however, is based on the assumption that a molecule has a fixed shape which remains unchanged on excitation. In practice, there are always some vibrations in a polyatomic molecule which periodically change the shape of that molecule, *e.g.,* bending vibrations. This enables a forbidden transition to become allowed (see also below).

It should also be noted that the molar absorptivity ($\varepsilon_{max}$) is a measure of the probability of a transition taking place, the higher the probability of the transition, the larger is the $\varepsilon_{max}$.

We now return to the problem of what happens to the molecule when it is in an excited state. In general, a molecule in the ground state will also be in the lowest vibrational state (rotational states, which lie between vibrational states, have been omitted in Figs. 4.1 *(b)* and *(c)).* Now suppose light is absorbed. If the energy of absorption is above that required for dissociation (> $E_1$) for a diatomic molecule, the molecule splits into free atoms (path A). If the absorption of energy raises the molecule to the first excited state, the transition is from ground state to the first excited singlet state, $S_1$ (path B), since $S_0 \rightarrow T_1$ is forbidden. It should also be noted that when the electron is promoted (from the ground state molecule), it moves from a given vibrational level (usually the lowest one) to some (allowable) vibrational level in $S_1$. When excitation occurs, it does so extremely rapidly ($10^{-13}$S) and consequently the nuclei do not have time to change their positions *(Franck-Condon principle), i.e.,* although the vibrational level has also been raised, the intel uclear distance remains unchanged *immediately* after the excitation. Thus, the excited molecule is formed in a 'compressed' condition and *then* begins to vibrate normally at this higher vibrational level.

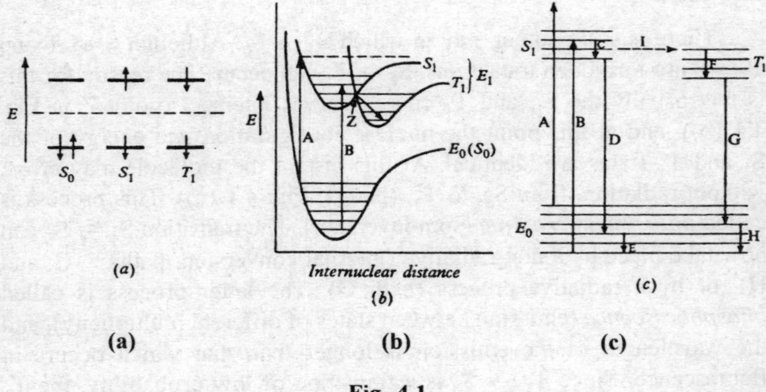


(a) (b) (c)

**Fig. 4.1**

In the excited state, the molecule can lose its excess of energy (*i.e.,* $S_1 \rightarrow E_0$) in a number of ways. If the excitation raises the promoted electron to a vibrational level in $S_1$ which is *above* the lowest vibrational level in $S_1$ the electron rapidly drops to this lowest level, the excess of vibrational energy being lost by collisions with other molecules (path C). Loss of energy by this process of *vibrational relaxation* occurs so rapidly ($10^{-12}$–$10^{-11}$ S) that the vibrationally relaxed excited state is reached before any other loss process can operate. This state can now lose energy by further collisions to reach the original ground state ($S_0$; paths D and E). When the molecule drops from $S_1$, to the ground state, the vibrational level in the latter is higher (in this level) than that in the former. In this series of processes, *all* the absorbed energy is lost through collisions, and the molecule is said to return from its excited state to the ground state by a *non-radiative process,* the 'lost' energy appearing as heat. This type of return is called *internal conversion.* On the other hand, the vibrationally relaxed excited state can lose energy by emission of radiation *before* collisions occur (path D). Loss of energy by this *radiative process,* however, results in the wavelength of the emitted light being longer than that of the absorbed light (ΔE for the former is smaller than that of the latter). This type of transition is known as *fluorescence* (emission between two states of the *same* multiplicity). When fluorescence occurs, it does so about $10^{-8}$ s after the light source is shut off (this is the life-time of a singlet state). Furthermore, had the original excitation been to the $S_2$ state, this decays extremely rapidly to the $S_1$ state, which then returns to $S_0$ by either of the two processes described.

There is still another way in which $S_1 \rightarrow S_0$. Although $S \rightarrow T$ and $T \rightarrow S$ are forbidden transitions, $S_1 \rightarrow T_1$ can occur. The reason for this is that usually the $S_1$, and $T_1$ energy curves intersect (point Z in Fig. 4.1 (b)), and at this point the nuclear configuration and energy of the $S_1$ and $T_1$ states are identical. At this instant the molecule may cross, without radiation, from $S_1$, to $T_1$, (path I; Fig. 4.1 (c)). This process is called *intersystem crossing* (spin inversion). The transition $S_1 \rightarrow S_0$ can now take place by a non-radiative (internal conversion; paths F, G, and H), or by a radiative process (path G). The latter process is called *phosphorescence* (emission between states of different multiplicity), and the wavelength of the emission is longer than that which occurs in fluorescence. Since $T_1 \rightarrow S_0$ is a transition of low probability, the $T_1$ state has a life-time of $10^{-5} - 10^{-3}$ S, and so phosphorescence lasts longer than fluorescence.

*Chemiluminescence* is the emission of light from chemical reactions at ordinary temperatures in the absence of flame. This light is emitted from an excited molecule which has been produced in the chemical reaction, and it is believed that most chemiluminescent reactions involve the singlet excited state, which decays to the ground state by fluorescing. The energy required for the excitation is obtained as a result of an *exothermic* reaction. A common type of reaction that exhibits chemiluminescence is the oxidation by oxygen of many organic compounds in solution.

## Valence Bond Approach to Colour

An extremely important difference between V.B. and M.O. theories is that in the former, electrons are dealt with in *pairs,* whereas in the latter they can be dealt with *singly.*

According to the V.B. theory, the electron pairs of a molecule in the ground state are in a state of oscillation, and when placed in the path of a beam of light, the amplitude of the oscillation is increased when a photon of the appropriate energy is absorbed. The molecule is now in an excited state, and the wavelength of the photon absorbed depends on the energy difference between the excited and ground states, the smaller the difference, the longer being the wavelength.

Let us first consider the case of ethylene. This, from the point of view of the V.B. method, may be regarded as a resonance hybrid of (I) and (II).

$$\underset{(I)}{CH_2=CH_2} \longleftrightarrow \underset{(II)}{\overset{+}{C}H_2-\overset{-}{C}H_2}$$

Also, to a first approximation, the ground state is represented predominantly by (I) and the excited state by (II). The energy difference between these two is very large and so the energy of the photon required to excite ethylene is very high, *i.e.,* its wavelength is very short. This is the general situation for *short* molecules, and it has been shown that: (i) Resonance among *charged* structures *lowers* the energies of both ground and excited states; (ii) Charged structures contribute more to the excited state than to the ground state; (iii) The larger the number of electrons involved in resonance, the smaller is the energy difference between the ground and excited states. Hence, the more extended the conjugation in a molecule and the greater the contribution of charged structures, the longer is the wavelength of the photon required to excite the molecule, *e.g.,*

| | $CH_2=CH_2$ | $CH_2=CH-CH=CH_2$ | $CH_2=CH-CH=CH-CH=CH_2$ |
|---|---|---|---|
| $\lambda_{nm}$ | 175 | 217 | 258 |

From what has been said, it can be expected that the presence of + I/ + R and/or – I/ – R groups at the *ends* of a conjugated system will both *extend* the conjugation and *increase* the contributions of charged structures to the resonance hybrid. Therefore, in these circumstances, the wavelengths should become progressively longer, *e.g.,*

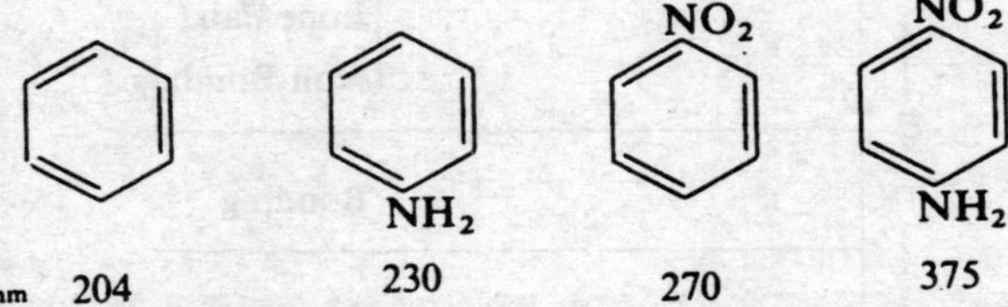


Benzene itself is a chromophore; it is considered to be a resonance hybrid of two Kekule structures and a small amount of charged canonical structures (dipolar ions).

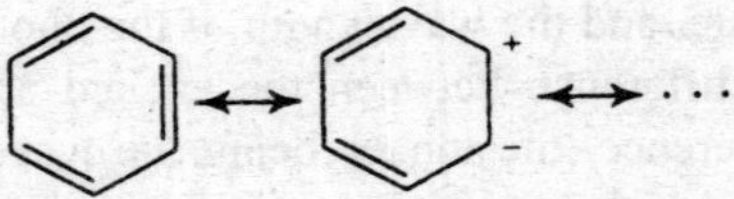

*Steric effects and colour.* As we have seen, resonance in a conjugated system is a maximum when the system is completely (or almost completely) planar. Hence, if resonance is restricted or inhibited, *i.e.,* steric inhibition of resonance is operating, depth of colour will diminish or the compound may even become colourless. It should be remembered that the colour of a compound is that of the complementary colour (this excludes reflection, etc.).

## Molecular Orbital Approach to Colour

In M.O. theory, an atom or molecule is excited when *one* electron is transferred from a *bonding* to an *anti-bonding* orbital. Electronic transitions, however, can occur in different ways. A transition in which a bonding σ-electron is excited to an anti-bonding σ -orbital is referred to as a $\sigma \rightarrow \sigma^*$ transition. In the same way, $\pi \rightarrow \pi^*$ represents the transition of a bonding n-electron to an anti-bonding *n*-orbital. An $n \rightarrow \pi^*$ transition represents the transition of *one* electron of a *lone pair, i.e.,* a non-bonding pair of electrons, to an anti-bonding n-orbital. This type of transition occurs with compounds containing double bonds involving hetero-atoms, *e.g.,* $>$C=O, $>$C=S, $>$C=N— etc., and may be represented as follows :

| | | |
|---|---|---|
| | $\sigma^*$ | **Anti-Bonding** |
| | $\pi^*$ | **Anti-Bonding** |
| ↑ *E* | $n$ | **Lone Pair; Non-Bonding** |
| | $\pi$ | **Bonding** |
| | $\sigma$ | **Bonding** |

**Fig. 4.2**

$$>C=\ddot{O}: \longleftarrow >C=\ddot{O}: \longrightarrow >C=\ddot{O}:$$

$(\pi \rightarrow \pi^*)$ $\qquad$ $(n \rightarrow \pi^*)$

Fig. 4.2 shows diagrammatically the general pattern of the energy levels, and it can be seen that the transitions are brought about by the absorption of different amounts of energy. Of the large number of possibilities, only the following transitions are allowed, and their general order of energy difference is.

$$\sigma \rightarrow \sigma^* > n \rightarrow \sigma^* > \pi \rightarrow \pi^* > n \rightarrow \pi^*$$

The principles discussed so far are illustrated in Fig. 4.3, which represents the relative energies of the various types of orbitals for formaldehyde. This compound shows three bands; 285 (n → π*), 190 (π → π*), and 160 nm (this appears to be n → σ*).

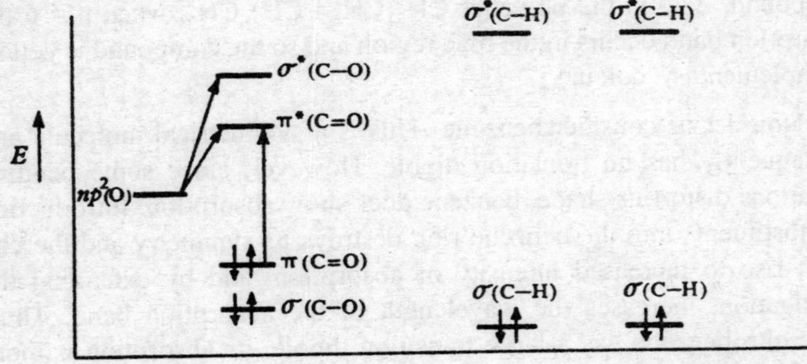


**Fig. 4.3**

Now let us consider some examples. In alkanes, only σ → σ* transitions (C — H and C — C) are possible. The energy of a photon required to effect this transition is very high, *e.g.,* ethane has a $\lambda_{max}$ 135 nm (C — C), and since this is in the far ultra-violet region, ethane is colourless. In ethylene, two types of electronic transitions are now possibles → σ* and π → π*, *e.g.,* ethylene has $\lambda_{max}$ 175 nm and e (molar absorptivity), 5 000. This absorption band is due to a π → π* transition, and again, since it is in the far ultra-violet region, ethylene is colourless. Butadiene has $\lambda_{max}$ 217 nm *(e,* 21000); this is also a π → π* transition. Thus, conjugation has shifted the absorption band to a longer wavelength (but still not into the visible region), and the intensity has greatly increased.

These results can be explained as follows. If we refer to we see that the two bonding orbitals *(b)* and *(c)* represent the molecule in the ground state; *(d)* and *(e)* represent excited states, that of *(e)* being the higher state. Excitation of butadiene can therefore cause transfer of one electron from *(b)* or *(c)* to *(d)* or *(e)*. If all of these occurred, *four* absorption bands would be produced. However, since the energy difference between the highest occupied π-orbital level and lowest unoccupied level *(i.e., c → d)* is *smaller* than any other electronic transition, the *longest* wavelength in the absorption spectrum corresponds to this transition, and the shorter wavelengths correspond to the other transitions.

Calculation (and experimental work) has shown that as conjugation increases, the energy difference between the highest occupied and lowest unoccupied π-orbitals decreases *(cf.* V.B. theory). Thus, as conjugation increases, the wavelength of the absorption band increases, and when it reaches the visible region, colour (complementary) will appear in the compound, *e.g.,* in the polyenes $CH_3(CH = CH)_nCH_3$, when n = 6 the absorption band occurs in the blue region and so the compound is yellow (complementary colour).

Now let us consider benzene. This is a symmetrical molecule and consequently has no transition dipole. However, since some bending vibrations distort its shape, benzene does show absorption. Introduction of substituents into the benzene ring destroys its symmetry and thereby gives rise to increased intensity of absorption, and by extending the conjugation, increases the wavelength of the absorption band. Thus, since nitrobenzene has a large transition dipole, its absorption is more intense than that of benzene, and because of extended conjugation the wavelength is shifted to the longer region.

So far, we have discussed $\sigma \rightarrow \sigma^*$ and $\pi \rightarrow \pi^*$ transitions. Now let us consider the case of methanol; this has $\lambda_{max}$ 150 and 183 nm. As we have seen earlier, ethane has $\lambda_{max}$ 135 nm. Hence, the introduction of the auxochrome, OH, has produced a bathochromic shift. The shorter wavelength is attributed to a $\sigma \rightarrow \sigma^*$ transition (C — H) and the longer one to an $n \rightarrow \sigma^*$ transition (this involves one electron of a lone pair on the oxygen atom). Crotonaldehyde has $\lambda_{max}$ 220 (ε, 16000) and 321 nm (ε, 20); the former is due to a $\pi \rightarrow \pi^*$ transition, and the latter to an $n \rightarrow \sigma^*$ transition.

Steric effects on colour may be explained as in the V.B. approach, *i.e.,* decreased orbital overlap raises the energy difference between the highest occupied and lowest unoccupied M.O.s, thereby causing the absorption to shift to a shorter wavelength.

The u.v. (and i.r.) spectra of free radicals differ from those of the corresponding neutral molecules. This is due to the fact that the presence of the unpaired electron changes the electronic energy levels from those in the parent molecule. In general, the energy difference between the highest occupied and lowest unoccupied M.O.s is lowered, and consequently the absorption wavelength is lengthened and may enter the visible region. Hence, most free radicals are coloured.

## ULTRA-VIOLET AND VISIBLE SPECTROSCOPY

The principles of electronic absorption have been described. However, before discussing their applications, we shall restate some of the terms previously mentioned, but with their modifications. A *chromophore* is any structural feature which produces light absorption in the-ultra-violet region or colour in the visible region. An *auxochrome* is any group which, although not a chromophore, brings about a red shift when attached to a chromophore. Thus, the combination of chromophore and auxochrome behaves as a *new* chromophore. A *bathochromic effect (red shift)* and a *hypsochromic effect (blue shift)* are the shifting of the absorption band to the longer and shorter wavelengths, respectively. A *hyperchromic effect* and *hypochromic effect* are those which respectively increase and decrease the intensity of absorption.

As we have seen, isolated double bonds do not give strong bands, but when conjugated systems are present, the bands are usually strong and in the longer wavelength region. Thus, one particularly important application of ultra-violet (and visible) spectroscopy is the detection and elucidation of the nature of conjugated systems (including aromatics). This is often carried out with the use of 'model' molecules, *i.e.,* a 'simple' molecule that differs from the compound under investigation in a way that should have no effect on the chromophore.

Alkanes absorb in the region of 140–150 nm ($\sigma \rightarrow \sigma^*$), and when combined with various auxochromes, new absorption bands ($n \rightarrow \sigma^*$) are produced at longer wavelengths, *e.g.,* RCl, 170–175, RBr, 200–210, RI, 255–260; ROH and $R_2O$, 180-185; $RNH_2$, 190–200 nm. On the other hand, the carbonyl chromophore (in various functional groups) absorbs, in general, above 200 nm, *e.g.,* aldehydes, 160 nm ($n \rightarrow \pi^*$) and 290–295 ($\pi \rightarrow \pi^*$); ketones 190 ($\pi \rightarrow \pi^*$) and 270–280 ($n \rightarrow \pi^*$); saturated monocarboxylic acids, 200–210 ($n \longrightarrow \pi^*$), esters, 200–205 ($n \rightarrow \pi^*$); amides, 205–220 ($n \rightarrow \pi^*$). It is this absorption at wavelengths longer than 200 nm that permits the identification of many chromophores in compounds.

When ethylenic double bonds are in conjugation or conjugated with a carbonyl group, the absorption moves to longer wavelengths, *e.g.,* for crotonaldehyde, there is one band at 220 nm ($\pi \rightarrow \pi^*$) and another at 321 nm ($n \rightarrow \pi^*$).

An interesting point about conjugated systems is that the geometry of conjugated dienes and trienes affects both $\lambda_{max}$ and e. In general, the acyclic compound absorbs at a shorter wavelength and has a greater intensity than the corresponding cyclic compound, *e.g.,* butadiene, 217 (e, 21000) and cyclohexa-l, 3-diene, 257 (ε, 8000). $\lambda_{max}$ and e are also affected by strain in a molecule, the greater the strain, the shorter the wavelength, *e.g.,* cyclobutanone, 281 nm; cyclohexanone, 290 nm. Steric effects, when operating, decrease conjugation, and so the *trans*-isomer will absorb at a longer wavelength than the *cis.*

Aromatic compounds show a number of bands, *e.g.,* benzene absorbs at 184 (ε, 60000), 204 (ε, 7400) and 254 (ε, 200) nm. All are $\pi \longrightarrow \pi^*$ transitions, and the 254 nm band is called the *benzenoid band* and is characterised by a large degree of fine structure (Fig. 4.4). For benzene derivatives, this benzenoid band generally occurs between 250 and 280 nm, but for polynuclear aromatics it moves to the longer wavelengths as the number of rings increases (see Table 4.2).

All substituents in benzene have a bathochromic effect (Table 4.2). For disubstituted benzenes, the positions of the absorption maxima depend on their orientation, and for para disubstitution, whether the substituents electronically assist, *e.g.,* $NH_2$ and $NO_2$, or whether they electronically oppose each other, *e.g.,* $NH_2$ and OMe. In the latter case, the absorption maximum is usually close to that of the 'stronger' chromophore.

The case of aniline is worth further consideration. In ethanol, $\lambda_{max}$ is 230 nm (Fig. 4.5), but in dilute aqueous acid, $\lambda_{max}$ is 203 nm. In the free base, the nitrogen lone pair of electrons can enter into conjugation with the benzene ring.

Thus, there is increased delocalisation in aniline and consequently the absorption maximum is shifted to the longer wavelength. In the anilinium cation, the lone pair is no longer available for conjugation with the ring, and so the molecule now behaves like benzene itself.

**Table 4.2**

| *Compound** | $\lambda_{max}$ *nm* *(mμ) (ε)* | *Compound* *(mμ) (ε)* | $\lambda_{max}$ |
|---|---|---|---|
| Ethylene | 175(5000) | Resorcinol | 277 (2200) |
| Butadiene | 217(21000) | Quinol | 225(5000) |
| Hexatriene | 258 (35 000) | | 293(2700) |
| Acetaldehyde | 180(10000)<br>290(15) | o-Nitroaninile | 222 (16000)<br>275 (5000) |
| Acetone | 190 (900)<br>280(12) | *m*– ,, | 235 (16000)<br>373 (1 500). |
| Crotonaldehyde | 220(16000<br>321 (20) | *p*- ,, | 229 (5000)<br>375 (16000) |
| Benzene | 204(7400)<br>254 (200) | Naphthalene | 220 (100000)<br>275 (5700) |
| Toluene | 206.5(7000)<br>261 (225) | Anthracene | 253 (200000)<br>375 (8000) |
| Chlorobenzene | 210(7400)<br>264 (200) | Phenanthrene | 252 (50000)<br>293 (16000) |
| Aniline | 230(8600)<br>280(1400) | Furan | 205 (6000)<br>250 (2) |
| Nitrobenzene | 270 (7 800) | Thiophen | 235 (4500) |
| Phenol | 210(6200)<br>271 (1 450) | Pyrrole | 210 (10000)<br>240 (400) |
| Catechol | 214(6000)<br>278(2600) | Pyridine<br>Quinoline | 252 (2000)<br>313 (2500) |
| Stilbene (trans) | 295(27000) | Stilbene (cis) | 280 (10500) |

*In most cases, ethanol is the solvent.

Fig. 4.6–8 are the u.v. spectra of acetophenone, acetone, and m-nitrophenol, respectively.

Heterocyclic compounds, to a large extent, have u.v. spectra similar to those of the analogous benzenoid compounds.

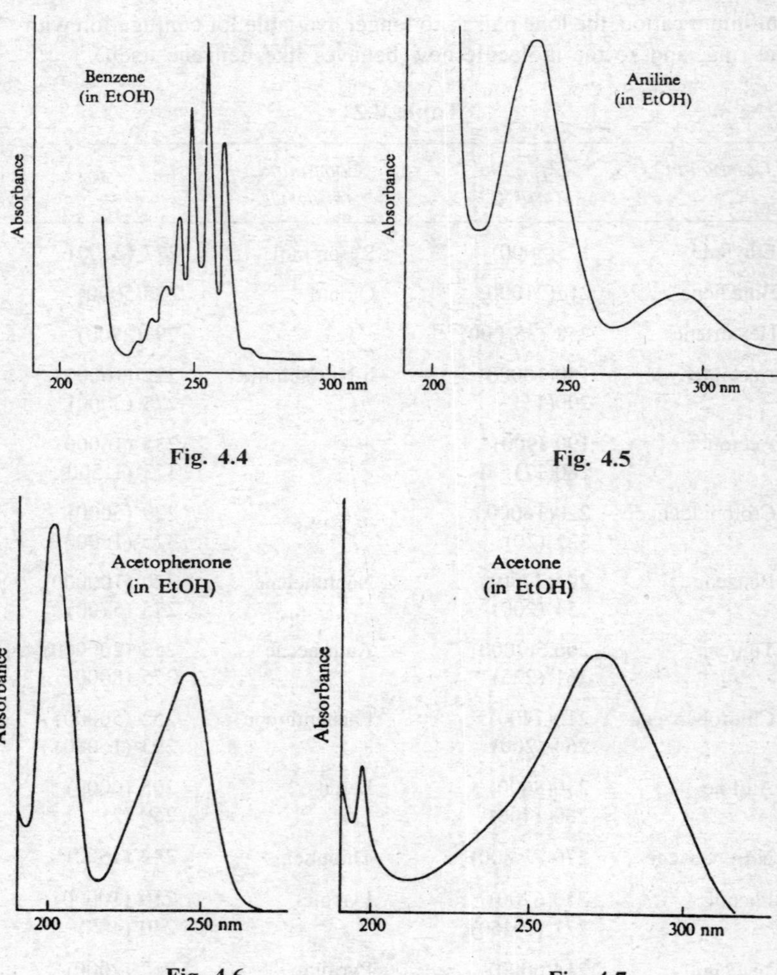


**Fig. 4.4**

**Fig. 4.5**

**Fig. 4.6**

**Fig. 4.7**

Woodward and Fieser nave developed empirical rules for calculating $\pi \rightarrow \pi^*$ maxima of a given chromophore associated with unsaturation in conjugation and the type and position of substituents in the conjugated system.

The final point we shall make here is the effect of solvents. Ethanol is most commonly used since it is a good solvent for many organic compounds and is transparent above 200 nm. However, polar solvents (and those which can form hydrogen bonds) tend to interact electrostatically (and form hydrogen bonds) with various chromophores, *e.g.*, the carbonyl group. This changes the charge distribution in the molecule and results, in effect, in increased delocalisation. For $\pi \rightarrow \pi^*$ transitions, both ground and excited states are stabilised, and the absorption moves to longer wavelengths. On the other hand, for n $\pi \rightarrow \pi^*$ transitions, the ground state is, *e.g.*, hydrogen-bonded to a lone pair of electrons, whereas in the excited state, hydrogen bonding involves only one electron of the lone pair (the other having been promoted to an upper energy state).

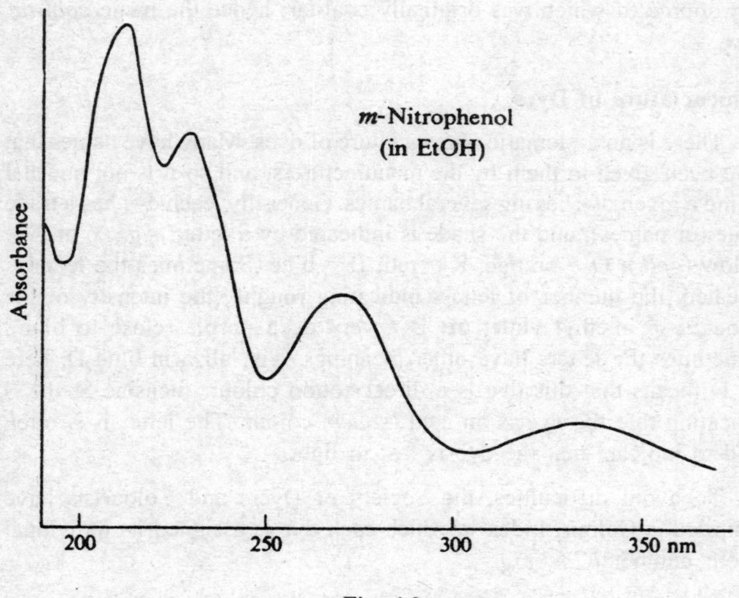


**Fig. 4.8**

In these circumstances, the ground state is more stabilised than the excited state, and consequently absorption shifts to the shorter wavelengths. This blue shift with increasing polarity of solvent, *e.g.*, cyclohexane → EtOH → $H_2O$, is a useful means of recognising n → $\pi^*$ transitions.

**Dyes**

For a substance to act as a dye, certain conditions must be fulfilled.

(i) It must have a suitable colour.

(ii) It must be able to 'fix' itself or be capable of being 'fixed' to the fabric.

(iii) The fixed dye must have fastness properties, (a) fastness to light; (6) resistance to the action of water, dilute acids and alkalis, various organic solvents used in dry cleaning, etc.

Many natural dyes have been known for a long time. These were obtained from animal and vegetable sources. Today, however, practically all dyes are synthetic and are prepared from aromatic compounds, the only source of which was originally coal-tar; hence the name *coal-tar dyes.*

**Nomenclature of Dyes**

There is no systematic nomenclature of dyes. Many have names that have been given to them by the manufacturers, and so it is not unusual to find a given dye having several names. Generally, each dye has a trade name (or names), and the shade is indicated by a letter, *e.g.*, Y or G = yellow (*gelb*); O = orange; R = red; B = blue. Sometimes the letter is repeated, the number of letters indicating roughly the intensity of the colour, *e.g.*, methyl violet 6B is a very deep purple (close to blue). Sometimes the letters have other meanings, *e.g.*, alizarin blue D; here the D means that this dye is a direct cotton colour; fuchsine S, the S indicating that the dye is an acid (*sauer*) colour. The letter F is often used to indicate that the dye is fast to light.

To avoid difficulties, the Society of Dyers and Colourists have compiled .a Colour. Index in which each dye is assigned its individual colour number *(C.I. no.).*

*Classification of dyes.* Dyes are classified according to their chemical constitution or by their application to the fibre. The former is of theoretical value to the chemist but of little importance to the dyer who is mainly concerned with the reaction of dyes towards the fibre being used.

*Chemical classification.* The chemical constitutions of dyes are so varied that it is difficult to classify them into distinct groups. The Colour Index classifies dyes as shown; in some cases a particular dye could be placed in one or other group.

Nitroso; nitro; azo (monazo, etc.); azoic; stilbene; diarylmethane; triarylmeihane; xanthen; quinoline; methine; acridine; sulphur; thiazole;

thiazine; indamine; azine; oxazine; lactone; anthraquinone; indigoid; phthalocyanine.

**Classification According to Application**

The method of application of a dye depends on the nature of both the fibre and the dye. *Natural fibres* are of two types: vegetable fibres are cellulosic, *e.g.*, cotton, linen, flax, hemp, and jute; animal fibres are proteins, *e.g.*, wool, silk, leather, and fur. *Synthetic fibres* are high polymers built up of various types of monomers, *e.g.*, polyamides, polyesters, polyacrylonitrile, etc. Apart from the nature of the fibre, there is also the problem of the treatment of the fibre before dyeing, *e.g.*, natural and synthetic fibres are treated chemically to make them crease-resistant, shrink-resistant, water-repellent, etc. This chemical treatment alters the structure of the fibre, and these alterations can be detected by spectroscopy (u.v., visible, and particularly i.r.).

Dyeing has always been carried out in aqueous solutions, but recent work appears to show that various solvents, *e.g.*, 1, 1, 1-trichloroethane, may be used.

As pointed out above, dyes have diverse chemical structures, but from the point of view of their application, there are four ways in which the dye molecule may be bound to the fibre:

(i) covalent bonds;

(ii) hydrogen bonds;

(iii) ionic bonds;

(iv) van der Waals forces.

The type of binding for a given dye will thus depend largely on the chemical nature of the fibre.

1. *Acid dyes* are the sodium salts of sulphonic and carboxylic acids, and are applied to wool, silk, polyamides and acrylic fibres. *Basic dyes* are salts of the colour bases.

2. *Azoic dyes* are insoluble azo-dyes which are prepared in situ; they are mainly applied to cellulosic fibres. The term *ingrain dyes* has also been used with the same meaning, *i.e.*, produced on the fibre.

3. *Mordant dyes* do not dye a fibre directly; they require a mordant. This is, for acidic dyes, a metal hydroxide, and for basic dyes,

tannin (tannic acid). The metal salts have been referred to as *lakes.*

4. *Metal complex dyes* differ from mordant dyes in that they are prepared as complexes which are *then* applied to the fibre.
5. *Vat dyes* are those which are used in their reduced state *(leuco-compounds),* and after application to the fibre, are oxidised to the dye. They are used mainly with cotton fibres.
6. *Sulphur dyes* are used mainly with cellulosic fibres.
7. *Disperse dyes* are water-insoluble dyes which are dispersed by suitable reagents before application to synthetic fibres.
8. *Reactive dyes* contain a reactive group which combines directly with cellulosic fibres.
9. *Organic pigments* are water-insoluble compounds which are used for colouring paints, etc., and when made water-soluble, are used as dyes.

Owing to the very large number of dyes in use, it is only possible to give a bare outline of dye chemistry. (For further information, see the appropriate references.)

*NITRO-DYES* Two examples of this class are *Naphthol Yellow S, (I),* and *Amido Yellow E,* (II).

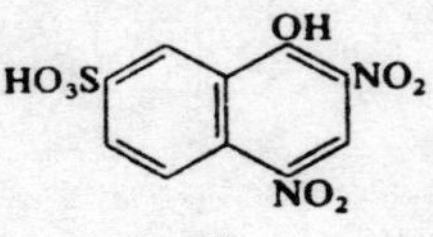

OH, $HO_3S$, $NO_2$, $NO_2$

(I)

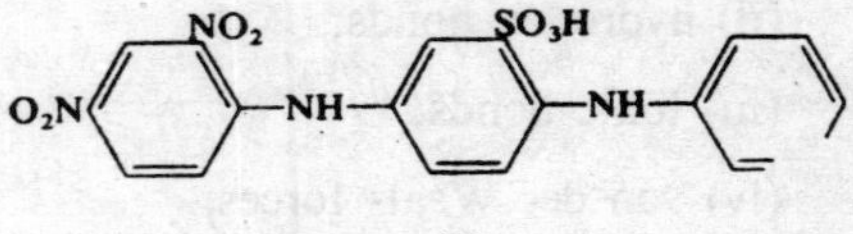

$NO_2$, $SO_3H$, $O_2N$, NH, NH

(II)

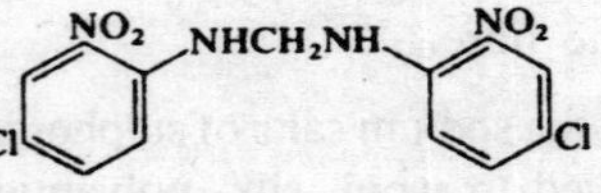

$NO_2$ $NHCH_2NH$ $NO_2$, Cl, Cl

(III)

(I) is made by nitrating l-naphthol-2, 4, 7-trisulphonic acid, and (II) by reaction between 1-chloro -2, 4-dinitrobenzene and 4-aminodiphenylamine-2-sulphonic acid.

*Lithol Yellow G,* (III), is a pigment, and is used as a non-poisonous substitute for *Chrome Yellow* (lead chromate). It is prepared by condensing/π-chloro-o-nitroaniline with formaldehyde. *NITROSO-DYES*

Most nitroso-dyes have one hydroxyl group *ortho* to the nitroso-group, and their iron salts (lakes) are green, *e.g.*, the iron complex of *Gambine Y* (1-nitroso-2-naphthol).

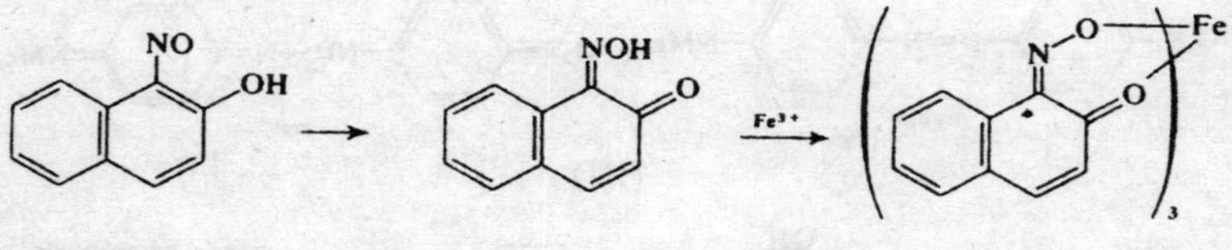

## Azo-dyes

Azo-dyes (and pigments) are the largest class of synthetic dyes. Their chromophore is at aromatic system joined to the azo-group, and the common auxochromes are $NH_2$, $NR_2$, OH The usual method of preparation of azo-dyes is direct coupling between a diazonium salt and 2 phenol or an amine.

The structure of an azo-dye is readily found by reduction with stannous chloride and hydrochloric acid, or with sodium hyposulphite (dithionite), whereupon the azo-group is ruptured with the formation of primary amines which are then identified.

$$Ar^1N = NAr^2 \xrightarrow{e;H^+} Ar^1NH_2 + Ar^2NH_2$$

Azo-dyes are classified according to the number of azo-groups in the molecule, monazo, disazo, trisazo, etc., and are also sub-classified as basic (cationic) dyes, acid (anionic) dyes, etc.

## Monazo Dyes

*Cationic dyes. Chrysoidine,* (I), is a cationic dye, and is prepared by coupling benzenediazonium chloride with m-phenylenediamine. It is used for dyeing paper, leather, and jute. Another cationic dye is *Acid Red,* (II; from diazotised p–nitroaniline and naphthionic acid).

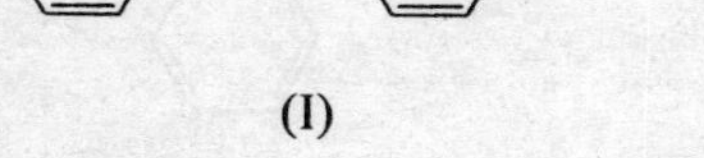

(I) 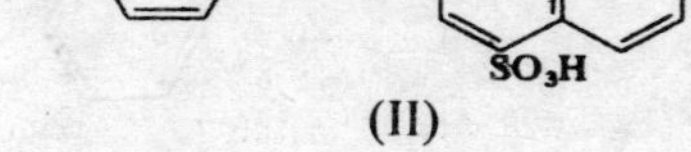 (II)

## Anionic Dyes

These contain a sulphonic or a carboxylic group, and also belong to the class of *soluble dyes. Methyl Orange (Helianthin),* (III), prepared by coupling diazotised sulphanilic acid with dimethylaniline, is not used as a dye. It is used as an indicator, being orange in alkaline solution and red in acid solution *(cf.* Congo Red; see later).

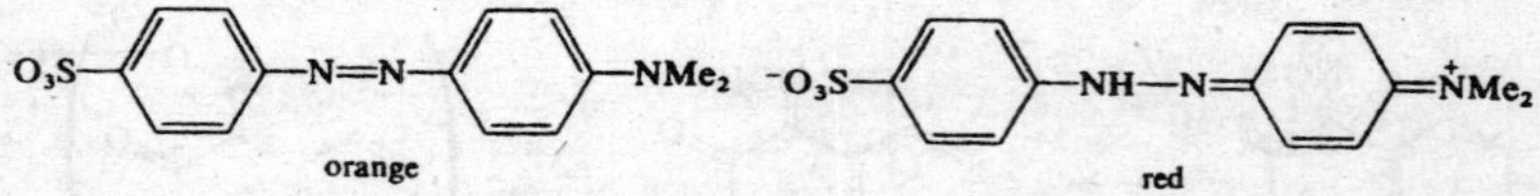

orange red

(III)

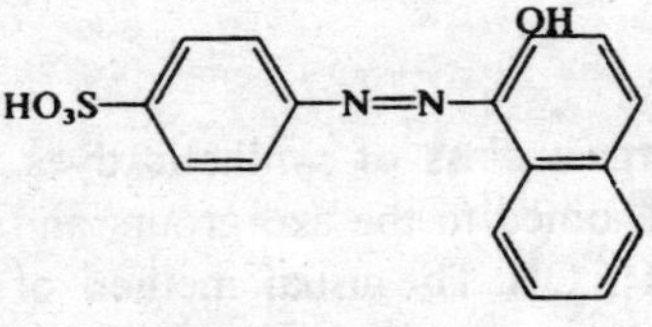

(IV) 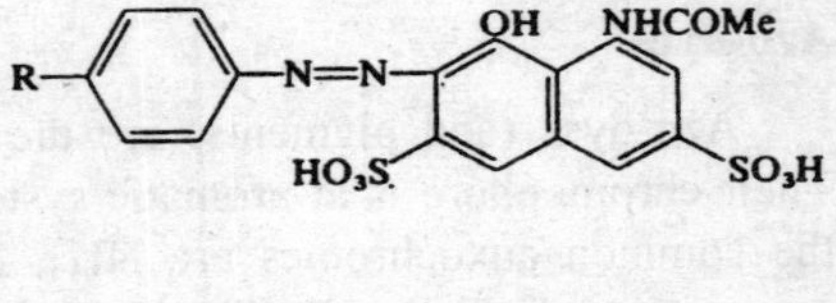 (V)

*Orange II,* (IV) (from diazotised sulphanilic acid and 2-naphthol) is used for dyeing paper.

A serious disadvantage with anionic azo-dyes is their relative easy removal from the fibre by water, This has been overcome by the introduction of the *Carbolan dyes,* (V). R is a long-chain alkyl group, *e.g.,* $C_{12}H_{25}$–. Because of their hydrophobic nature, these large groups prevent the removal of the dye by water.

*Azoic dyes (ingrain azo-dyes)*

These are water-insoluble azo-dyes which are formed on the fibre. They are prepared by 'padding' the fibre with an alkaline solution of 2-naphthol containing Turkey-red oil, drying, and then immersing in a solution of the diazotised amine, *e.g., Para Red,* (VI) (diazotised p-nitroaniline).

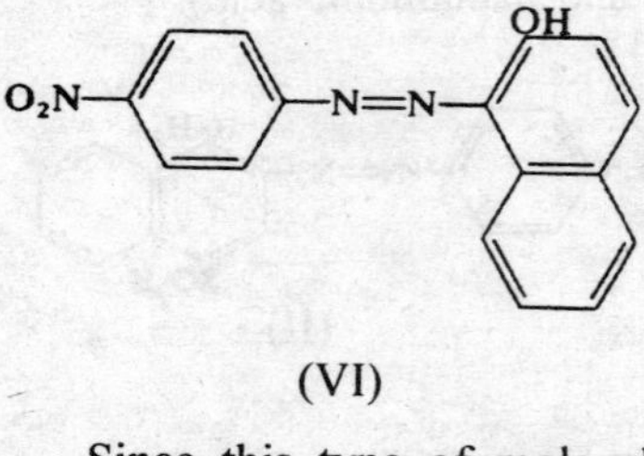

(VI) 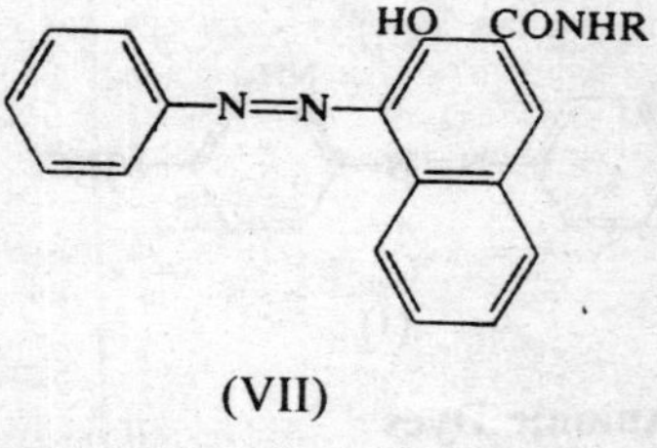 (VII)

Since this type of molecule has poor affinity for the fibre, 3-carbonamido-2-naphthols were introduced, *e.g., Naphthol AS,* (VII,

R = Ph). By changing the diazotised molecule and also R, the colour of the dye can be varied.

**Mordant Azo-dyes**

Their characteristic structural feature is the presence of a hydroxyl group *ortho* to the azo-group. The most important mordanting metal is chromium. The fibre is mordanted by boiling with potassium dichromate solution, usually with a reducing agent such as formic, lactic, or oxalic acid (the dichromate is converted into chromium hydroxide). Metal salts (lakes) are then formed by reaction with various azo-dyes, *e.g., Mordant Brown* (VIII; formed by coupling diazotised p-aminophenol with pyrogallol).

*Metal azo-dye complexes.* One of the difficulties with mordant dyes is that the fibre tends to be damaged under the conditions of chroming. This has been overcome by using neutral-dyeing premetallised dyes. One group of such dyes is the *Irgalans,* which are 2 : 1 complexes (2 dye molecules co-ordinated with the chromium atom to give a negative ion), *e.g.,* (IX):

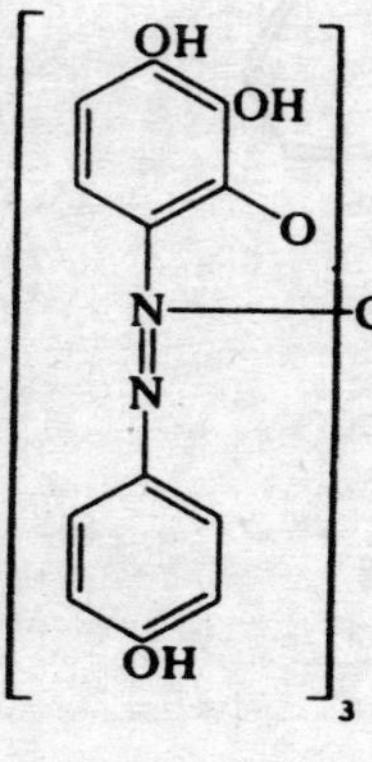

(VIII) 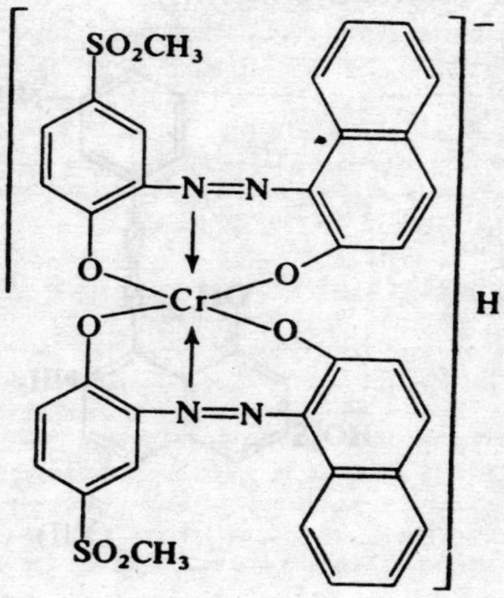 (IX)

These complexes are anionic, and dye wool in neutral solution. They have excellent fastness to light and to washing.

o-Amino- or carboxyl groups can be used instead of o-hydroxyl groups, and cobalt can replace chromium. Furthermore, a wider range of shades can be obtained by using two different azo-compounds to form the 2 : 1 complex.

**Disperse Azo-dyes**

These are used for synthetic fibres, and the method of application is their preparation, by means of dispersing reagents, in a state of uniform fine dispersions in the dye bath. Terylene is dyed mainly with disperse dyes, *e.g.,* (X; a pyrazolone derivative; yellow), and (XI; red).

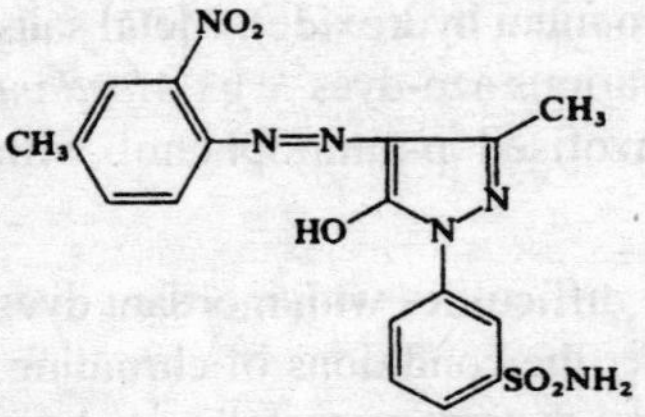

(X) 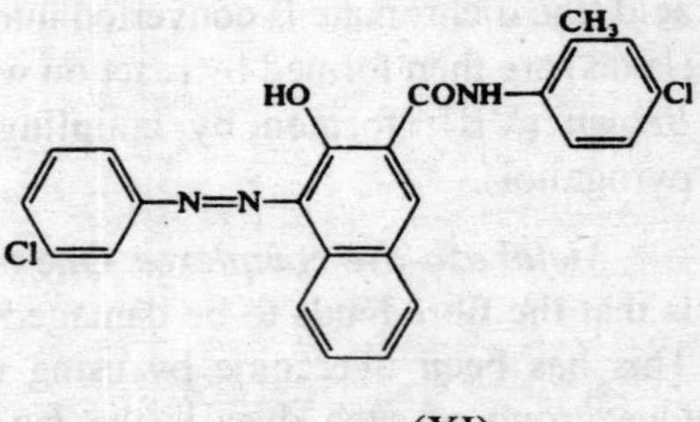 (XI)

Nylon may be dyed with acid dyes (Nylon is basic), with 2 : 1 metal complexes, and with disperse dyes. A typical acid dye is (XII), and a typical disperse dye is (XIII).

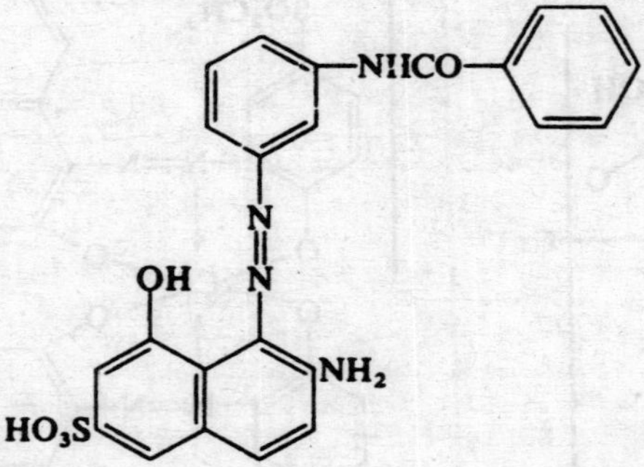

(XII)

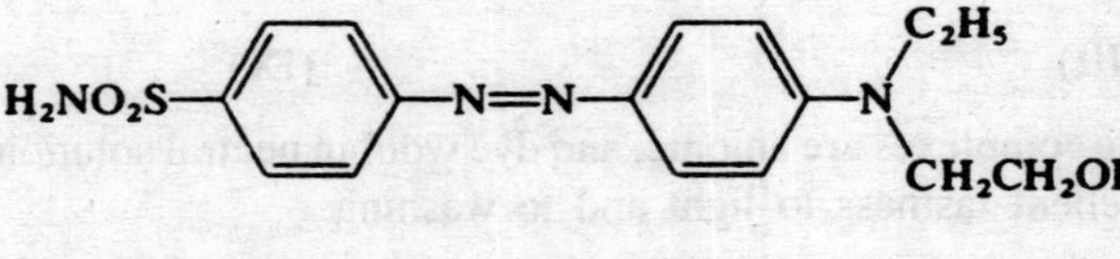

(XIII)

Polyacrylonitrile may be dyed with basic dyes (since it contains free carboxyl groups), with cationic dyes, *e.g.*, *Astrazone Red GTL,* (XIV), and with disperse dyes.

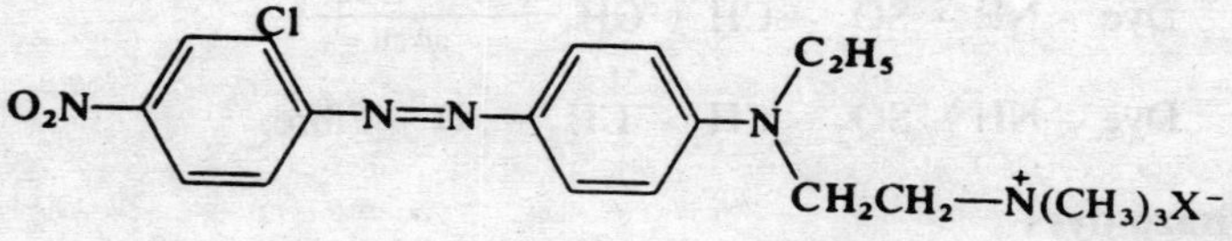

$$O_2N-C_6H_3(Cl)-N{=}N-C_6H_4-N(C_2H_5)CH_2CH_2-\overset{+}{N}(CH_3)_3X^-$$

(XIV)

**Reactive Azo-dyes**

The contains groups which react directly with the fibre. Thus, *Procion dyes* (water-soluble dyes used for cellulosic fibres), the dye is attached to a choloro-1, 3, 5-triazine, *e.g.*,

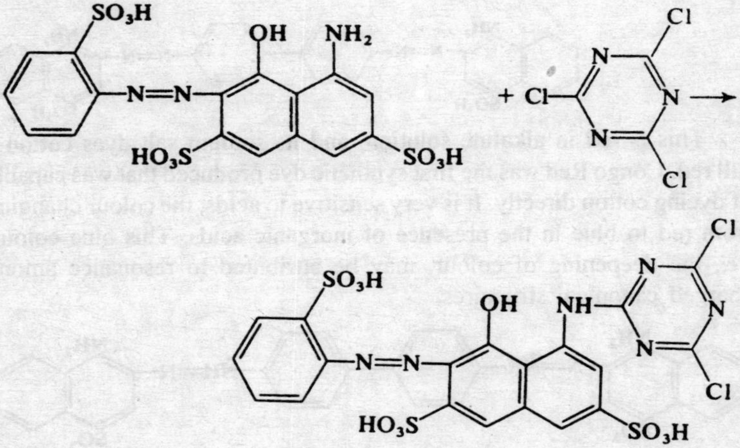

SO₃H, OH, NH₂, N=N, HO₃S, SO₃H + Cl, Cl, Cl, N, N, N →

SO₃H, OH, NH, N=N, HO₃S, SO₃H, Cl, Cl, N, N, N

These reactive dyes are stable in water, and become attached to the cellulose fibre via hydroxyl groups in weakly alkaline solution.

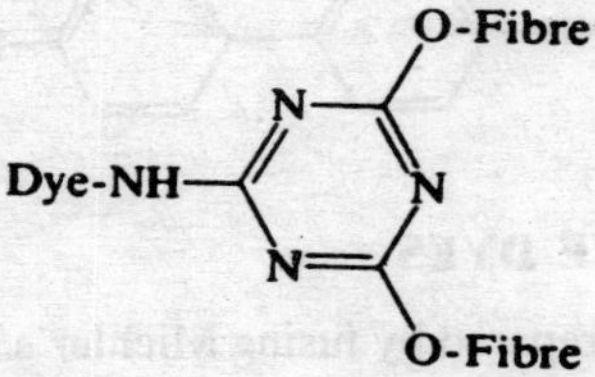

Dye-NH, N, N, N, O-Fibre, O-Fibre

There are also other reactive dyes, *e.g.*, *Ramazol*.

$$\text{Dye} - \text{NH} - \text{SO}_2\text{CH}_2\text{CH}_2\text{OSO}_3\text{H} \xrightarrow{-\text{H}_2\text{SO}_4}$$

$$\text{Dye} - \text{NH} - \text{SO}_2 - \text{CH} = \text{CH}_2 \xrightarrow[\text{alkali}]{\text{HO-Fibre}}$$

$$\text{Dye} - \text{NH} - \text{SO}_2 - \text{CH}_2 - \text{CH}_2 - \text{O} - \text{Fibre}$$

**Disazo-dyes**

*Congo Red* is an example of this group; it is prepared by coupling tetrazotised benzidine with two molecules of naphthionic acid.

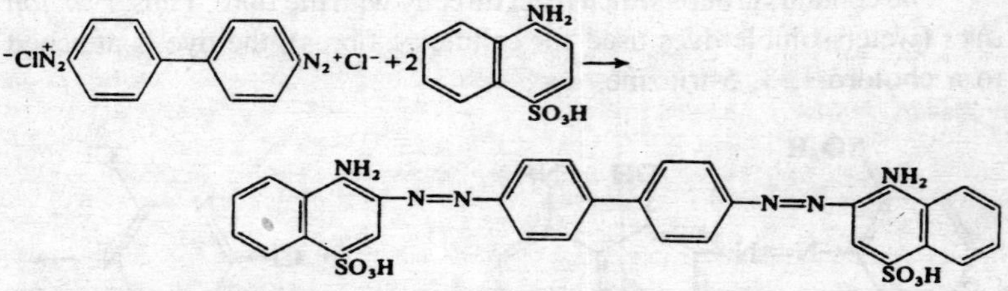

This is red in alkaline solution, and its sodium salt dyes cotton a full red. Congo Red was the first synthetic dye produced that was capable of dyeing cotton directly. It is very sensitive to acids, the colour changing from red to blue in the presence of inorganic acids. This blue colour, *i.e.*, the deepening of colour, may be attributed to resonance among charged canonical structures.

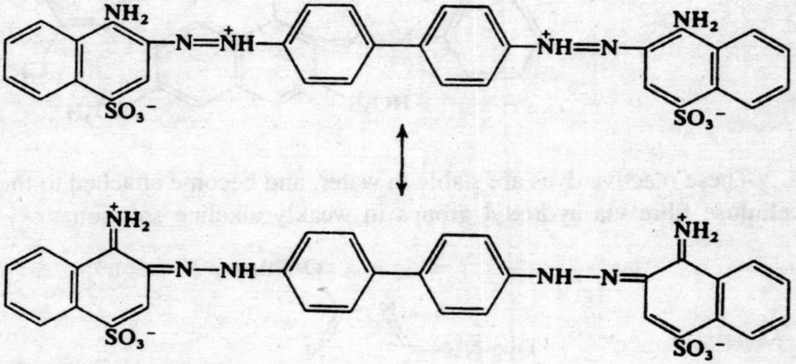

## DIARYLMETHANE DYES

*Auramine O* is prepared by fusing Michler's ketone with ammonium and zinc chlorides at 150-160°C, or by heating p,p'-tetramethyl-diaminodiphenylmethane with sulphur, ammonium chloride and a large excess of sodium chloride in a current of ammonia at about 200°C. This produces *Auramine Base* which, with hydrochloric acid forms the hydrochloride, *Auramine* O, a yellow basic dye.

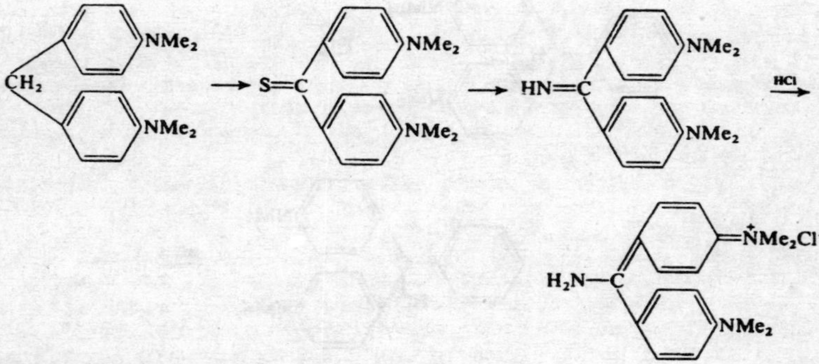

$CH_2$ ($NMe_2$, $NMe_2$) → $S{=}C$ ($NMe_2$, $NMe_2$) → $HN{=}C$ ($NMe_2$, $NMe_2$) $\xrightarrow{HCl}$ $H_2N{-}C$ ($={\overset{+}{N}}Me_2Cl^-$, $NMe_2$)

## Triarylmethane Dyes

Triarylmethane dyes are obtained by the introduction of $NH_2$, $NR_2$ or OH groups into the rings. The compounds so obtained are colourless–the *leuco-compounds*–and these, on oxidation, are converted into the corresponding tertiary alcohols, the *colour bases,* which readily change from the colourless benzenoid forms to the quinonoid dyes in the presence of acid, due to salt formation. The salts are easily reconverted into the leuco-base.

$$\underset{\text{(colourless)}}{\text{leuco-base}} \underset{\text{reduction}}{\overset{\text{oxidation}}{\rightleftharpoons}} \underset{\text{(colourless)}}{\text{colour-base}} \underset{\text{alkali}}{\overset{\text{acid}}{\rightleftharpoons}} \underset{\text{(coloured)}}{\text{dye}}$$

*Malachite Green* is prepared by condensing dimethylaniline (2 molecules) with benzaldehyde (1 molecule) at 100°C in the presence of concentrated sulphuric acid.

The leuco-base produced is oxidised with lead dioxide in a solution of acetic acid containing hydrochloric acid; the resulting colour base gives Malachite Green with excess of hydrochloric acid.

Malachite Green dyes wool and silk directly, and cotton mordanted with tannin.

*Rosaniline, Magenta, Fuchsine* is produced by oxidising an equimolecular mixture of aniline, *o*- and toluidines, and their hydrochlorides, with nitrobenzene in the presence of iron filings.

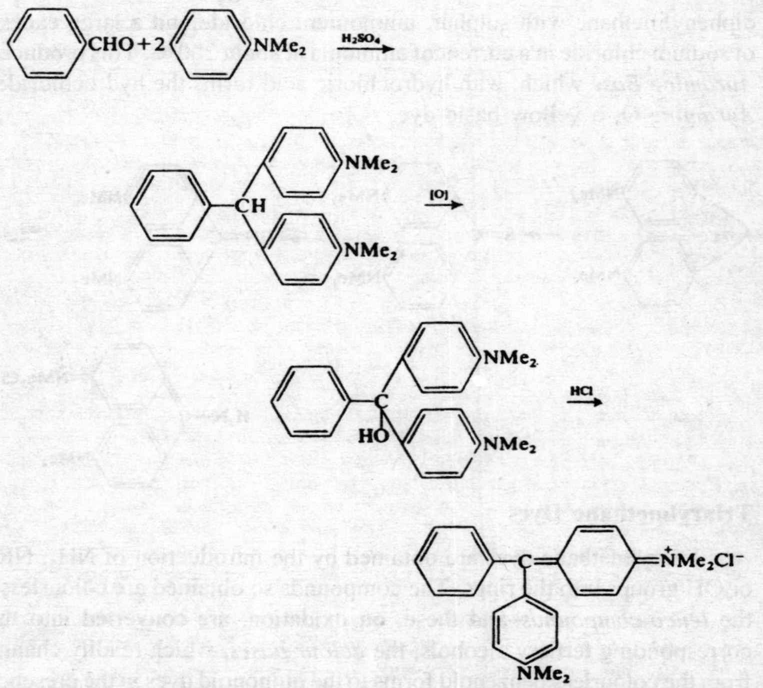

The product is a mixture of rosaniline and pararosaniline (no methyl group present), in which the former predominates.

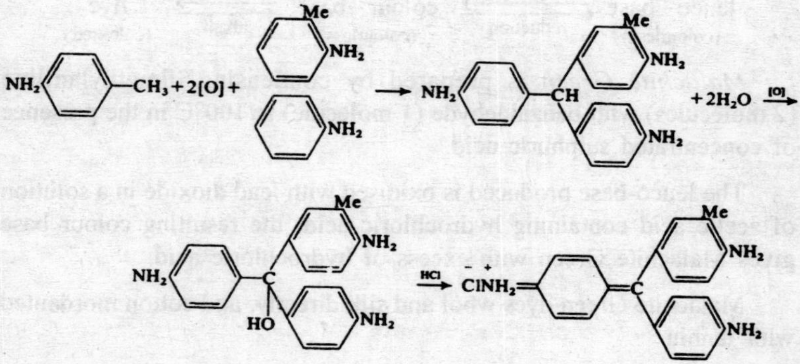

Crystals of rosaniline show a green metallic lustre, and dissolve in water to form a deep-red solution. This solution is decolorised by sulphur dioxide and is then known as Schiff's reagent, which is used as a test for aldehydes.

Rosaniline (and pararosaniline) dyes wool and silk directly, producing a violet-red colour; cotton must first be mordanted with tannin.

*Crystal Violet* may be prepared by heating *Michler's ketone* with dimethylaniline in the presence of phosphoryl chloride or carbonyl chloride. If the latter compound is used, then Crystal Violet may be prepared directly by heating carbonyl chloride and dimethylaniline.

$$COCl_2 + 2C_6H_5NMe_2 \longrightarrow CO(C_6H_4NMe_2)_2$$

$$CO(C_6H_4NMe_2)_2 + C_6H_5NMe_2 + COCl_2 \longrightarrow Me_2NC_6H_4{-}C(C_6H_4NMe_2){=}C_6H_4{=}\overset{+}{N}Me_2Cl^- + HCl + CO_2$$

A Weakly acid solution of Crystal Violet is purple. In *strongly* acid solution the colour is green, and in still more strongly acid solution, the colour is yellow. The explanation of these colour changes may be as follows. In weakly acid solution Crystal Violet exists as the singly charged ion (I). In this state two-thirds of the charge can oscillate in the horizontal direction. In strongly acid solution, if Crystal Violet exists as the *doubly* charged ion (II), the whole unit of charge can oscillate in the horizontal direction, and consequently the colour *deepens* (the vertical direction of oscillation is inhibited due to the fixation of the lone pair by proton addition). In very strongly acid solution another proton is added to form the ion (III) with three charges. In this ion relatively little resonance (with oscillation of charge) is possible, and consequently the colour lightens.

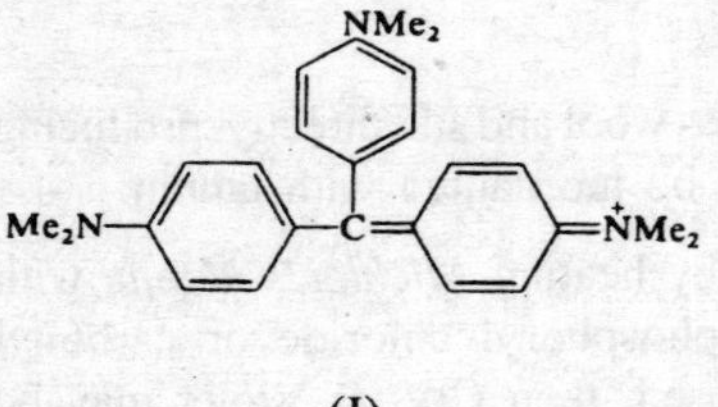

(I)

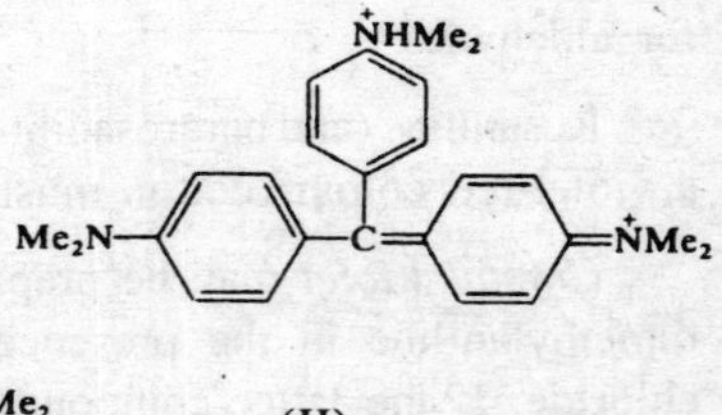

(II)

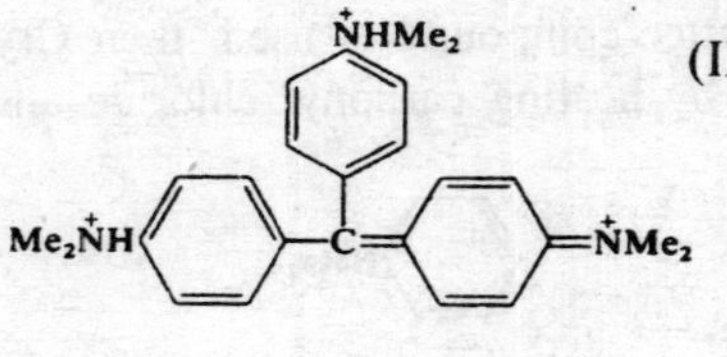

(III)

## XANTHEN DYES

The parent substance of this group of dyes is xanthen *(dibenzo*-1 ,4-*pyran)*. This, on oxidation, yields *xanthone* (*9-ketoxanthen)* which, on reduction, yields *xanth-hydrol* (*9-hydroxyxanthen)*. This forms *oxonium salts* with inorganic acids.

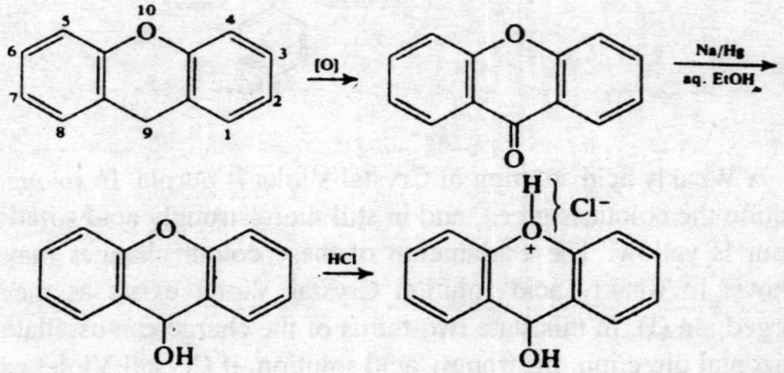

Dyes are obtained from xanthen by the introduction of auxochromes into positions 3 and 6 (*i.e.,* the p-positions with respect to the carbon atom linking the two benzene nuclei).

*Pyronines. Pyronine G* is prepared by condensing formaldehyde (1 molecule) with m-dimethyl-aminophenol (2 molecules) in the presence of concentrated sulphuric acid, and then oxidising the leuco-compound with ferric chloride. Pyronine G dyes silk and cotton mordanted with tannin a crimson red.

Since two electronic structures are possible in which the conjugation is totally different, more information is necessary to say which form predominates, or even whether both forms are present. However, because nitrogen is less electronegative than oxygen, the former can support a positive charge more easily than the latter, and therefore the structure with positive nitrogen is the more likely one.

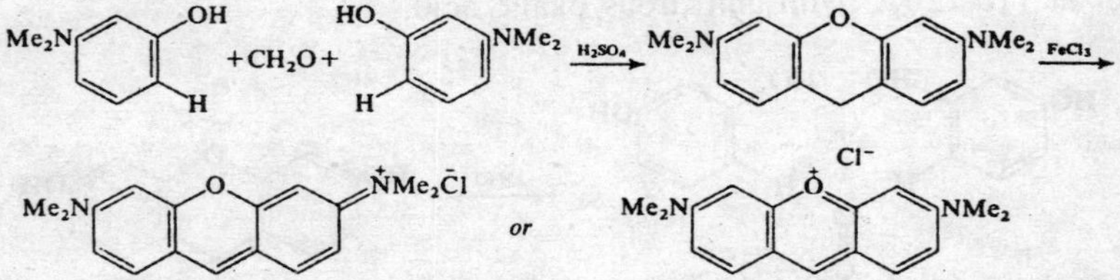

$Me_2N$ … OH + $CH_2O$ + HO … $NMe_2$ $\xrightarrow{H_2SO_4}$ … $\xrightarrow{FeCl_3}$

$\overset{+}{N}Me_2\overset{-}{Cl}$ *or* $Cl^-$ $\overset{+}{O}$

*Phthaleins.* These are obtained by condensing phenols with phthalic anhydride in the presence of certain dehydrating agents, *viz.*, concentrated sulphuric acid, zinc chloride or anhydrous oxalic acid.

*Phenolphthalein* is not a xanthen derivative but a triphenylmethane derivative. Since, however, it is prepared from phenol and phthalic anhydride, it is more closely allied to the phthaleins (in preparation and properties), and hence is here considered in the phthalein group. Phenolphthalein is prepared by heating phthalic anhydride (1 molecule) with phenol (2 molecules) in the presence of concentrated sulphuric acid.

$+H_2O$

(colourless)

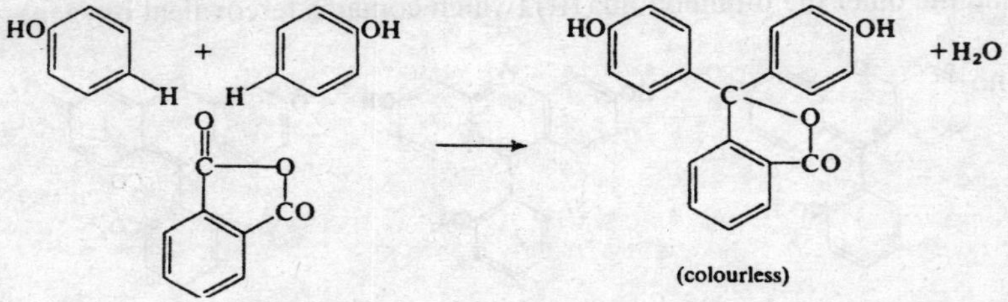

It is a white crystalline solid, insoluble in water, but soluble in alkalis to form deep red solutions.

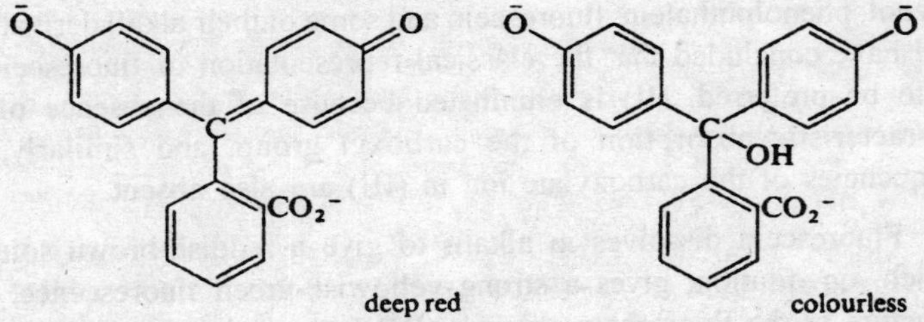

deep red colourless

In the presence of *excess* of strong alkali, the solution of phenol-phthalein becomes colourless again due to the loss of the quinonoid structure and resonance (but see fluorescein, below).

*Fluorescein* is a xanthen derivative, and is prepared by heating phthalic anhydride (1 molecule) with resorcinol (2 molecules) at 200°C, or at 110-120°C with anhydrous oxalic acid.

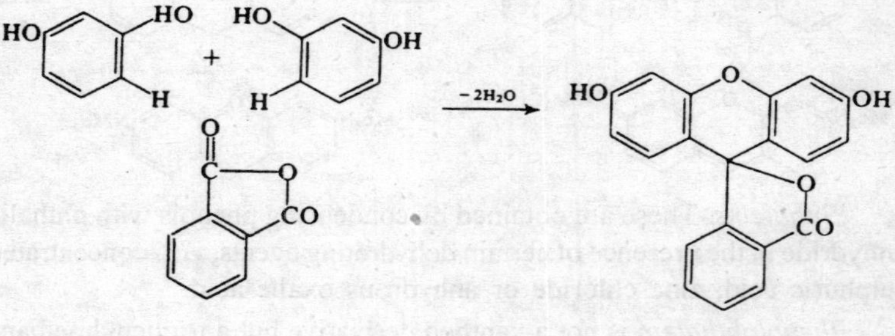

$\xrightarrow{-2H_2O}$

(I)

Fluorescein is a red powder insoluble in water. Since it is coloured, the non-quinonoid uncharged structure (I) has been considered unsatisfactory. Two quinonoid structures are possible in which the conjugation is totally different, one having the p-quinonoid structure (II), and the other the o-quinonoid (III) (which contains tercovalent oxygen).

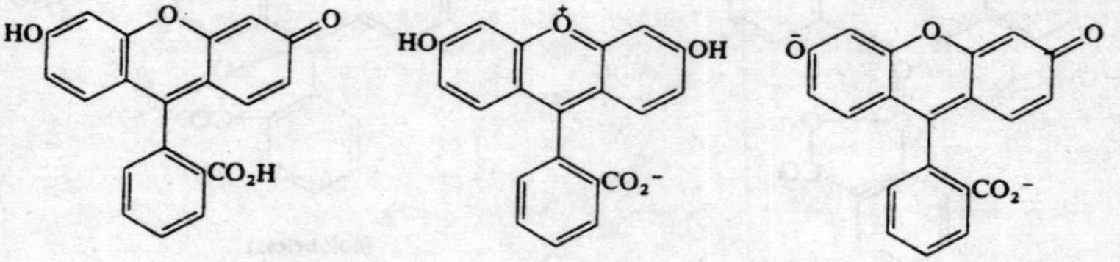

(II) (III) (IV)

Davies *et al.,* (1954) have examined the infra-red absorptions of phenol, phenolphthalein, fluorescein and some of their alkali derivatives, and have concluded that the classical representation of fluorescein (I) is to be preferred. (II) is eliminated because of the absence of the characteristic absorption of the carboxyl group, and similarly, the frequencies of the carboxylate ion in (III) are also absent.

Fluorescein dissolves in alkalis to give a reddish-brown solution which, on dilution, gives a strong yellowish-green fluorescence. The structure of the fluorescein anion is (IV) (oxygen becomes tercovalent unielectrovalent only in acid solution). The sodium salt of fluorescein is known as *Uranine,* and dyes wool and silk yellow from an acid bath; the colours are fugitive.

*Eosin* (Caro, 1871) is tetrabromofluorescein, and is prepared by the action of bromine on fluorescein in glacial acetic acid solution.

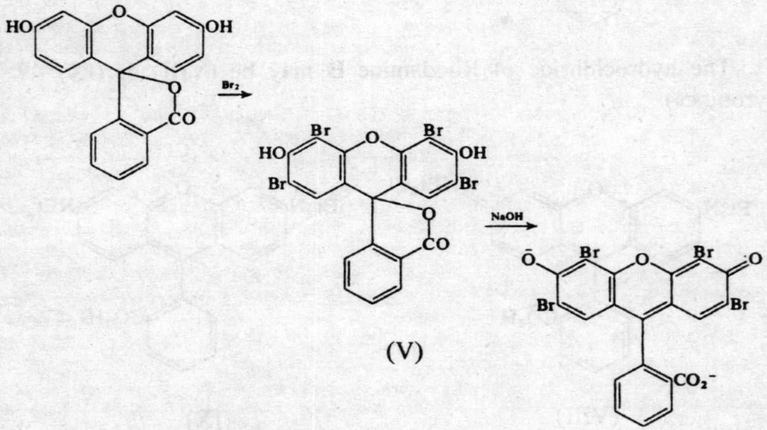

HO OH O CO $\xrightarrow{Br_2}$ Br Br HO OH Br Br O CO $\xrightarrow{NaOH}$ $\bar{O}$ Br O Br O Br Br $CO_2^-$

(V)

Eosin itself is a red powder and so its structure may not be (V), but may be (VI) or (VII) *(cf.* fluorescein).

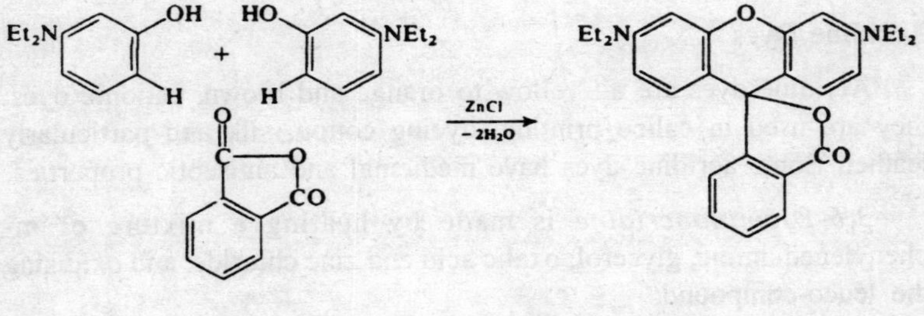

$Et_2N$ OH + HO $NEt_2$ H H O C O CO $\xrightarrow[-2H_2O]{ZnCl}$ $Et_2N$ O $NEt_2$ O CO

(VI) (VII)

Eosin dyes wool and silk a pure red, with a yellow fluorescence in the case of silk. Eosin is also used as the lead lake *Vermilionette* for poster printing. Most red inks are dilute solutions of eosin.

*Rhodamines* are prepared by condensing phthalic anhydride (1 molecule) with m-aminophenol (2 molecules) or its derivatives; *e.g., Rhodamine B.*

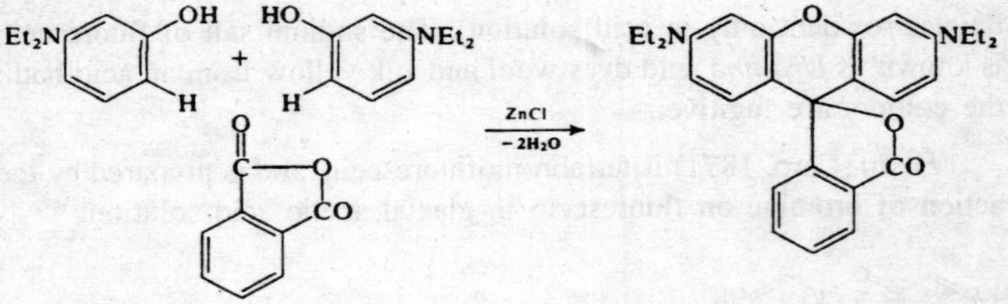

The hydrochloride of Rhodamine B may be (VIII) or (IX) *(cf.* pyronines).

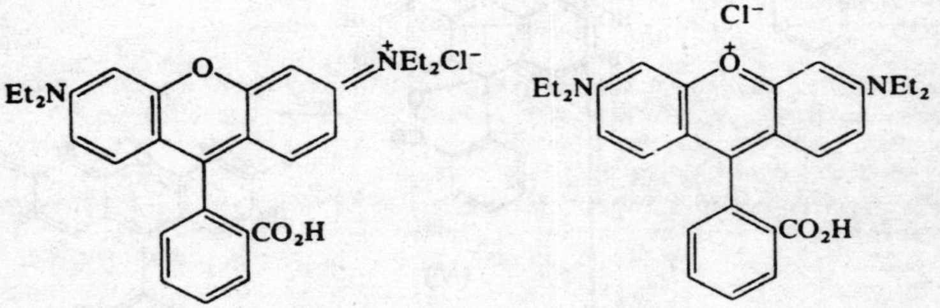

(VIII) (IX)

The rhodamines are red cationic dyes, the N-alkylated derivatives being more highly coloured than the unsubstituted compounds. They dye wool and silk directly, and cotton mordanted with tannin.

**Acridine Dyes**

Acridine dyes are all yellow to orange and brown, cationic dyes; they are used in calico printing, dyeing cotton, silk and particularly leather. Some acridine dyes have medicinal and antiseptic properties.

*3,6-Diaminoacridine* is made by heating a mixture of m-phenylenediamine, glycerol, oxalic acid and zinc chloride, and oxidising the leuco-compound.

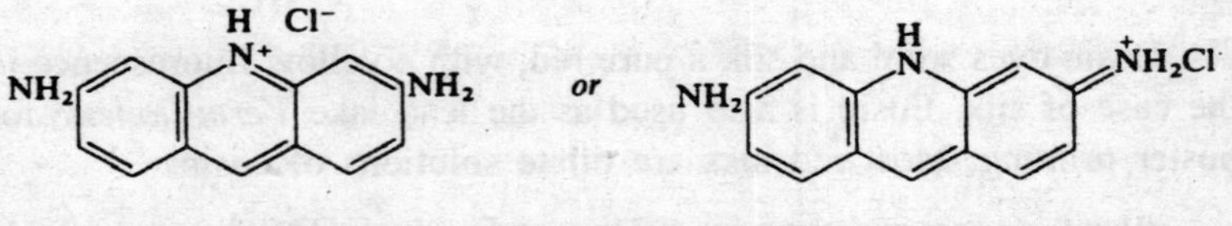

The corresponding sulphate is the antiseptic *proflavine.* On acetylating 3,6-diaminoacridine, methylating the 10-N atom by means of methyl p-toluenesulphonate, and then hydrolysing the product with dilute hydrochloric acid, 3,6-diamino-10-methylacridinium chloride is produced. This is known as *acriflavine,* and possesses trypanocidal action *(i.e.,* the power to kill trypanosomes which are micro-organisms causing sleeping sickness and other diseases). It has now been replaced by more potent trypanocides, but it is still used as an antiseptic.

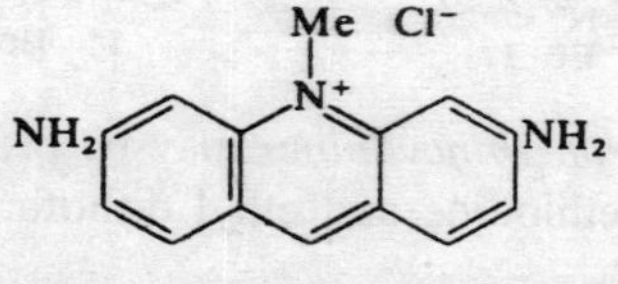

*Acridine Yellow G* is one of the more important acridine dyes, and is prepared from 2,4-diamino-toluene and formaldehyde.

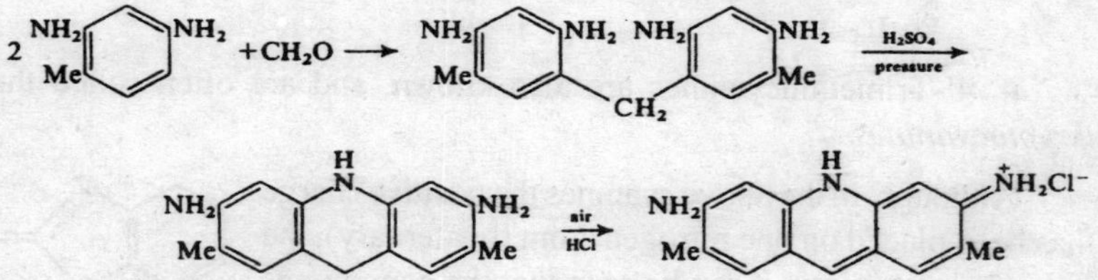

## Quinoline Dyes

Very few of these have any value as dyes because of their lack of fastness to light. However, most of them are very important photographic sensitisers. *Cyanines* consist of two heterocyclic nuclei (containing nitrogen) linked by an *odd-numbered* conjugated carbon chain, and are in the form of a mono-acid salt. The term *cyanine,* when not preceded by a prefix, is used for the compounds in which *two quinoline nuclei* are linked through the 4, 4′-positions. In other cases, numbers are used to designate other positions of attachment. Since the odd-conjugated chain can be of different lengths, a common practice is to indicate this by the prefix methin (= CH —), trimethin (= CH — CH = CH —), etc.

*Cyanine (methincyanine; blue)* may be prepared by heating a mixture of 1-ethylquinolinium iodide and 1-ethyl-4-methylquinolinium iodide in ethanolic sodium hydroxide solution.

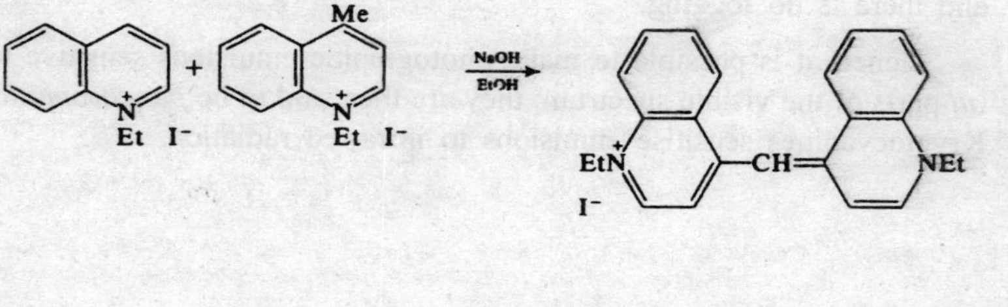

*Ethyl Red* (2, 4′-*methincyanine)* is prepared in a similar manner from quinoline and quinaldine ethiodides.

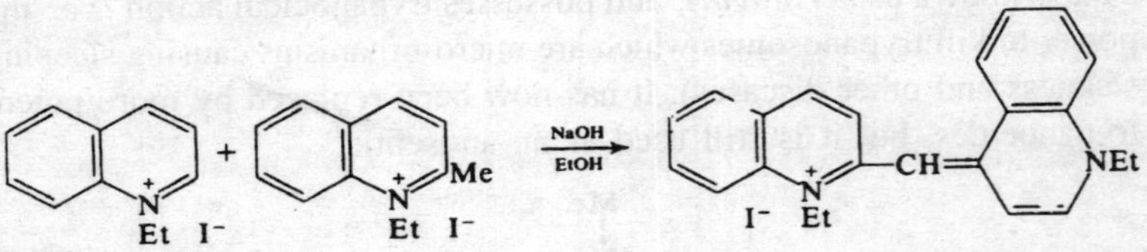

Pinacyanol (2,2′-*trimethincyanine)* may be prepared by heating a mixture of quinaldine ethiodide and ethyl orthoformate in pyridine.

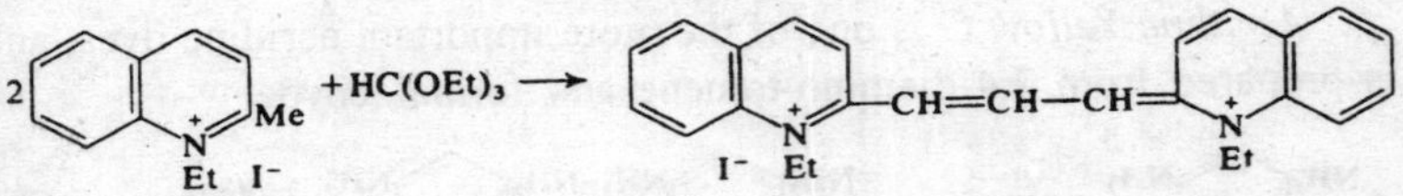

4, 4′-Trimethincyanines are also known, and are often called the *kryptocyanines.*

Although, in the above cyanines the positive charge has been placed on one nitrogen atom (quaternary), the other being tertiary, it can be seen that the compounds are resonance hybrids, the other 'extreme' canonical structure having the valencies of the two nitrogen atoms reversed. Also, extending the conjugation between the rings produces a red shift.

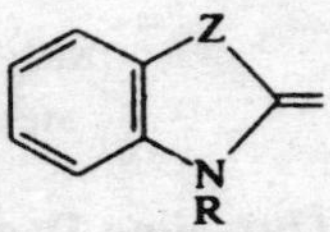

Z=S; benzthiazole
Z=O; benzoxazole

Other nuclei besides quinoline may be used to prepare 'cyanines', *e.g.,* benzthiazole and benzoxazole, etc. One or both quinoline nuclei may be replaced, and the absorption maximum depends on the nature of the nucleus present.

Photographic plates of silver chloride emulsions are sensitive from to 350 nm; those of silver bromide, from 350 to 530 nm. On the other hand, silver bromide emulsions are sensitised to the green region (350 to 600 nm) by cyanine, but the emulsion is fogged.

Pinacyanol sensitises the emulsion to the red region (350-710 nm), and there is no fogging.

Hence, it is possible to make photographic emulsions sensitive to *all* parts of the visible spectrum; they are then said to be *panchromatic.* Kryptocyanines sensitise emulsions to infra-red radiation.

## Azine Dyes

Perkin (1856), in search for the synthesis of the alkaloid quinine, oxidised crude aniline sulphate with potassium dichromate, and obtained a black precipitate from which he extracted a mauve colouring matter with methylated spirit. Perkin found that this material had the properties of a dye. This was the first synthetic dye to be used, and was used for dyeing silk under the name of *Aniline Purple* or *Tyrian Purple.* Aniline Purple was the first basic dye to be used, but no method was known (when if was discovered) for applying it to cotton. Perkin and Pullar, independently in 1857, discovered the general method of applying basic dyes to cotton mordanted with insoluble inorganic compounds of tannin. Aniline Purple was used in France under the name of *Mauve,* and the dye afterwards became known by this name; it is also known as *Mauveine.*

Perkin found that his material contained two products, one from pure aniline–*pseudo-Mauveine,* and the other from aniline and o-toluidine (the latter was an impurity in crude aniline)–*Mauveine.* The structure of pseudo-mauveine was established by Nietzki who prepared it by oxidising a mixture of p-phenylenediamine and diphenyl-m-phenylenediamine.

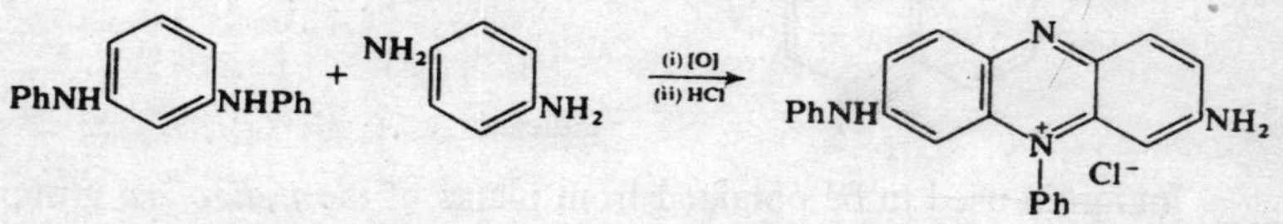

The structure may be the p-quinonoid instead of the *o*- shown above (*cf.* fluorescein, etc.). Mauveine is a mixture of compounds of the type

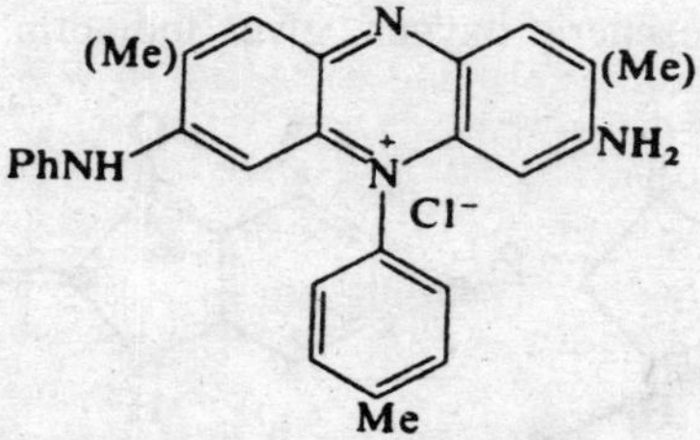

More useful dyes than Mauveine are the *Safranines*–these are more strongly basic than the mauveines. *Safranine T* is prepared by oxidising a mixture of 2, 5-diaminotoluene, o-toluidine and aniline with sodium dichromate and hydrochloric acid (an indamine is formed as an intermediate).

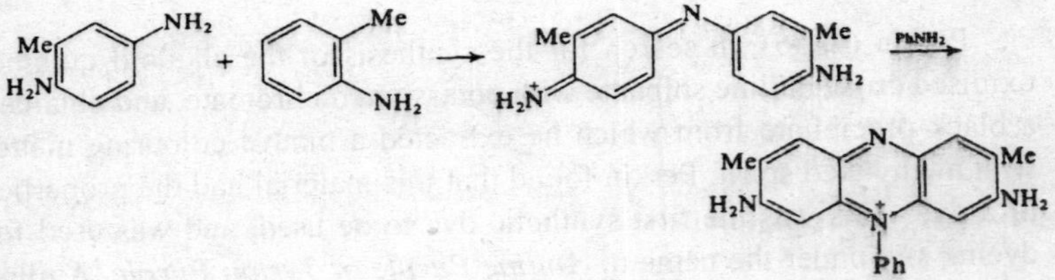

It is a red basic dye, dyeing wool, silk and 'tanned' red shades.

## Vat Dyes

*Indigoid Group.* India appears to be the birth-place of indigo—indigotin is its official name. It is the oldest known dye–mummy cloths, 5 000 years old, have been found that were dyed with indigotin. *Wood,* which is an impure form of indigotin (and is obtained from the plant *Isatis tinctoria),* contains a small amount of *indican* which is the glucoside of indoxyl.

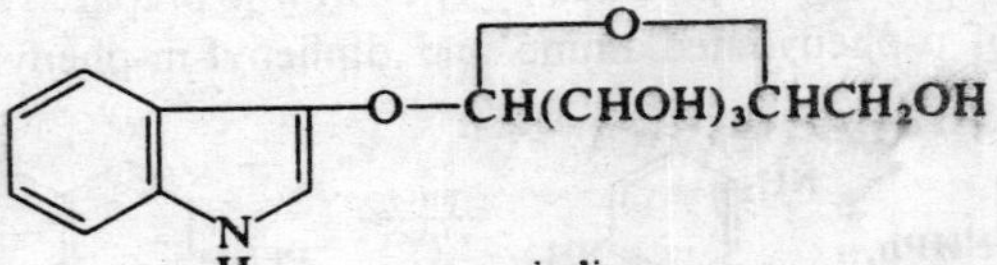

indican

Indigotin used to be obtained from plants of the *indigofera* group, but is now prepared synthetically. These plants do not contain indigotin but indican which, on hydrolysis with hydrochloric acid or by enzymes which occur in the crushed plant, is converted into indoxyl and this, on oxidation by atmospheric oxygen, gives indigotin.

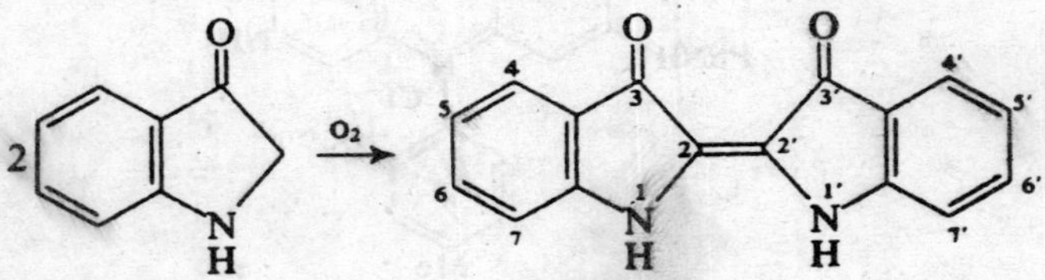

*Synthesis of indigotin.* Many syntheses of indigotin are known; only the commercial methods are given here.

(i) Anthranilic acid is heated with chloroacetic acid and the product, phenylglycine-o-carboxylic acid, is heated with a mixture of potassium hydroxide and sodamide. Indoxylic acid is thereby produced, and this decarboxylates into indoxyl which, on exposure to air. is oxidised to indigotin.

$$C_6H_4(CO_2H)(NH_2) + ClCH_2CO_2H \longrightarrow C_6H_4(CO_2H)(NHCH_2CO_2H) \xrightarrow[NaNH_2]{KOH}$$

$$\text{(3-hydroxyindole-2-carboxylic acid)} \xrightarrow{-CO_2} \text{(indoxyl)} \xrightarrow{O_2} \text{indigotin}$$

This method is due to Heumann (1896).

(ii) Aniline is heated with chloroacetic acid, the product, phenylglycine, converted into a mixture of its sodium and potassium salts, and these fused with sodamide and a mixture of sodium and potassium hydroxides at 220-240°C. Indoxyl which is thereby obtained is converted into indigotin by atmospheric oxygen (the 'mixed salt' requires a lower temperature of fusion, and consequently the yield of indoxyl is better).

$$C_6H_5NH_2 + ClCH_2CO_2H \longrightarrow C_6H_5NHCH_2CO_2H \xrightarrow[KOH]{NaOH}$$

$$C_6H_5NHCH_2CO_2 \begin{cases} Na \\ K \end{cases} \xrightarrow[NaOH/KOH]{NaNH_2} \text{(indoxyl)} \xrightarrow{O_2} \text{indigotin}$$

This method was first used in 1890, the fusion, however, being carried out with potassium hydroxide alone (Heumann). In 1896 this method was abandoned, and method (i) was used due to the cheap cost of phthalic anhydride from naphthalene. This oxidation was successful due to the accidental breaking of a thermometer, an accident which led to the discovery in 1893 that mercury catalyses the oxidation of naphthalene with sulphuric acid. Furthermore, method (i) gave better yields than (ii). About 1898, however, it was found that the addition of sodamide in the fusion in method (ii) increased the yield, and so (ii) was used again. Still later the preparation of phenylglycine was modified as follows. Formaldehyde is converted into its bisulphite compound on treatment with an aqueous saturated solution of sodium hydrogen sulphite at 50 – 80°C. This compound is warmed with aniline at 50–70°C, then treated with aqueous sodium cyanide, and the product hydrolysed with water at 70°C.

$$CH_2O + NaHSO_3 \longrightarrow CH_2(OH)SO_3Na$$
$$\xrightarrow{C_6H_5NH_2} C_6H_5NHCH_2SO_3Na$$
$$\xrightarrow{NaCN} C_6H_5NHCH_2CN$$
$$\xrightarrow{H_2O} C_6H_5NHCH_2CO_2H$$

(iii) Aniline, on treatment with ethylene oxide, forms N-phenyl-2-hydroxyethylamine and this, on fusion with a mixture of sodium and potassium hydroxides at 200°C, gives the sodium (and potassium) alkoxide.

When this sodium salt is heated with sodamide and a mixture of sodium and potassium hydroxides at 200°C, the N-sodio-derivative is produced. This is dehydrogenated on heating rapidly to 300°C, and on rapid cooling to 235°C, ring-closure takes place with the formation of the sodium salt of indoxyl. This, when treated with water and allowed to stand exposed to the air, gives indigotin.

$$C_6H_5NH_2 + \underset{\text{(O bridge)}}{CH_2\text{—}CH_2} \longrightarrow C_6H_5NHCH_2CH_2OH$$
$$\xrightarrow{NaOH/KOH} C_6H_5NHCH_2CH_2ONa$$
$$\xrightarrow[NaNH_2]{NaOH/KOH} C_6H_5N(Na)CH_2CH_2ONa \xrightarrow[(ii)\ 235°C]{(i)\ 300°C}$$

ONa / N / H (sodium indoxylate) $\xrightarrow{air/H_2O}$ indigotin

*Properties of indigotin.* Indigotin is a dark blue powder with a coopered lustre. It is insoluble in water, but when its paste is agitated with alkaline sodium hyposulphite in large vats, the insoluble indigotin is reduced to the soluble leuco-compound, *indigotin-white.*

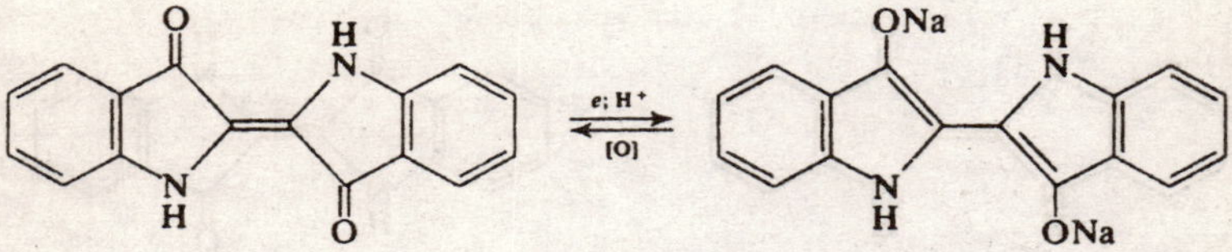

The material to be dyed is soaked in this alkaline solution and then exposed to the air, whereupon the original blue dye is regenerated in the cloth.

In vat-colour printing (of calico), 'local dyeing' is achieved by making the dye pigment into a thick paste (by means of a 'thickener', *e.g.*, starch), and applying this locally in the presence of alkali and a reducing agent, *e.*

Sodium hydroxymethanesulphinate *(Rongalite, Formosul),* $CH_2(OH)SO_2Na$, which is prepared by reducing formaldehyde bisulphite with zinc and acetic acid. The colour is finally developed by means of an oxidising agent, *e.g.,* sodium nitrite.

In textile printing, *e.g.,* shirtings, dress materials, etc., indigotin-white is not sufficiently stable to be used. Hence textile printing (and ordinary dyeing) may be effected by first converting indigotin into *Indigosol O* by treating indigotin-white with chlorosulphonic acid in the presence of pyridine.

This produces the disulphuric ester which, by means of sodium hydroxide, is converted into its sodium salt, Indigosol O. This is readily applied to animal or vegetable fibres by soaking the fabric in the solution, and then oxidising the Indigosol O in *acid* solution (with sodium nitrite) to the original insoluble vat dye.

Here, we have an example of a vat dye being applied without the vatting process. Indigosol O is stable in neutral or alkaline solution, but is hydrolysed to indigotin-white in acid solution.

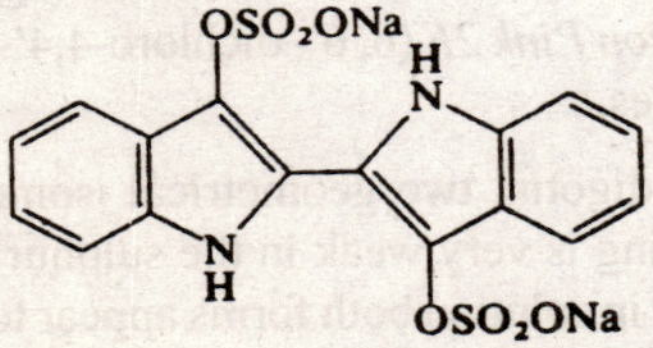

Two geometrical isomers are possible for indigotin.

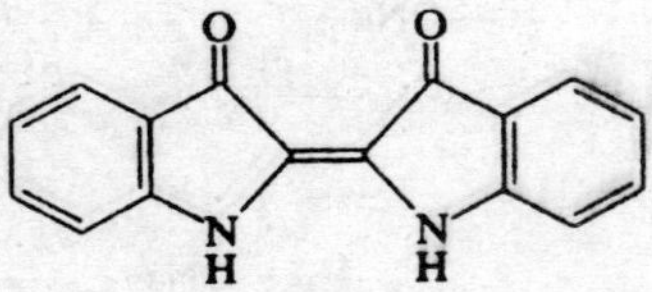

*cis*-form

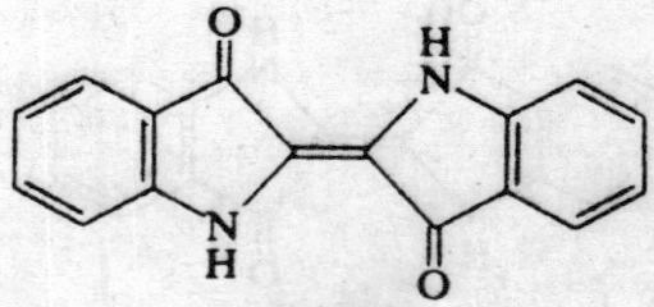

*trans*-form

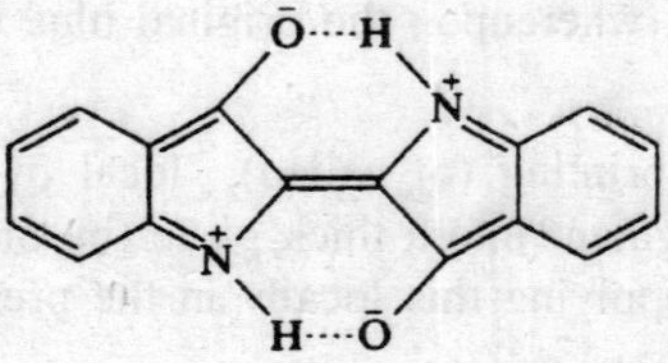

(I)

Derivatives of both are known, but the stable form is the *trans,* X-ray analysis shows that the indigotin molecule contains a centre of symmetry. There is, however, some evidence to show that it is the m-form which is produced in the cloth, and that it slowly changes into the stable *irons-form.*

The colour of indigotin requires explanation by charged structures, *e.g.,* the o-quinonoid charged structure (I) is a possibility; at the same time, this structure would be stabilised by hydrogen bonding.

*Derivatives of indigotin.* Indigotin can be chlorinated or brominated directly in the 5-, 5′, 7- and 7′- positions. The tetrachloro-derivative, *Brilliant Indigo B.* and the tetrabromo-derivative, *Brilliant Indigo 2B,* are both used commercially, both are brighter than indigotin.

*Tyrian Purple ('Purple of the Ancients')* is 6, 6′- dibromoindigotin, and was originally obtained from a species of molluscs in the Mediterranean. It is not commericially important since much cheaper dyes of similar colour are available. *Indigo Carmine* is the sodium salt of indigotin 5, 5′- disulphonic a acid. *Thioindigo* and many of its derivatives, *e.g., Hydron Pink 2F* (6, 6′- dichloro-4,4′-dimethylthioindigo), are important vat dyes.

In the case of indigotin, two geometrical isomers are possible, but since hydrogen bonding is very weak in the sulphur compounds, neither form is stabilised, and in solution both forms appear to exist in equilibrium (Wyman *et al.,* 1951).

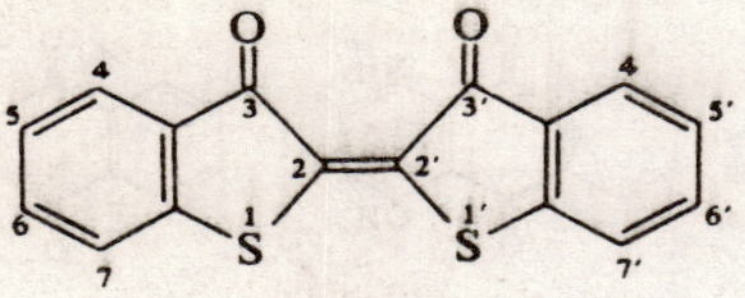

thioindigo

## ANTHRAQUINOID DYES

*Alizarin* is one of the most important anthraquinoid dyes; it is the chief constituent of the madder root. Alizarin is 1, 2-dihydroxyanthraquinone, and is now manufactured synthetically by sulphonating anthraquinone with fuming sulphuric acid at high temperatures, and fusing the sodium salt of the product, anthraquinone-2-sulphonic acid, with sodium hydroxide and the calculated quantity of potassium chlorate at 180–200°C under pressure.

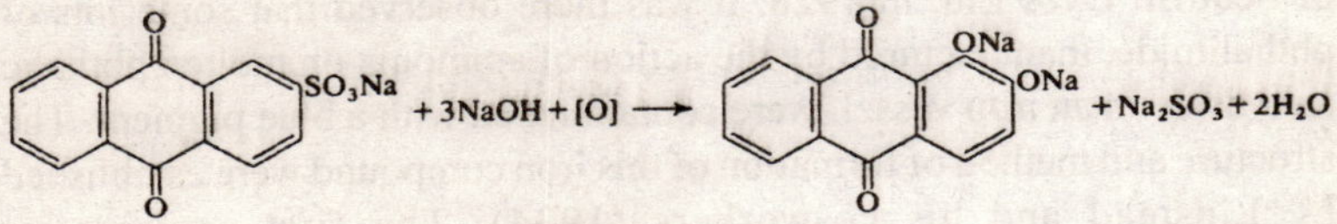

$$\text{(anthraquinone-}SO_3Na) + 3NaOH + [O] \longrightarrow \text{(1,2-di-}ONa\text{ anthraquinone)} + Na_2SO_3 + 2H_2O$$

The function of the potassium chlorate is to provide oxygen for the oxidation of the 1-carbon atom to COH.

*Properties of alizarin.* Alizarin forms ruby red crystals, m.p. 290°C, and dissolves in alkalis to give purple solutions. It is a mordant dye, and the colour of the lake depends on the metal used.

Aluminium gives a red lake *(Turkey red),* iron (ferric) a violet-black, and chromium a brown-violet. Aluminium and iron lakes are usually employed for cotton dyeing and for printing, and aluminium and chromium lakes for wood dyeing.

In addition to the aluminium compound, the lake contains calcium salts, Turkey-red oil and sodium phosphate. These additional compounds affect the brightness of the lake and increase its stability towards acids and alkalis.

It appears that aluminium hydroxide alone does not yield a lake with alizarin, but only after the addition of a calcium salt.

Some anthraquinoid dyes of various types are.

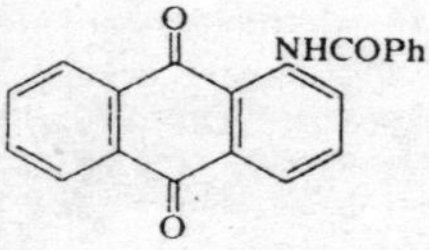

NHCOPh

**Indanthrene Yellow GK**
(vat dye)

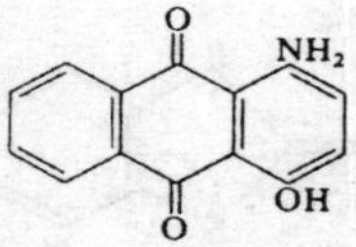

$NH_2$ OH

**Duranol Red 2B**
(disperse dye)

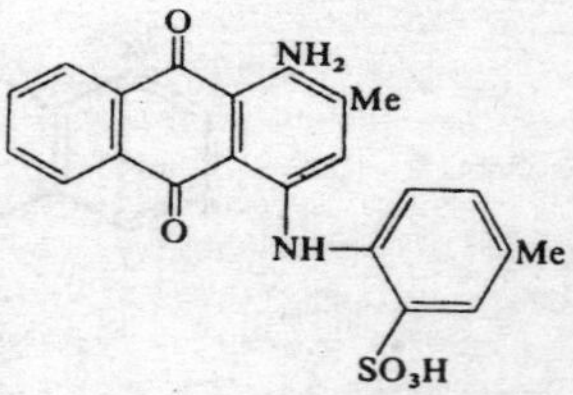

$NH_2$ Me NH Me $SO_3H$

**Kiton Fast Blue CR**
(acid dye)

## PHTHALOCYANINE PIGMENTS AND DYES

The *phthalocyanines* are a very important class of organic dyes and pigments; they are coloured blue to green. They are very fast to light, heat, acids and alkalis, and are very useful for paints, printing inks, synthetic plastics and fibres, rubber, etc.

The phthalocyanine dyes were discovered by accident at the works of Scottish Dyes Ltd. in 1928. It was there observed that some lots of phthalimide, manufactured by the action of ammonia on molten phthalic anhydride in an iron vessel, were contaminated with a blue pigment. The structure and method of formation of this iron compound were established by Linstead and his co-workers (1934). The first commercial phthalocyanine dye was *Monastral Fast Blue BS.* This is copper phthalocyanine, the copper atom taking place of the hydrogen atoms in the two NH groups of phthalocyanine itself and being co-ordinated with the other two nitrogen atoms. The colour depends on the metal (copper, magnesium, lead, etc.), and the greener shades are obtained by direct chlorination or bromination.

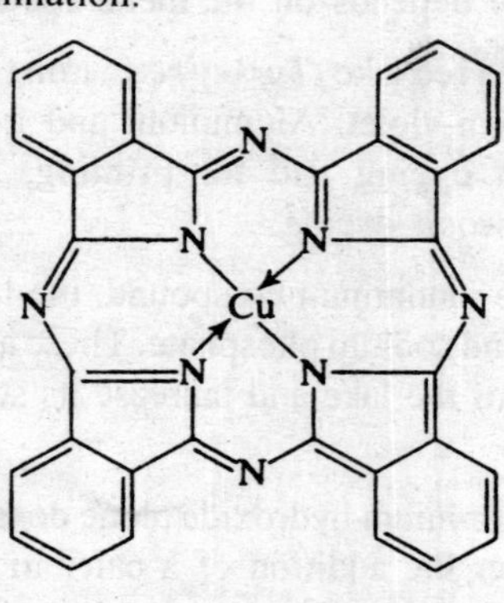

N N N N Cu N N N N

Metal phthalocyanines may be prepared.

(i) By passing ammonia into molten phthalic anhydride or phthalimide in the presence of a metal salt.

(ii) By heating o-cyanoarlamides or phthalimide with metals or metallic salts.

(iii) By heating phthalic anhydride or phthalimide with urea and a metallic salt, preferably in the presence of a catalyst such as boric acid.

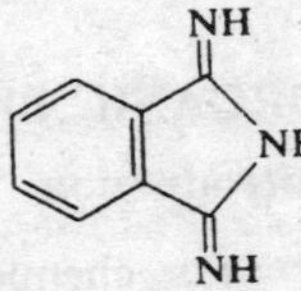

Phthalocyanine may be prepared by heating phthalonitrile with a little triethanolamine.

Phthalocyanines are made water soluble by soluble, and these soluble salts are used as direct dyes. The most successful water-soluble copper phthalocyanines, however, are the *Alcian dyes.* The pigment is chloromethylated, and the product rendered soluble by formation of the isothiouronium salt. Decomposition of this salt on the fibre generates insoluble substituted phthal-ocyanines.

A more recent method is to prepare copper phthalocyanine on the fibre itself. A compound such as (I; see (i) above) is printed on to a cotton cloth as a paste, together with a suitable copper salt. On gentle heating, copper phthalocyanine is formed in the cloth. These precursors are sold under the trade name *Phthalogens.*

*Fluorescent brightening agents.* These compounds are often called *optical bleaches,* and their use is based on their ability to absorb in the region 330-390 nm (hence they are 'colourless') and to fluoresce in the blue region of the spectrum (430–490 nm). Fabrics, after repeated washing, tend to develop a yellow colour.

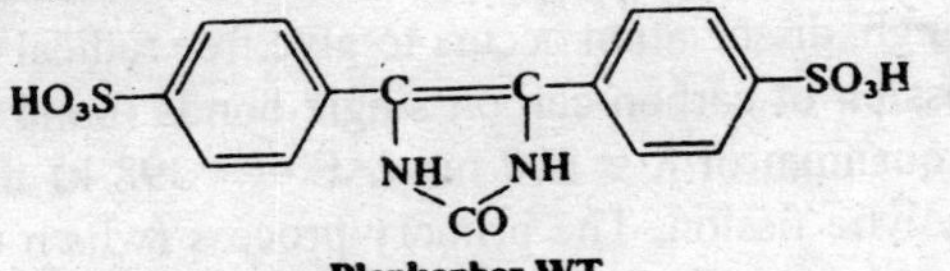

**Blankophor WT**

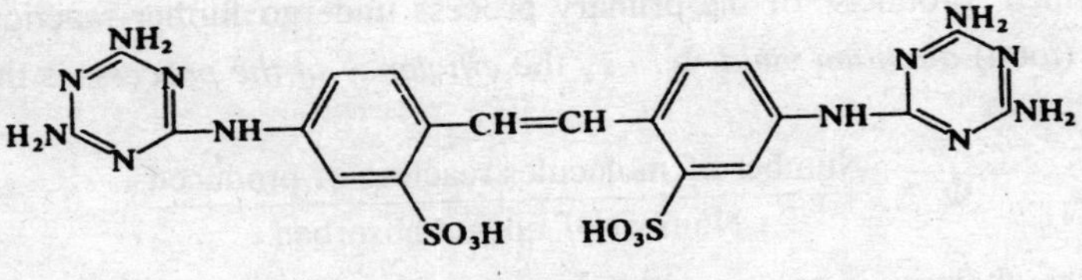

**Tinopal BV**

Since the complementary colour of yellow is blue (Table 4.1), 'addition' of blue light will cause the fabric to appear white. Hence the use of detergents containing 'whiteners', many of which are stilbene derivatives (as sodium salts), *e.g.*,

## ORGANIC PHOTOCHEMISTRY

### Introduction

For chemical reaction to occur, the reacting molecules must be activated, and as we have seen, molecules usually acquire this activation energy by increasing their kinetic energy. In *photochemical reactions,* however, the activation energy is obtained by the absorption of radiation (light energy). In order to study the mechanisms of photochemical reactions, a necessary requirement is the nature of the electronic transition, *i.e.,* whether it is $\sigma \rightarrow \sigma^*$, $\pi \rightarrow \pi$ or $n \rightarrow \pi^*$, and also whether the state is singlet or triplet. As we have seen, a molecule in the ground state is raised to an excited singlet state by absorption of a quantum of the appropriate energy. We have also seen how an excited molecule may lose this excess of energy to return to the ground state (loss by collision, fluorescence, and phosphorescence). However, because of their excitation, the molecules are more reactive (than ground-state molecules) and so can undergo chemical reaction.

In view of what has been said, it might be expected that *one* molecule would react for *each* quantum absorbed. This is the *law of photochemical equivalence* (Einstein, 1912), but is rarely observed in practice. The reason for this has been explained as follows. Photochemical reactions consist of *two* processes. The first is the *primary process,* in which the light is absorbed to produce the excited molecule. If the energy is high enough, dissociation occurs to give free radicals. Since this often involves fission of carbon-carbon single bonds (bond energy = 348 kJ $mol^{-1}$), a quantum of $\lambda$ = 300 nm ($\Delta E$ = ~ 398 kJ $mol^{-1}$) will bring about homolytic fission. The primary process (which obeys the law of photochemical equivalence) is then followed by *secondary processes,* in which 'products' of the primary process undergo further reactions. The (total) *quantum yield,* $\Phi$ , *i.e.,* the *efficiency of the process,* is then denned as

$$\Phi = \frac{\text{Number of molecules reacting or produced}}{\text{Number of quanta absorbed}}$$

$\Phi$ is rarely unity in practice; its value varies from $10^{-2}$ to $10^4$ (or more).

Photochemical secondary processes are of various types, *e.g.*, substitution, addition, dimerisation, elimination, rearrangement, etc.

Photochemical reactions are usually divided into two main types, *direct light-induced or non-sensitised reactions, and photosensitised reactions.* In non-sensitised reactions, the molecule that actually absorbs the light undergoes various secondary processes to give the final products of the reaction. On the other hand, in photosensitised reactions, one type of molecule, the *donor* (D) that is present absorbs the light (which is of the wavelength characteristic of the donor), becomes excited, and then transfers its excitation energy to another type of molecule, the *acceptor* (A), that is also present. In this way, the donor molecule returns to the ground state and the acceptor molecule is raised to an excited state, which then undergoes emission or photochemical reaction. This transfer of energy from D to A, followed by emission of light or reaction of A is called *sensitisation.*

For the donor-acceptor relationship to function, certain conditions must be fulfilled, (i) The donor-excited state must have a sufficiently long life-time to be able to transfer its energy to the acceptor molecule. Since the triplet state has a longer life–time than the singlet state, sensitisation involving the former is much more favourable than the latter. It therefore follows that sensitisers will be more effective the more efficient is their intersystem crossing, (ii) The transfer of energy from D to A will occur only if the excitation energy of the former is higher than that of the latter. A common sensitiser is benzophenone; the energies of its $S_1$, and $T_1$ states are, respectively, 346 kJ and 291 kJ above the 50 state, (iii) In general, the multiplicities of the D and A excited states are the same.

Since molecules excited above the $S_1$ state rapidly decay to this state, it is the first excited triplet state that is usually involved in photochemical reactions. The actual transfer of energy from the donor to the acceptor molecule occurs through collisions, and occurs on *every* collision if the above conditions are fulfilled. The most usual types of energy transfer are (the asterisk represents the excited state).

$$D(S_0) \xrightarrow{h\nu} D(S_1) \longrightarrow D^*(T_1) \xrightarrow{A} D(S_0) + A^*(T_1)$$

$$D(S_1) \xrightarrow{A} D(S_0) + A^*(S_1)$$

Photochemical reactions are being increasingly used for the synthesis of organic compounds, and in many cases, for the preparation of new compounds that cannot be obtained by the usual methods.

Many photochemical reactions have been mentioned in the text (Index), but no details of mechanisms have been described. These will now be discussed, and since the most widely studied reactant molecules are the carbonyl compounds, they will be largely used to describe the various types of photochemical reactions. However, before proceeding further, let us first define two additional terms commonly used in photochemistry. *Photolysis* is the term used to describe the breaking of chemical bonds by light energy. The bonds are broken homolytically to form free radicals, and sometimes small molecules are eliminated *(cf.* mass spectrometry). In relatively few cases, photolysis produces ions. *Flash photolysis* is the process in which a flash of light of very high intensity, lasting a number of microseconds ($10^{-6}$ S), irradiates the reaction mixture. In this way, reactive intermediates are formed in high concentration and their rate of decay may be measured by various means, particularly spectroscopically. Porter *et al.,* (1968), by means of a laser pulse, have shortened the flash duration to 18 nanoseconds ($10^{-6}$S).

*Photochemical elimination reactions.* For ketones, these reactions usually involve the loss of carbon monoxide. An extensively studied reaction is the photolysis of acetone in the vapour phase (v.p.), and it has been found that the products depend on the temperature of the experiment:

*(a) Elevated temperature:* the products are carbon monoxide and ethane.

$$CH_3\overset{\overset{O}{\|}}{C}CH_3 \xrightarrow[\text{(v.p.)}]{h\nu} [CH_3\overset{\overset{O}{\|}}{C}CH_3]^* \longrightarrow CH_3\overset{\overset{O}{\|}}{C}\cdot + CH_3\cdot$$

$$CH_3\overset{\overset{O}{\|}}{C}\cdot + CH_3\cdot + CO$$

$$2CH_3\cdot \longrightarrow C_2H_6$$

The excited state of acetone could be $S_1$, or $T_1$, but it has been shown that the decomposition via the triplet state is the more efficient, the quantum yield $\Phi$ being almost unity. It should also be noted that of all the bonds present in acetone, the C—C bond is the weakest.

(b) *Room temperature:* the products are biacetyl, acetaldehyde, methane, and possibly hexane -2,5-dione.

$$CH_3\overset{O}{\overset{\|}{C}}CH_3 \xrightarrow{hv} [CH_3\overset{O}{\overset{\|}{C}}CH_3]^* \longrightarrow CH_3\overset{O}{\overset{\|}{C}}\cdot + CH_3\cdot$$

$$CH_3\overset{O}{\overset{\|}{C}}\cdot + CH_3\overset{O}{\overset{\|}{C}}CH_3 \longrightarrow CH_3\overset{O}{\overset{\|}{C}}H + \cdot CH_2\overset{O}{\overset{\|}{C}}CH_3$$

$$CH_3\cdot + CH_3\overset{O}{\overset{\|}{C}}CH_3 \longrightarrow CH_4 + CH_3\overset{O}{\overset{\|}{C}}CH_2$$

$$2CH_3\cdot\overset{O}{\overset{\|}{C}} \longrightarrow CH_3—\overset{O}{\overset{\|}{C}}—\overset{O}{\overset{\|}{C}}—CH_3$$

$$2CH_3\cdot\overset{O}{\overset{\|}{C}}CH_2 \longrightarrow CH_3—\overset{O}{\overset{\|}{C}}CH_2CH_2\overset{O}{\overset{\|}{C}}CH_3$$

Acetone is a symmetrical molecule, but when the ketone is unsymmetrical, preferential fission occurs to give the more stable alkyl radical, *e.g.,* photodissociation of ethyl methyl ketone gives ethyl and methyl radicals, with the former predominating.

$$CH_3\overset{O}{\overset{\|}{C}}\cdot + C_2H_5 \xleftarrow{hv} CH_3\overset{O}{\overset{\|}{C}}C_2H_5 \xrightarrow{hv} CH_3\cdot + \overset{O}{\overset{\|}{C}}C_2H_5$$

Photochemical reactions of ketones (and aldehydes) which result in fission of the C—CO bond fallowed by elimination of carbon monoxide, are classified as the *Norrish type I process.*

As the length and complexity of the alkyl groups increase, the products of photolysis become more complicated.

Thus, when a hydrogen atom in the alkyl group can enter into the formation of a six-membered cyclic T.S. involving the oxygen atom as one member of the ring, *$\gamma$ -hydrogen abstraction* and fission of the molecule occur to give an alkene and a ketone (via the enol form), *e.g.,* hexan-2-one gives mainly propane and acetone, the elimination of carbon monoxide occurring to a lesser extent *(cf.* mass spectra of ketones).

$$CH_3CH(H)-CH_2-CH_2-C(=O)CH_3 \xrightarrow[\text{(v.p.)}]{h\nu} \left[\begin{matrix} CH_3CH & O \\ \| & \| \\ CH_2 & CCH_2 \\ & CH_2 \end{matrix}\right] \longrightarrow$$

$$\begin{matrix} CH_3CH \\ \| \\ CH_2 \end{matrix} + \left[\begin{matrix} HO \\ | \\ C \\ CH_2 \;\; CH_3 \end{matrix}\right] \longrightarrow CH_3COCH_3$$

This type of photolysis undergone by ketones (and aldehydes) is called the *Norrish type II process.*

α, β–Unsaturated ketones, provided the light energy is high enough, undergo photolysis with elimination of carbon monoxide (see also later).

Cyclic ketones undergo elimination of carbon monoxide and also ring opening with the formation of an unsaturated aldehyde, etc., *e.g.*, cyclopentanone.

$$\text{cyclopentanone} \xrightarrow[\text{(v.p.)}]{h\nu} [\text{diradical}] \longrightarrow CO + [\text{tetramethylene diradical}] \longrightarrow \text{cyclobutane}$$

$$[\text{diradical}] \xrightarrow{-H \text{ abstraction}} CH_2{=}CH{-}CH_2{-}CH_2{-}CHO \qquad [\text{tetramethylene diradical}] \longrightarrow 2CH_2{=}CH_2$$

The photolysis of aldehydes is more complicated than that of ketones, the usual primary reaction resulting in the formation of the *formyl radical, i.e.,*

$$RCHO \xrightarrow[\text{(v.p)}]{h\nu} R. + .CHO$$

This is then generally followed by the secondary process.

$$R. + .CHO \longrightarrow RH + CO$$

All of the foregoing decarbonylations are vapour phase reactions, and the products are the same if the photochemical reactions are carried out in a *chemically inert solvent.* In protic solvents, the products are usually different.

Nitrogen is another type of molecule that is readily eliminated from compounds containing nitrogen-nitrogen double bonds, particularly azo-compounds, *e.g.*, azomethane in the gas phase (g.p.) ($\Phi$ = 1).

$$CH_3 - N = N - CH_3 \xrightarrow[(g.p)]{h\nu} N_2 + 2CH_3\cdot \longrightarrow C_2H_6$$

In the same way, diazomethane gives nitrogen and methylene.

The photolysis of diazonium salts in alcohols gives aromatic hydrocarbons and aryl ethers, and the mechanism is believed to be via aryl free-radicals.

The photolysis of nitrites which do not contain $\gamma$-hydrogen atoms usually results in the elimination of nitric oxide and the formation of hydroxy and carbonyl compounds, *e.g.*, [NO ≡ NO.].

$$\overset{\beta}{C}H_3\overset{\alpha}{C}H_2CH_2ONO \xrightarrow{h\nu} NO + CH_3CH_2CH_2O\cdot$$

$$\begin{matrix} CH_3CH_2CH{-}O \\ H\ \cdot OCH_2CH_2CH_3 \end{matrix} \longrightarrow CH_3CH_2CH{=}O + HOCH_2CH_2CH_3$$

When $\gamma$-hydrogen atoms are present (for the formation of a six-membered cyclic T.S.), then the product is an oxime or the corresponding nitroso dimer. This reaction, which is very important synthetically, is known as the *Barton reaction* (1960) and may be formulated in general terms as shown (note the nitroso-compound intermediate).

ONO, H–CHR, RCH, CH₂, CH₂ $\xrightarrow{h\nu}$ O·, (NO) H–CHR, RCH, CH₂, CH₂ ⟶ OH, ·CHR (NO), RCH, CH₂, CH₂ ⟶ OH, ON–CHR, RCH, CH₂, CH₂ ⟶ dimer

OH, ON–CHR, RCH, CH₂, CH₂ ⟶ OH, HON=CR, RCH, CH₂, CH₂

*Photochemical reductions.* In these reactions, a hydrogen atom is abstracted from the solvent or from another reactant to produce two free radicals. The most notable example of this type is the irradiation of benzophenone in alcoholic solvent to give benzopinacol and oxidation products of the alcohol; isopropanol is generally the alcohol used. Several mechanisms have been proposed, a widely accepted one is.

$$\text{(i) } Ph_2C = O \xrightarrow{h\nu} [Ph_2C = O]^*(S_1) \longrightarrow [Ph_2C = O]^*(T_1)$$

$$\text{(ii) } [Ph_2C = O]^*(T_1) + Me_2CHOH \longrightarrow Ph_2\dot{C} - OH + Me_2\dot{C} - OH$$

There is much evidence for this mechanism. *Two* molecules of benzophenone are consumed, one *excited* molecule (step i) is converted into the benzhydryl radical (step ii), and another *ground-state* benzophenone molecule reacts with the hydroxyispropyl radical to give a benzhydryl radical (step iii). Thus, the quantum yield would be expected to be 2, and this is the value obtained in practice (under ideal experimental conditions). There is also much evidence to show that it is the triplet state of benzophenone that is involved. Benzophenone, in the presence of benzhydrol, also gives benzopinacol on irradiation.

$$\text{(v) } [Ph_2C = O]^*(T_1) + Ph_2CHOH \longrightarrow$$

$$2Ph_2\dot{C}OH \longrightarrow Ph_2C(OH)C(OH)Ph_2$$

When the photoreduction is carried out in the presence of a low concentration of benzhydrol, the quantum yield is still appreciable, thereby implying that the excited state has a relatively long life-time (hence triplet rather than singlet). Also, fluorescence has never-been observed for benzophenone solutions, and this shows that the singlet state decays rapidly, too rapidly to react with benzhydrol molecules in solution. Calculations have shown that benzophenone singlets undergo intersystem crossing to benzophenone triplets extremely rapidly, this explains the failure of benzophenone to fluoresce and supports the argument that it is benzophenone triplets which are the excited molecules involved in hydrogen abstraction.

At first sight it might have been expected from step (ii) that pinacol would be produced by dimerisation of the hydroxyisopropyl radicals. The absence of this product may be explained on the basis that the life-time of the hydroxyisopropyl radical is too short for it to combine with another radical to form the dimer; it transfers its hydroxylic proton too rapidly. On the other hand, the benzhydryl radical has a relative long life-time due to resonance stabilisation, and is sufficiently stable not to attack $Me_2CHOH$.

One other point about this reduction is the structure of the benzophenone triplet state. There are two possibilities, (n, π*) and (π, π*) triplets. The former, however, has more free-radical character than the latter, and can therefore abstract hydrogen atoms relatively easily. On the other hand, aromatic ketones which are not readily reduced are believed to be in the (π, π*) triplet state.

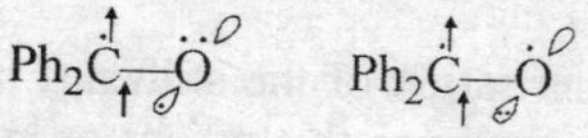

$Ph_2C—O$ (n, π*) $Ph_2C—O$ (π, π*)

*Photochemical oxidations.* Many photochemical reactions, when carried out in the presence of oxygen, result in the formation of peroxides, hydroperoxides, oxidation of a functional group, etc. Although oxygen is a triplet (diradical) in its ground state, it does not add to cyclic dienes. In the presence of a sensitiser, *e.g.,* fluorescein, eosin, the oxygen is raised to its S, state (all electrons paired), and it is the oxygen in this state that is the reactive species. Many linear polynuclear aromatic compounds, however, add oxygen in the absence of a sensitiser. In these cases, the substrate molecule is believed to behave as the sensitiser. Thus, the photo-oxidation of anthracene to form the transannular (9, 10-) peroxide may possibly proceed as follows.

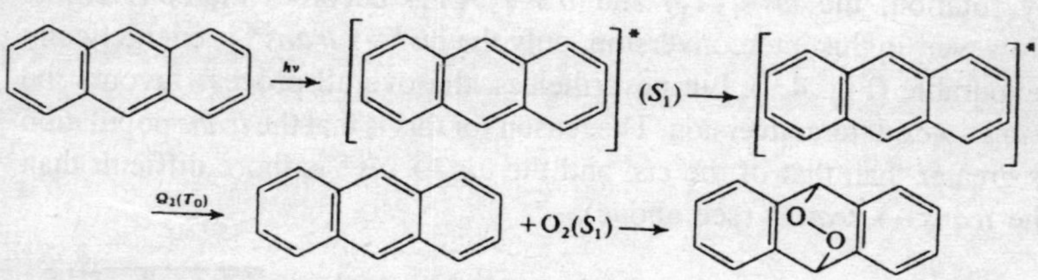

$h\nu$ $[\ ]^*$ $(S_1)$ ⟶ $[\ ]^*$

$\xrightarrow{O_2(T_0)}$ $+O_2(S_1)$ ⟶

Alkenes which possess an allylic hydrogen atom tend to produce the 'allylic isomerised' hydro-peroxide when they undergo photosensitised oxidation. This involves hydrogen abstraction, and it appears that a cyclic T.S. is involved, *e.g.,* 4-methylpent-2-ene.

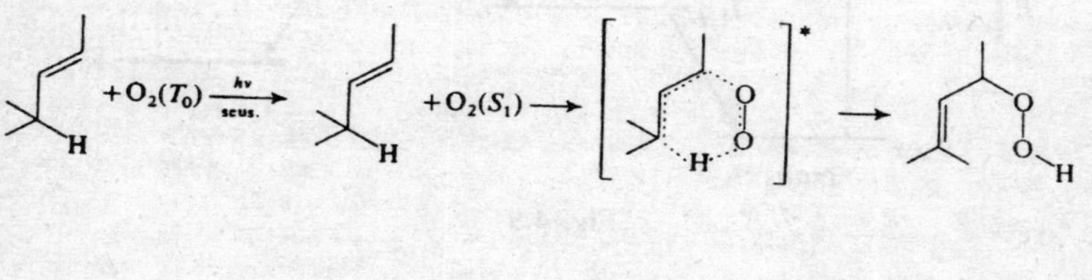

H $+O_2(T_0)\xrightarrow[\text{sens.}]{h\nu}$ H $+O_2(S_1)$ ⟶ $[\ ]^*$ ⟶ O–O–H

*Photochemical cis-trans isomerisations. cis-trans* Isomerisation can be carried out (usually in solution) by irradiation alone or in the presence of a sensitiser or a catalyst. In general, an equilibrium mixture is reached, the *cis-trans* ratio remaining constant no matter how much longer the irradiation is continued. This condition, called a *photostationary state,* is independent of which isomer is the starting material and always contains predominantly the cis-isomer. The actual ratio of the *cis-trans* forms, however, depends on a number of factors, *e.g.*, solvent, temperature, nature of the sensitiser, etc.

The *cis-trans* isomerisation of the stilbenes has been examined in great detail, and so we shall use this as our example. The evidence is strongly in favour that the reaction proceeds via a triplet state (Ph – CH— CH–Ph), but since the *cis-trans* ratio depends on the temperature, several theories have been proposed to explain the isomerisation. According to one theory, the *cis-* and *transforms* give *different* excited states, and the temperature effect is due to the existence of an energy barrier between these excited states (Fig. 4.9). Because of steric effects, the cis-isomer has a higher energy content than the *trans* in the ground state, and this is also (usually) the case for the corresponding excited states. Also, because of this larger steric effect, the molar absorptivity of the cis-isomer is (usually) lower than that of the *trans-isomer.* Hence, the population of the *trans* excited state is greater than that of the cis. These excited states can return to their respective ground states (Fig. 4.9) or, by rotation, the *cis** ($T_1$) and *trans** ($T_1$) become interconvertible. However, in this interconversion, only the *cis** → *trans** is energetically favourable (Fig. 4.9), but nevertheless, the overall process favours the *trans* → *cis* interconversion. The reason for this is that the *trans* population is greater than that of the *cis,* and the *cis* → *cis** is more difficult than the *trans* → *trans** (see above).

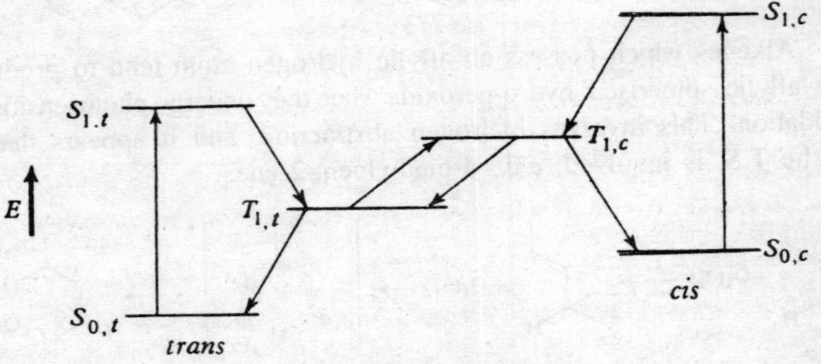


**Fig. 4.9**

In *photosensitised cis-trans* isomerisation, the sensitiser excitation energy (of the *T* state) must be higher than that of the cis- and trans-isomers for energy transfer to occur. Hence, in practice, as long as the sensitiser has a triplet energy state above a certain minimum, the *cis-trans* ratio is very little affected by the nature of the sensitiser. Thus, if benzophenone is the sensitiser (donor), the mechanisms for the isomerisation of stilbene (acceptor) may be written.

(i) $Ph_2CO\ (S_0) \xrightarrow{h\nu} [Ph_2CO]^*\ (S_1) \longrightarrow [Ph_2CO]^*\ (T_1)$

(ii) $[Ph_2CO]^*\ (T_1) + \text{cis-}Ph_2C_2H_2 \longrightarrow$

$$Ph_2CO(S_0) + \left[\begin{matrix} H & & H \\ & \dot{C}-\dot{C} & \\ Ph & & Ph \end{matrix}\right] (T_1)$$

(iii) $[Ph_2CO]^*\ (T_1) + \text{trans-}Ph_2C_2H_5 \longrightarrow$

$$Ph_2CO\ (S_0) + \left[\begin{matrix} Ph & & H \\ & \dot{C}-\dot{C} & \\ H & & Ph \end{matrix}\right] (T_1)$$

(iv) $$\left[\begin{matrix} H & & H \\ & \dot{C}-\dot{C} & \\ Ph & & Ph \end{matrix}\right] (T_1) \rightleftharpoons \left[\begin{matrix} Ph & & H \\ & \dot{C}-\dot{C} & \\ H & & Ph \end{matrix}\right] (T_1)$$

$$\downarrow \text{spin inversion} \qquad\qquad \downarrow \text{spin inversion}$$

$$\begin{matrix} H & & H \\ & C=C & \\ Ph & & Ph \end{matrix}\ (S_0) \qquad \begin{matrix} Ph & & H \\ & C=C & \\ H & & Ph \end{matrix}\ (S_0)$$

Actually, the mechanism given is an over-simplification; there is evidence to show that another triplet state of stilbene (intermediate in energy content between the *cis* and *trans* triplet states) may also be produced. This triplet state, because it can be reached *directly* from the ground state (normally a forbidden transition) has been called a *phantom triplet* (Hammond, 1960).

The structures of the *cis* and *trans* triplet states of stilbene are not clear. Both would be expected to be resonance hybrids (of the canonical forms given in equations ii and iii, above), but because rotation can occur, the CH – CH bond in both would be expected to be predominantly single in character. However, because of the different steric effects operating, the double bond character in the two forms will be different, presumably greater in the *trans* than in the *cis,* since the former is more stable than the latter.

Photochemical *cis-trans* isomerisation can also be effected in the presence of, *e.g.,* bromine or iodine. A probable mechanism is.

$$Br_2 \xrightarrow{h\nu} 2Br\cdot$$

Me(H)C=C(H)Me —Br·→ Me(H)Ċ—C(Me)(H)Br ⇌ H(Me)Ċ—C(Me)(H)Br

↓ –Br· (from each radical)

Me(H)C=C(H)Me  H(Me)C=C(Me)H

In addition to photochemical *cis-trans* isomerisations of alkenes, other systems, *e.g.,* oximes, azo-compounds, can be isomerised under similar conditions.

*Photochemical rearrangements.* α,β Unsaturated ketones (acyclic and cyclic enones and dienones) undergo various types of rearrangements when irradiated with light. Acyclic α,β-unsaturated ketones containing a γ-hydrogen atom often undergo double-bond migration provided the light energy is not high enough to cause dissociation. The presence of the γ-hydrogen is necessary for the formation of a six-membered cyclic T.S., leading to hydrogen abstraction and transfer to the oxygen atom.

[scheme: enone —hν→ [cyclic T.S.] → enol (OH) → β,γ-unsaturated ketone]

*e.g.*, 5-methylhex-3-en-2-one.

It should be noted that the system changes from conjugation to non-conjugation, and there is *no* equilibrium (as is the case with *cis-trans* isomerisation). The reason for this is that irradiation is carried out with a wavelength absorbed by the *conjugated* substrate, and since this is longer than that absorbed by the non-conjugated product, the latter remains unaffected once it is formed. Thus, isomerisation gives completely the less stable isomer. This type of isomerisation is an example *of optical pumping.*

The most widely studied compounds that undergo photochemical rearrangements are the dienones. These undergo the dienone-phenol rearrangement in acid solution, and also undergo this rearrangement under the influence of light, but in this case the reaction is more complex, *e.g.,* 4,4-diphenylcyclohexa-2,5-dienone (a cross-conjugated dienone), in aqueous dioxan solution, gives four products.

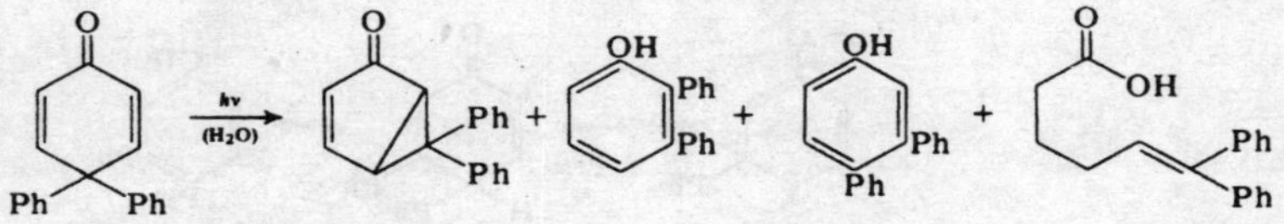

One mechanism that has been proposed (Zimmerman *et al.,* 1964) is given here to illustrate the variety of steps that can be involved in a photochemical reaction. (I) is the compound that leads to the formation of all other products, and it should be noted that the steps involve first a series of free radicals and then a series of carbonium ions, the latter undergoing rearrangements characteristic of such ions.

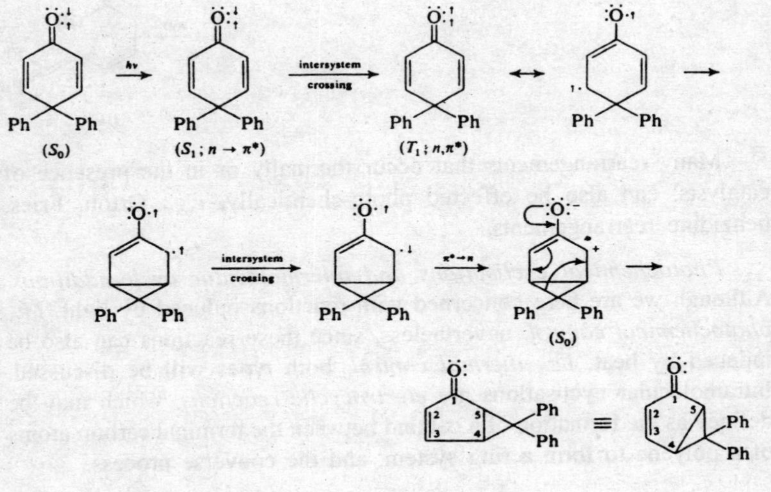

On the assumption of a similar mechanism, the formation of the other products can be explained as follows.

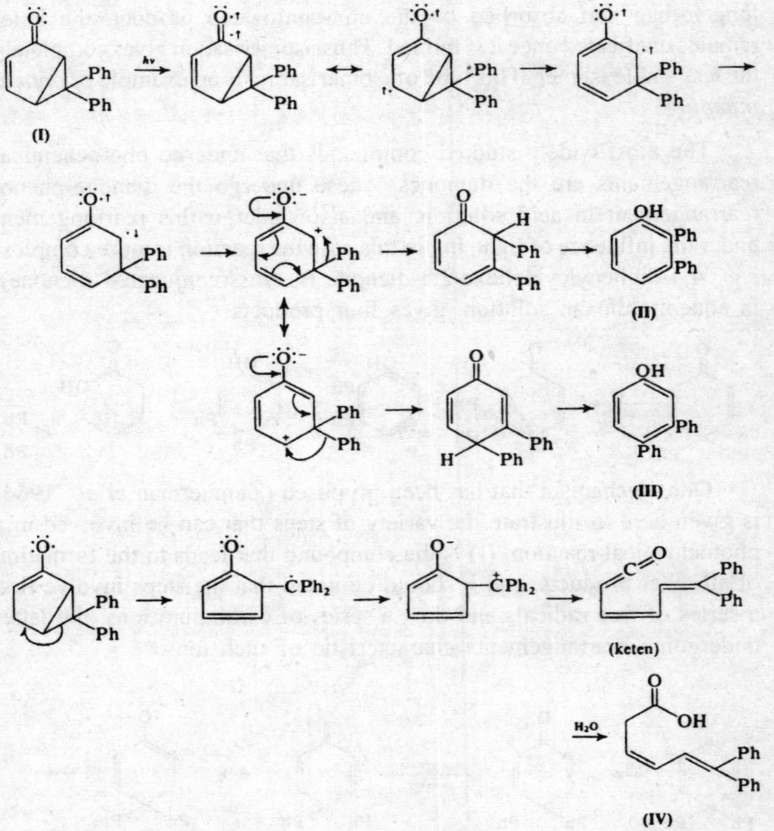

(I)

(II)

(III)

(keten)

(IV)

Many rearrangements that occur thermally or in the presence of catalysts, can also be effected photo-chemically, *e.g.*. Orton, Fries, benzidine rearrangements.

*Photochemical cyclisations and intermolecular cycle-additions.* Although we are here concerned with reactions induced by light, *i.e., photochemical control,* nevertheless, since these reactions can also be induced by heat, *i.e., thermal control,* both types will be discussed. Intramolecular cyclisations are *electrocyclic reactions,* which may be defined as the formation of a σ-bond between the terminal carbon atoms of a polyene to form a ring system, and the converse process.

Intermolecular cyclo-additions are of two types, *concerted* and *non-concerted.* Here we shall deal with only the concerted additions, which may be defined as a reaction in which *both new σ-bonds are formed simultaneously.* The Diels-Alder reaction is one example of this type, and another is the dimerisation of alkenes (see below).

Both intramolecular cyclisation and concerted intermolecular addition are characterised by having a negative entropy of activation, which is in keeping with formation of a cyclic T.S. These reactions may also be stereospecific, and this stereospecificity has been explained by means of the *Woodward-Hoffmann selection rules* (1965). These rules are based on the *principle of conservation of orbital symmetry,* according to which combination (involving $p_z$ orbitals) to form the σ-bond occurs by overlap of *like-sign orbitals, i.e.,* a positive lobe overlaps with another positive lobe, or a negative lobe with another negative lobe. The way in which this mode of overlap is achieved, however, depends on the number of n-electrons involved in the reaction and also on whether the reaction is under thermal or photochemical control. For *thermally controlled* intramolecular cyclisations, the reactant molecule is assumed to be in a *vibrationally excited ground state,* and that the two π-electrons in the highest occupied M.O. are involved in the formation of the new σ-bond, and vice versa. For *photochemically controlled* intramolecular cyclisations, the reactant molecule is in *is first electronic excited state, i.e.,* the (π,π*) excited state, and the two electrons, one in the highest (occupied) bonding M.O. and the other in the lowest (unoccupied) anti-bonding M.O. are involved in the formation of the new σ-bond. In this case it is the symmetry of the *anti-bonding* M.O. that is involved in the conservation of orbital symmetry, and because this symmetry differs from that of the highest occupied M.O. (in the ground state), the stereochemistry of the products is different for thermal and photochemical control.

For thermally controlled concerted intermolecular cycle-additions, it is assumed that the two π-electrons in the highest occupied M.O. overlap with the two π-electrons of the other reactant molecule in the lowest unoccupied M.O., *i.e..* in the first electronic excited state. For the photochemically controlled reaction, however, *both* reactant molecules are in their first electronic excited states, *i.e.,* the highest occupied M.O. for thermal control is now replaced by the M.O. in the first electronic excited state. Once again, the stereochemistry of the products is different for the two processes.

We are now in a position to apply the Woodward-Hoffmann selection rules and, at the same time, introduce some other terms necessary for our discussion. We shall be concerned mainly with ethylene and butadiene as linear reactant molecules and so, for ease of reference, the M.O.s of these two molecules are given here in Fig. 4.10. In Fig. 4.10, positive lobes are indicated by blacked circles, negative lobes by circles.

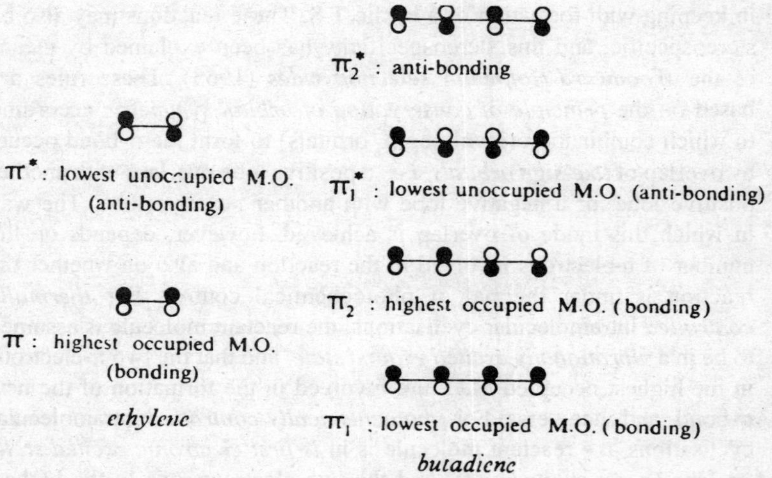


Fig. 4.10

*Intramolecular electrocyclic reactions. (i) Thermal control.* The most common examples are butadiene-cyclobutene intercoversions, but ring fission is far more common than ring closure. We shall first consider the ring fission of 3,4-dimethylcyclobutene, and it should be noted that the terms *cis* or *trans* refer to the positions of the two hydrogen atoms with respect to a given double bond, *e.g.,* a diene described as *cis, trans* has the *cis* configuration for one double bond and the *trans* configuration for the *other* double bond. The product obtained from cis-3,4-dimethylcyclobutene, (I), on heating, is cis, trans-1,4-dimethylbutadiene, (II).

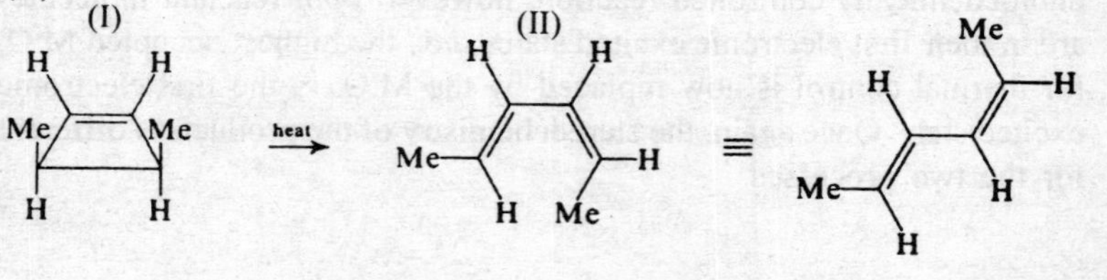

The formation of this stereospecific product is a consequence of conservation of orbital symmetry, and is explained as follows. When the ring in (I) opens, (II) contains four $\pi$-electrons, two of which occupy the lowest occupied M.O. (orbital $\pi_1$, see Fig. 4.10) and the other two the highest occupied M.O. (orbital $\pi_2$). On the assumption that the two $\pi$-electrons derived from the $\pi$-bond enter the highest occupied M.O., these enter orbital $\pi_2$, whereas the two $\pi$-electrons in the w-bond originally present in (I) enter orbital $\pi_1$. This sequence may be represented as (la) $\rightarrow \pi_1 + \pi_2$ (the combination of these being the ground state of II).

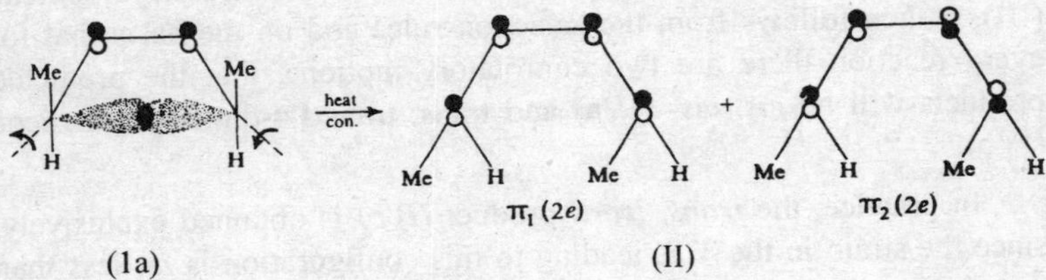


(1a) (II)

However, when the $\sigma$-bond undergoes fission, in order that the two individual $\pi$-electrons may enter orbital $\pi_2$ (in which the terminal lobes have opposite signs), it is necessary that in the 'de-overlapping' process, one lobe must rotate upwards and the other downwards. This is achieved by *both* rotating simultaneously in the same direction (anti-clockwise) as shown in (la). This mode of rotation, *i.e.,* both rotating in the *same* direction, is said to be *conrotatory.* It should be noted that if the mechanism was not concerted, *i.e.,* did not proceed via a cyclic T.S., *e.g.,* proceeded via a diradical, it is then very difficult to explain the stereospecificity of the reaction. It should also be noted that a *clockwise* conrotatory motion would still give the same product, *cis, trans* (Ib $\rightarrow \pi_2$ (a) = $\pi_2$, this is so, since + and – signs are purely relative).

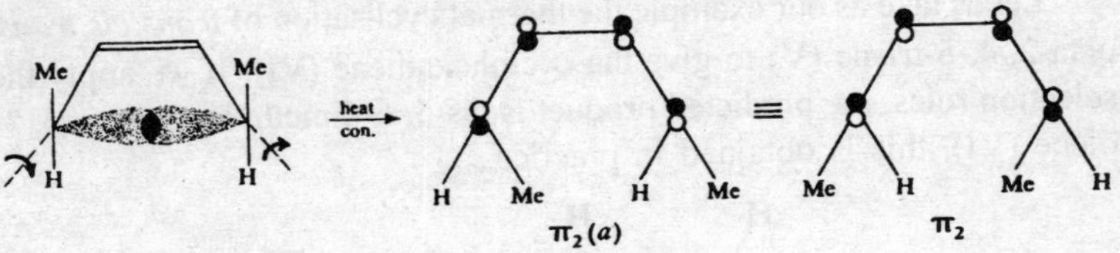

(1b) (II)

If the conversion of cis, trans -l,4-dimethylbutadiene into cis-3,4-dimethylcyclobutene were thermally possible, then this reaction would also occur by a conrotatory motion, since both positive (or negative) lobes must be aligned to give maximum overlap for the formation of the σ-bond. This, of course, could also have been anticipated from the principle of microscopic reversibility.

(III) —con. heat→ *cis, cis* (IVa) ; *trans, trans* (IVb) ←con. heat— (III)

If we now consider the ring fission of trans -3,4-dimethylcyclobutene (III), it then follows from the selection rules and on the basis that for every reaction there are two conrotatory motions, that the predicted products will be *cis, cis-* (IVa) and trans, trans l,4-dimethylbutadiene (IVb).

In practice, the *trans, trans* product *(IVb)* is obtained exclusively, since the strain in the T.S. leading to this configuration is far less than that leading to the *cis,cis* configuration (steric effects).

Now let us consider the thermal conversion of hexa-l, 3, 5-triene into cyclohexa-l, 3-diene. The triene contains three bonding and three anti-bonding M.O.s, but since we shall be concerned with only the highest occupied M.O. ($\pi_3$, bonding) and the lowest unoccupied M.O. ( $\pi_1^*$ anti-bonding), these two are given here (Fig. 4.11).

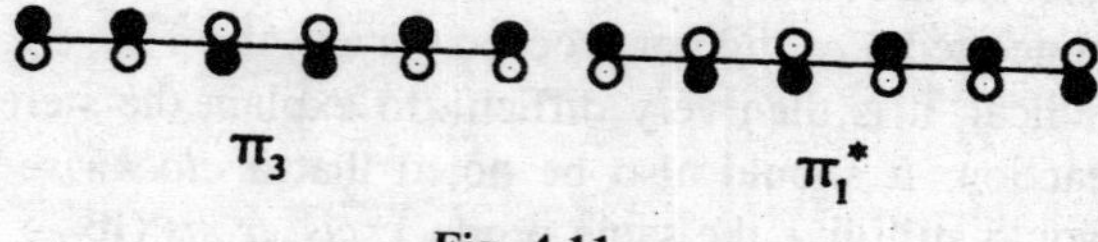


**Fig. 4.11**

Let us take as our example the thermal cyclisation of *trans,cis,trans.* octa-2, 4, 6-triene (V) to give the cyclohexadiene (VI). If we apply the selection rules, the predicted product is cis-5, 6-dimethylcyclohexa-l, 3-diene (VI), this is obtained in practice.

(V) —heat→ (VI)

In this case, however, to obtain maximum overlap in the formation of the a-bond, the terminal groups must rotate in opposite directions (see Va), *i.e.,* they undergo a *disrotatory* motion. This is represented by (Va) → (Via). In the same way, (Via) is converted into (Va) by thermal control, also by a disrotatory motion.

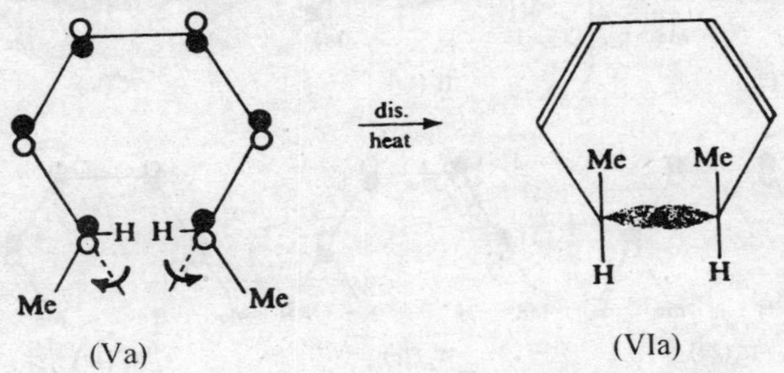


(Va) (VIa)

(II) *Photochemical control.* As we have seen, in thermally controlled cyclisations, the π–electrons forming the σ-bond are in the highest occupied M.O. For photochemically controlled reactions, the situation is different, the reactant molecule now being in its first excited electronic state. This also results in stereospecific reactions, but now the stereochemistry of the products is different, and the conrotatory and disrotatory motions are the reverse of those for thermal control.

As our first example, let us consider the photochemical conversion of trans-3,4-dimethylcyclobutene (VII) into the butadiene derivative (VIII). (VII), by thermal control, undergoes a conrotatory motion (in the ground state) to give the *trans, trans* product (IVb). Under photochemical control, however, the predicted product is the *cis, trans* isomer (VIII).

On irradiation, one π-electron in the reactant molecule is promoted to the π*-orbital (Vlla), then the excited molecule undergoes fission to give the butadiene derivative in its first excited state (VIII), in which two π-electrons are in orbital n, (Fig. 4.10), one in orbital $\pi_2$, and one in orbital $\pi_1^*$. Since it has been assumed that the two π-electrons derived from the σ-bond enter the higher energy M.O.s, to achieve this the σ-bond undergoes fission by a *disrotatory* motion, one π-electron entering the orbital $\pi_2$ and the other the orbital $\pi_1^*$.

The excited molecule now loses energy, the π-electron in the orbital $\pi_1^*$ returning to orbital $\pi_2$, thereby producing the butadiene derivative in its ground state [$\pi_1$ (2e) + $\pi_2$ (2e)]. This sequence is represented as shown in the diagram (VII) → (VIIa) → (VIIIa)

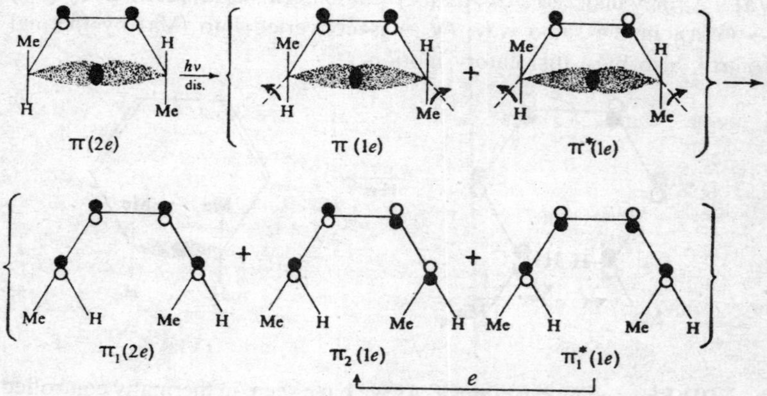

Me H H Me $\pi(2e)$ $\xrightarrow[\text{dis.}]{h\nu}$ { Me H H Me $\pi$ (1e) + Me H H Me $\pi^*$(1e) } →

{ Me H Me H $\pi_1$(2e) + Me H Me H $\pi_2$ (1e) + Me H Me H $\pi_1^*$ (1e) }

(VIII)

Now let us consider the photochemical cyclisation of *trains, cis, trans-octa-2, 4, 6-triene* (V). The predicted product is the *trans-* isomer, (IX) (cis, (VI), for the thermal cyclisation), and using similar arguments to those above, ring closure (and ring opening of the product) takes place by a conrotatory motion.

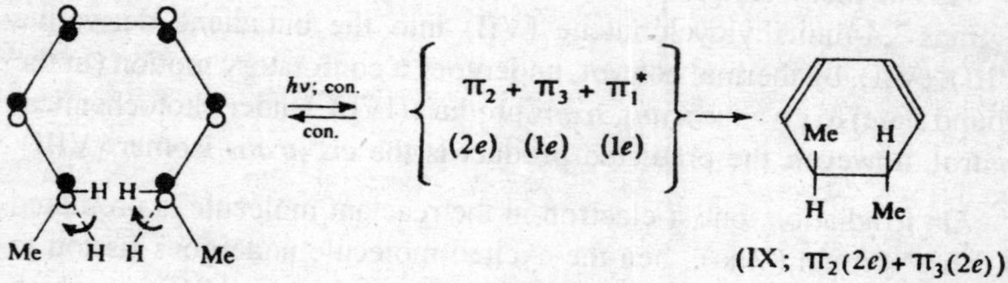

Me H H Me $\underset{\text{con.}}{\overset{h\nu;\ \text{con.}}{\rightleftharpoons}}$ { $\pi_2 + \pi_3 + \pi_1^*$ (2e) (1e) (1e) } → Me H H Me

(IX; $\pi_2(2e) + \pi_3(2e)$)

The foregoing discussion of the Woodward-Hoffmann rules for intramolecular electrocyclic reactions is summarised in Table 4.3, where $n$ is the number of $\pi$-electrons in the *acyclic polyene* (and hence $n - 2\pi$-eleetrons in the *cyclic isomer).*

It should also be noted that if the polyene *is part* of a ring system (usually a medium ring), then ring closure may occur to give a bicyclic product, the stereochemistry of which will, unless isomerisations have occurred, be in accordance with the selection rules.

**Table 4.3**

| *Acyclic polyene* (n = 0, 1, 2, ...) | *Thermal* | *Photochemical* |
|---|---|---|
| 4n | Conrotatory | Disrotatory |
| 4n + 2 | Disrotatory | Conrotatory |

*Concerted intermolecular cyclo-additions.* The most common type of reaction is *dimerisation,* but another important type is the Diels-Alder reaction. The selection rules also apply to these reactions which may, as before, be thermally or photochemically controlled. However, some complications arise in cyclo-additions owing to the fact that the reaction can proceed in four stereochemically distinct ways. Woodward and Hoffmann have introduced the following terms to indicate the geometrical relationship of the bonding (or fission) mode,

(i) *suprafacial process (supra;* subscript s) is the process in which bonds that are made or broken lie on the *same* face of the system undergoing reaction;

(ii) *antarafacial process (antara*; subscript a) is the process in which bonds that are made or broken lie on the *opposite* faces of the system. These terms are intended to replace the customary terms *cis (supra)* and *irons (antara).* The four combinations are illustrated by reactions leading to products (I-IV).

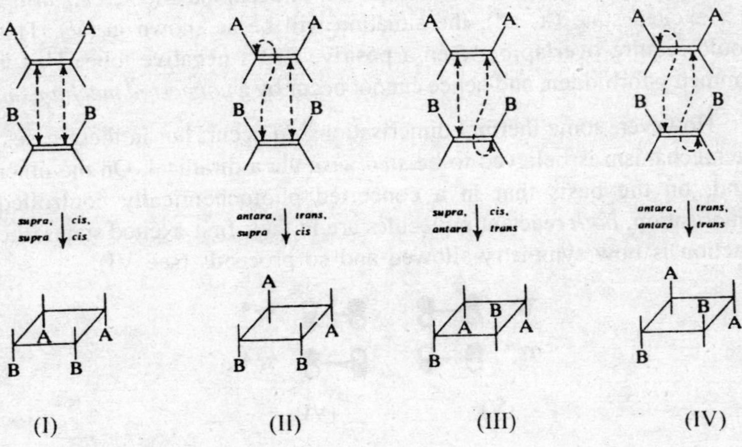

When the two reactant molecules are identical, *supra, antara* and *antara, supra* reactions give the same product. Also, for each process there are two possible cases, one producing the *endo* product, the other the *exo*.

Selection rules for predicting the geometry of the product have been developed, and these are given in Table 4.4, where n is the total number of n-electrons participating in the reaction. In practice, *supra, antara* and *antara, antara* cyclo-additions are uncommon, they appear to be sterically difficult.

**Table 4.4**

| n (1, 2,...) | *Thermal* | *Photochemical* |
|---|---|---|
| 4n (2 + 2, 4 + 4, | *supra, antara* | *supra, supra* |
| 6 + 2) | *antara, supra,* | *antara, supra* |
| 4n + 2 (4  2, | *supra, supra* | *supra, antara,* |
| 6 + 4, 8 + 2) | *antara, supra* | *antara, supra* |

*(2+2) Reactions.* According to the selection rules (Table 4.4), the $(2_5 + 2_5)$ thermal reaction is symmetry-forbidden. Let us consider ethylene as our example. On the basis that concerted dimerisation occurs by overlap between the two π-electrons in the highest occupied M.O. (orbital π*)* and the two π-electrons in the lowest unoccupied M.O., *i.e.,* in the first excited state (π, π*), the situation will be as shown in (V). This would require overlap between a positive and a negative lobe. This is symmetry-forbidden, and hence cannot occur by a *concerted mechanism.*

However, some thermal dimerisations *do* occur, but in these cases, the mechanism is believed to be *step wise* via a diradical. On the other hand, on the basis that in a concerted photochemically controlled dimerisation, *both* reactant molecules are in their first excited states, the reaction is now symmetry-allowed and so proceeds (see VI).

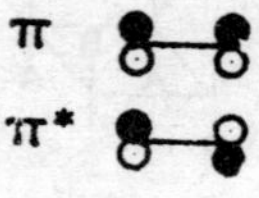

π π*

π* π*

(V) 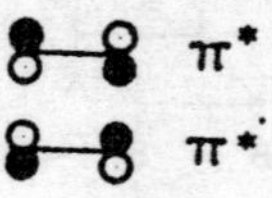 (VI)

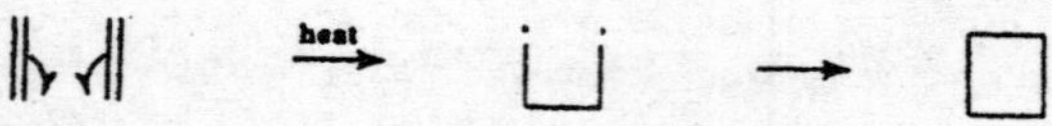

heat

An example of this type is the photodimerisation of tetramethylethylene. Photodimerisation of α, β-unsaturated carbonyl compounds is relatively common, and a well-studied example is *trans*-cinnamic acid. This acid exists in two crystalline forms, α and β, and when each is irradiated in the solid state, the α-form gives *α-truxillic acid* ('head-to-tail' dimerisation) and the β-form gives *β-traxinic acid* ('head-to-head' dimerisation).

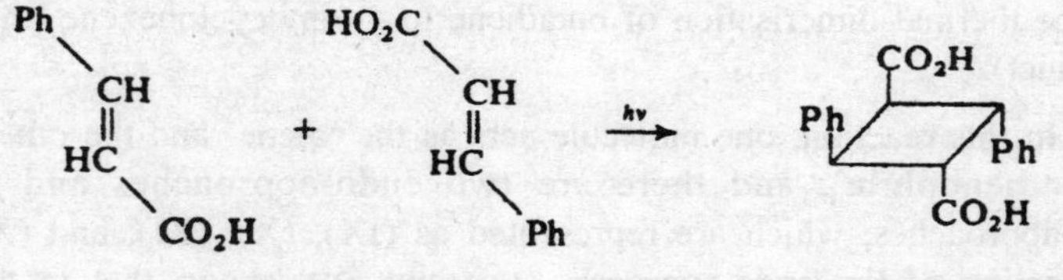

**α-form** **α-truxillic acid**

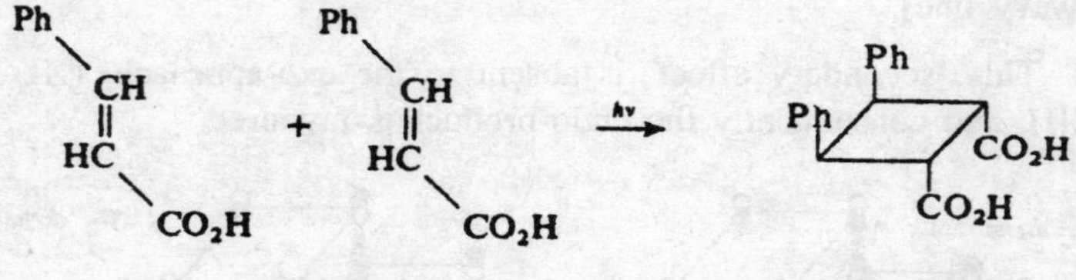

**β-form** **β-truxinic acid**

In these dimerisations, the crystalline lattice is the dominating factor in the preferential formation of the product; steric and electronic effects are less important.

*(4+2) Reactions.* The classical example of this type is the Diels-Alder reaction, and is carried out under thermal control. According to the selection rules (Table 4.4), if this reaction is concerted, the predicted bonding process is *supra, supra (cis,cis)* or *antara,antara (trans,trans).* As we have seen, the Diels-Alder reaction is a *cis,cis* cyclo-addition *(supra, supra),* and hence may be represented in two allowed ways :

(i) the diene in its first excited state and the dienophile in its highest occupied ground state, or

(ii) vice versa.

These alternatives may be represented (i) as (VII)-(VIII) and (ii) as (VIIa)-(VIII) [*cf.* with Fig. 4.10]. Thus, the. Diels-Alder reaction is a ($4_s$ + $2_s$) reaction and is *photochemically not allowed* (by a concerted mechanism).

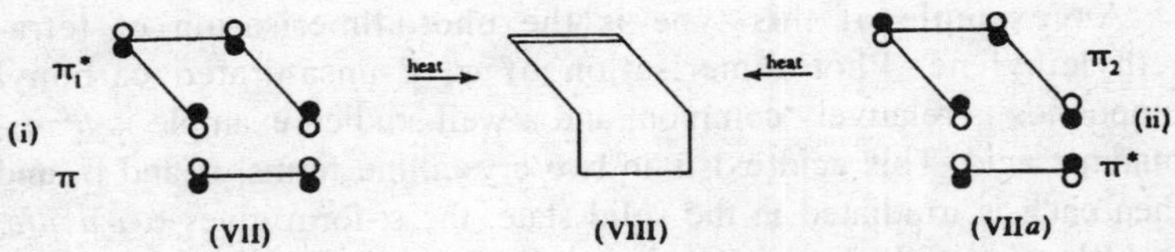


Another feature of the Diels-Alder reaction is that the product is predominantly the *endo* form. This may be explained by examination of the thermal dimerisation of butadiene to 4-vinylcyclohexene (major product).

In this reaction, one molecule acts as the 'diene' and the other as the 'dienophile', and there are two endo-approaches and two exo-approaches, which are represented as (IX), (X), (XI), and (XII). Inspection of the endo-approach, (IX) and (X), shows that in these configurations there is an interaction between like lobes (represented by a wavy line).

This 'secondary effect' is absent in the exo-approach, (XI) and (XII), and consequently the endo-product is favoured.

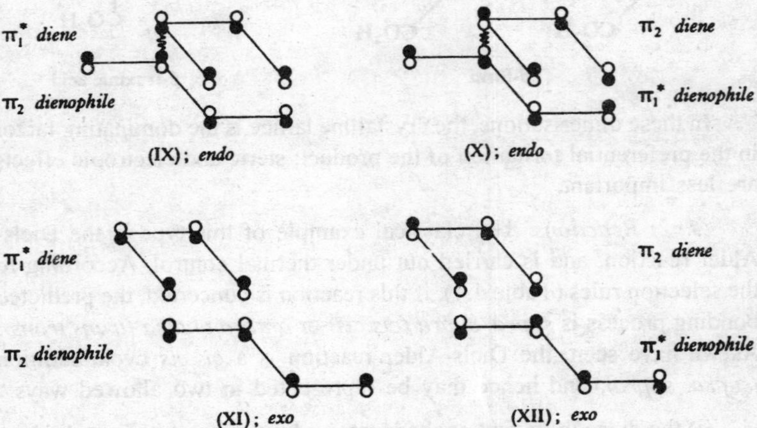


Now let us consider the photochemical dimerisation of butadiene. If this is carried out in ether solution in the absence of a photochemical, two products are obtained by intramolecular cyclisation, cyclobutene and bicyclo [1, 1, 0 ] butane.

$$\xrightarrow[Et_2O]{h\nu} \quad + $$

These arise from the excited singlet state. When irradiation of butadiene is carried out in *the presence* of a sensitiser, three products are obtained, *trans*-(XIII) and cis 1, 2-divinylcyclobutane (XIV), and 4-vinylcyclohexane (IX). (XIII) and (XIV) arise by 1, 2-cyclo-addition (2 + 2) and (IX) by 1,4-cyclo-addition (4 + 2), and their relative amounts depend on the nature of the sensitiser, *e.g.*,

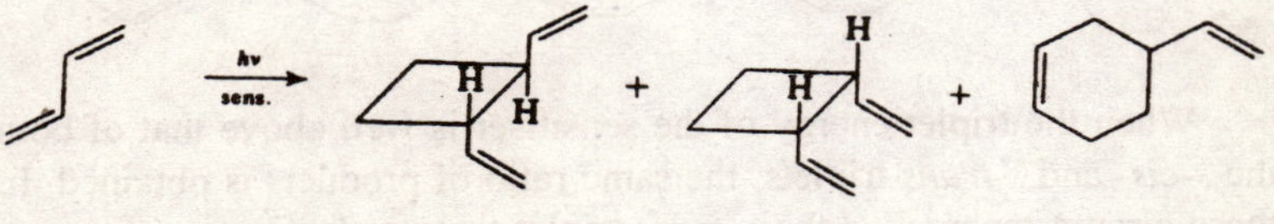

hv, sens. → (XIII) + (XIV) + (IX)

| | | (XIII) | (XIV) | (IX) |
|---|---|---|---|---|
| ***Sens.*** | acetophenone | 82 | 14 | 4% |
| | benzil | 50 | 8 | 42% |

Since there is dependence on the sensitiser, it is reasonable to accept that the excited triplet state is involved, thus.

$$[\text{sens.}]\,(S_0) \xrightarrow{h\nu} [\text{sens.}]^* \longrightarrow [\text{sens.}]^*\,(T_1)$$

$$\xrightarrow{CH_2=CH-CH=CH_2} [\text{sens.}]\,(S_0) + [CH_2=CH-CH=CH_2]^*\,T_1$$

If we also accept that butadiene in its triplet state can be represented as a resonance hybrid like butadiene in its ground state, then the central bond has double-bond character, with the *s-trans-form* being more stable than the *s-cis*.

$$\dot{C}H_2-\dot{C}H-CH=CH_2 \leftrightarrow \dot{C}H_2-CH=CH-\dot{C}H_2 \leftrightarrow CH_2=CH-\dot{C}H-\dot{C}H_2$$

*s-trans* *s-cis*

Furthermore, since one reacting species is in the triplet state, the cyclo-addition is not a concerted reaction, and hence the selection rules do not apply (diradicals are in the triplet state). In the case under discussion, the geometry of the reacting species plays a part in the course of the reaction, the *s-trains**- form undergoing 1,2-addition and the *s-cis**- form undergoing both 1,2- and 1,4-additions with ground-state *s-trans*-butadiene. This is represented as shown.

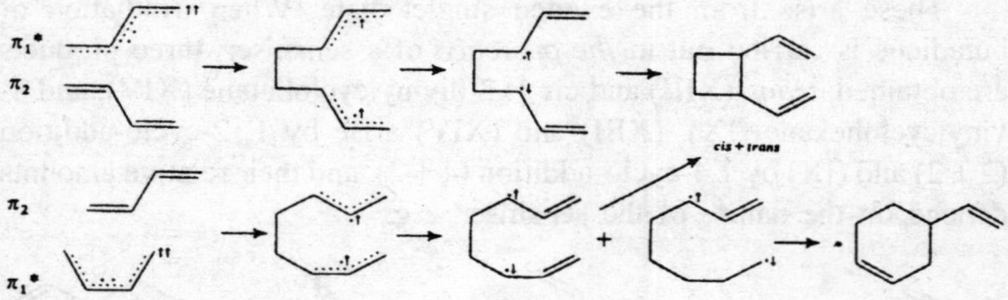

When the triplet energy of the sensitiser is well above that of both the *s-cis-* and *s-trans*-triplets, the same ratio of products is obtained. In these circumstances it is the *s-trans*-triplet that predominates (*cis-trans* isomerisation).

As the triplet energy of the sensitiser decreases, the amount of the *s-cis*-triplet increases, and consequently the amount of cyclohexene derivative increases.

## EXERCISES

1. The oxidation of cyclohexene with chromic acid gives a compound $C_6H_8O$ which shows a strong u.v. band at 225 nm. What is this compound and what is the 'expected' product that is also obtained?

2. This biphenyl derivative has been resolved and has $\lambda_{max}$ 249 nm ($\varepsilon$ 17 000). Biphenyl has $\lambda_{max}$ 251.5 nm ($\varepsilon$ 18 300). Discuss these observations.

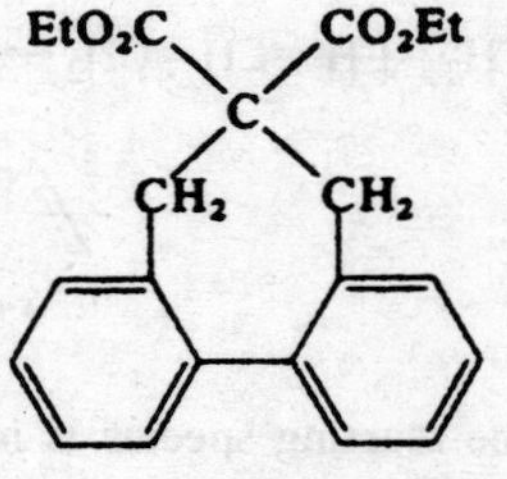

3. PhMe, $Ph_2CH_2$, and $Ph_3CH$ have similar u.v. spectra ($\lambda_{max}$ ~ 262 nm). Explain.

4. Discuss the following spectral data (nm).
   (i) PhOH (210), PhOMe (217), PhCT (235).

(ii) $C_6H_5-N{=}N-C_6H_4-NH_2$ and $C_6H_5-\overset{+}{N}H{=}N-C_6H_4-NH_2$

yellow violet

5. In 4-OH-2, 6-$Me_2$ – $C_6H_2$ – N = N – $C_6H_2$-2, 4, 6-$Me_3$, the *cis* isomer is colourless and the *irons* isomer is coloured. Offer an explanation.

6. What effect would you expect the polarity of the solvent to have on the position of the absorption maximum of alkenes ?

7. Offer an explanation for the following data. $\lambda_{max}$ (nm): methane, 125; ethane, 135; cyclopropane, 190.

8. Draw diagrams for the relative energies of the various types of π-orbitals of butadiene (which has two bonding and two anti-bonding π-orbitals) to illustrate the ground state and the $S_1$, $T_1$, $S_2$, and $T_2$ excited states.

9. Complete the equations.

   (i) diazotised picramic acid + p-OH-acetanilide ⟶ ?

   (ii) diazotised 3-$NH_2$-4-OH-benzenesulphonic acid + m-phenylenediamine ⟶ ?

   (iii) benzendiazonium chloride + resorcinol ⟶ ?

   (iv) (N-Me-pyridinium, $X^-$)–$C_6H_4$–$NH_2$ + $Me_2N$–$C_6H_4$–NO $\xrightarrow[EtOH]{KOH}$ ?

10. Suggest a synthesis of each of the following.

   (i) $HO_3S$–$C_6H_4$–N=N–(naphthyl)–OH from benzene and naphthalene;

   (ii) $O_2N$–$C_6H_4$–N=N–(naphthyl, OH) from benzene and naphthalene;

(iii) MeCONH–$C_6H_4$–N=N–$C_6H_3$(OH)(Me) from aniline and p-crecsol;

(iv) ($NO_2$)$C_6H_4$–N=N–$C_6H_3$(OH)($CO_2H$) from benzene;

(v) MeNH(Me)$C_6H_3$–C(=NH)–$C_6H_3$(Me)NHMe from N-Me-o toluidine;

(vi) (thioindigo: CO, C=C, S, S) from benzene;

11. What are the major products in the photochemical decomposition of methyl neopentyl ketone ? Suggest a mechanism.
12. Complete the following equations.

(i) (1-tetralone) $\xrightarrow[hv]{Me_2CHOH}$ ? (ii) $Me_2CHONO \xrightarrow{hv}$ ?

(iii) (1,3-diphenyl-1,3-dihydrobenzo[c]thiophene 2,2-dioxide: Ph, $SO_2$, Ph) $\xrightarrow{hv}$ ? (iv) (cyclohexylidenecyclohexane) $+ O_2 \xrightarrow[sens.]{hv}$ ?

(v) $Me_2CH_2CH_2COMe \xrightarrow{hv}$ ? a cyclobutane is major

(vi) (cyclodecanone, O) $\xrightarrow{hv}$ ? (a decalin)

(vii) (2-azidobiphenyl, $N_3$) $\xrightarrow{hv}$ ? (a heterocycle)

(viii) $CH_3CH_2CH_2CH_2N_3 \xrightarrow{hv}$ ? (a cyclic compound)

13. 5-Methylhex-3-en-2-one undergoes migration of the double bond when irradiated with light *of law* energy, whereas pent-3-en-2-one, under the same conditions, does not. Suggest a reason.

14. The photochemical decomposition of cyclohexanone gives five products: CO, $C_2H_4$, $C_3H_6$, $C_5H_{10}$, $CH_2=CH(CH_2)_3CHO$. Outline a mechanism for their formation.

15. The photochemical reaction between cyclohexane and pentane-2,4-dione gives a product A, $C_{11}H_{18}O_2$, A, on oxidation with alkaline hypobromite, gives B, $C_9H_{14}O_4$. B, on heating with soda-lime, gives C, $C_7H_{14}$, a saturated compound. What are A, B, and C ? Formulate the reactions.

16. Suggest a mechanism for the reaction.

$$2PhCH_2CH_2ONO \xrightarrow{hv} (PhCH_2NO)_2 + 2CH_2O$$

17. The Norrish type I process is not important for the photolysis of diaryl ketones. Suggest a reason.